The Value of Labor

The Value of Labor

The Science of Commodification in Hungary, 1920–1956

MARTHA LAMPLAND

The University of Chicago Press
Chicago and London

The University of Chicago Press, Chicago 60637
The University of Chicago Press, Ltd., London
© 2016 by The University of Chicago
All rights reserved. Published 2016.
Printed in the United States of America

25 24 23 22 21 20 19 18 17 16   1 2 3 4 5

ISBN-13: 978-0-226-31460-0 (paper)
ISBN-13: 978-0-226-31474-7 (e-book)
DOI: 10.7208/chicago/9780226314747.001.0001

Library of Congress Cataloging-in-Publication Data

Names: Lampland, Martha, 1952– author.
Title: The value of labor : the science of commodification in Hungary, 1920–1956 / Martha Lampland.
Description: Chicago ; London : The University of Chicago Press, 2016. | Includes bibliographical references and index.
Identifiers: LCCN 2015049562 | ISBN 9780226314600 (pbk. : alk. paper) | ISBN 9780226314747 (e-book)
Subjects: LCSH: Commodification—Hungary. | Hungary—Economic conditions—20th century. | Agriculture–Hungary.
Classification: LCC HD8420.5 .L357 2016 | DDC 331.01/30943909041—dc23 LC record available at http://lccn.loc.gov/2015049562

♾ This paper meets the requirements of ANSI/NISO Z39.48–1992 (Permanence of Paper).

Contents

Note on the Text vii
List of Abbreviations ix
Glossary xi
List of Illustrations xiii
Acknowledgments xv

Introduction 1

PART 1

1 Moral Imperatives, Political Objectives 29
2 Rationalizing the Economic Infrastructure 49
3 Formalizing Practices 77
4 The Problem with Money 109

PART 2

5 State Matters 139
6 A New Matrix of Labor Value 164
7 Administering Coercion 188
8 Fighting over Numbers 223
Conclusion 265

List of Archives 275
Notes 279
Bibliography 303
Index 327

Note on the Text

Hungarian law regarding archival materials, and common decency, require that the names of persons in government documents be redacted to protect their privacy. As a substitute, I have used the Hungarian shorthand for anonymity, "X. Y." This regulation does not apply to public figures.

Unless otherwise indicated, all the participants in this history have Hungarian names. Many names appear to be German, for example, but I make a point of identifying German citizens separately in the text.

All translations appearing in the text are my own, unless otherwise indicated.

Abbreviations

ÁEK	Állami Ellenörző Központ
	State Auditing Center
ÁVH	Államvédelmi Hivatal
	Office of State Security
DÉFOSZ	Dolgozó Parasztok és Földmunkások Országos Szövetsége
	National Union of Working Peasants and Agrarian Workers
DISZ	Dolgozó Ifjúság Szövetsége
	Union of Working Youth, the youth arm of the Communist Party
EMSZO	Egyházközségi Munkásszakosztályok
	Parish Workers' Departments
FÉKOSZ	Földmunkások és Kisbirtokosok Országos Szövetsége
	National Union of Agrarian Workers and Smallholders
FKgP	Független Kisgazdapárt
	Smallholders Party
FM	Földművelési Minisztérium
	Ministry of Agriculture
GF	Gazdasági Főtanács
	Economic Supreme Council
ha	hectare
ILO	International Labor Organization
KALOT	Katolikus Agrárifjúsági Legényegyletek Országos Testület
	National Boy of Catholic Agrarian Youth Men's Groups
kh	*katasztrális hold*
	cadastral acre (0.56 ha; 1.42 acres)
KTI	Közgazdaság-tudomány Intézet
	Institute for Economic Science (1951–1954)
	Közgazdaságtudományi Intézet
	Institute of Economics (1954–)

MDP	Magyar Dolgozók Pártja
	Hungarian Workers' Party (1948–1956)
MÉP	Magyar Élet Párt
	Hungarian Life Party
MGI	Magyar Gazdaságkutató Intézet
	Institute for Research on the Hungarian Economy
MGOE	Magyar Gazdatisztek Országos Egyesülete
	National Association of Hungarian Farm Managers
MKP	Magyar Kommunista Párt
	Hungarian Communist Party (1944–1948)
MMI	Mezőgazdasági Munkatudományi Intézet
	Agricultural Work Science Institute
NPP	Nemzeti Parasztpárt
	National Peasant Party
OMB	Országos Munkabérállapító Bizottság
	National Committee to Set Wages
OMGE	Országos Magyar Gazdasági Egyesülete
	National Hungarian Agricultural Association
OMKD	Országos Magyar Katolikus Diákszövetség
	(also known as Emericana szeniora)
	National Hungarian Catholic Student Association
OMMB	Országos Mezőgazdasági Munkabérmegallapító Bizottság
	National Committee to Set Agriculture Wages
PIL	Politikatudományi Intézet Levéltára
	Archive of the Political Sciences Institute
SZDP	Szociáldemokrata Párt
	Social Democratic Party
SZOT	Szakszervezetek Országos Tanácsa
	National Council of Trade Unions
UFOSZ	Ujonnan Földhöz Juttatottak Országos Szövetsége
	National Union of New Landowners

Glossary[1]

adó adópengő. tax pengő certificates used as currency
adópengő. tax pengő
beszolgáltatás. requisition or forced delivery of goods to the state
betétpengő. deposit pengő
Betriebswirtschaft. firm studies, industrial engineering, business economics
cseléd. servant; manorial servant if employed at a manorial estate
éberség. vigilance
egyenlősdi. egalitarian
egyke. only child
fegyelem. discipline
fejadag. per capita grain rations
fekete reakció. black reactionaries, specifically the priesthood
főiskola. technical college
forint. florin; currency as of 1946
gazdakönyv. farmer's book, designed for stewards at manorial estates
gazdaközönség. the farming community
gazdapengő. farmers' pengő
Gazdasági Főtanács. Economic Supreme Council
Gazdasági Lapok. Agricultural Papers
gazdasági rendőrség. police for economic transactions
gazdatiszt. estate manager, farm steward
gazdatitkár. farm advisor
Hajdú-Bihar Néplap. People's Newspaper of Hajdú-Bihar County
hangulatjelentés. mood report
háztáji. house plots
hitelpengő. credit pengő
járás. district (as in a subdivision of a county)
jegyző. village notary

1. Terms and names that only figure once in the text alongside the translation are not included.

kaloriapengő. monetary equivalent of a daily ration of calories
kommenció or *konvenció*. yearly payment package for manorial workers
Közellátási Minisztérium. Ministry of Supplies
Közigazgatási Racionalizálási Bizottság. Committee on Administrative Rationalization
közösen termelő szövetkezeti csoport. collectively producing cooperative group; type III cooperative farm
község. community
Köztelek. Commons
Magyar Függetlenségi Népfront. Hungarian Independence Popular Front
Magyar Gazdaszövetség. Hungarian Landowners' Association
Magyar Gazdaszövetség Fogyasztási Értékesítés Szövetkezete (Hangya). Consumer and Marketing Cooperative of the Hungarian Landowners' Association (Ant)
Magyar Közigazgatástudományi Intézet. Institute of Hungarian Administrative Science
megszilárdítás. consolidation
megváltási ár. conversion price
mezőgadász. agriculturalist
mezőgazdasági akadémia. high school (to study agriculture)
Mezőgazdasági Szervezési Intézet. Institute of Agricultural Organization
munkanap. work day
négyszögöl. square fathom
Néplap. People's Newspaper in Hajdú-Bihar County
népnevelő. person charging with enlightening the people, agitprop cadre
okleveles mezőgazda. licensed agriculturalist
Országos Mezőgazdasági Munkabérmegállapitó Bizottság. National Board to Set Agricultural Wages
Országos Mezőgazdasági Üzemi és Termelési Költségvizsgáló Intézet. National Institute for the Study of Agricultural Organization and Production Costs
Országos Zöldmező Szövetség. National Green Field Association
pengő. currency from 1918 to 1946
puszta. Hungarian steppe; in relation to manorial estates, sites interspersed across the property where manorial workers lived
robot. feudal service
Szabad Föld / Free Land. newspaper for the agrarian community
Szabad Nép / Free People. the Communist Party daily newspaper
szakember. expert, specialist
számvevő. accountant
szervezet. corporation
takarékpengő. savings pengő
társulat. association; type I cooperative farm
termelői szakcsoport. specialized producer group
termelő szövetkezeti csoport. type III cooperative farm
Terményhivatal. Produce Office
tervgazdálkodás. planned economy
törvényhatóság. municipal authority
üzem. firm
üzemgazdász. firm economist, industrial engineer
üzemtan. business economics, firm studies, industrial engineering

Illustrations

Figures

4.1	Rapacious pengő	133
4.2	Cartoon on the birth of the forint	134
7.1	Time is money	191
7.2	Winter courses on raising productivity	193
7.3	Hungarian emissaries to the Soviet Union	196
7.4	Heroes of the labor competition	199
7.5	Be the first in the competition	200
7.6	Kulaks as enemy of peace	206
7.7	Out with class enemies	210
7.8	Shaming laggards	214
8.1	Education for peace	240

Tables

1.1	The size and number of landed properties in Hungary	31
4.1	Daily price increases	127
6.1	Increases in requisitions between 1949 and 1951	175
7.1	Various prices for peasants' goods	204
8.1	Information on leading and mid-level cadres in the Ministry of Agriculture	232

Acknowledgments

I dread writing acknowledgments. Fruitful conversations forgotten, ongoing conversations taken for granted. Many, many people have patiently listened to me try to explain this project, and each time I learned how better to articulate my ideas. Thank you to all.

Archivists and librarians carried me throughout this project: vetting my numerous requests, hauling box after box to the reading room, dusting off files, and photocopying thousands of documents. I owe a special debt of gratitude to the hardworking staff of three national institutions in particular, where I spent countless hours immersed in bits of the past: the contemporary (post-1945) collection of the Hungarian National Archive, the Institute of Political History, and the National Széchényi Library. In addition, three county archives graciously hosted and helped me: Győr-Moson-Sopron County, Hajdú-Bihar, and Zala. The men, and one woman, I interviewed for this project were gracious with their time and their memories. Although they were often puzzled by my research project, they nonetheless humored me by spending hours answering questions and sharing their reflections of the world they once inhabited: Kálmán Eck, László Enese (formerly Steinzinger), András Hegedüs, Pál Izinger, Mihály Kalocsay, József Kecskés, Csaba Kovács, Miklós Kuzmiak, György Latkovics, Mrs. György Lonti, Pál Romány, István Szabó, Sándor Szakács, József Takács, and Zoltán Tószegi. During the first phase of research, three enterprising research assistants helped me to survey collections, run down sources, and plow through reams of documents: Zoltán Bellák, Edit Cserna, and Enikő Karácsony. At my request, Anna Hamar and Tibor Valuch wrote case studies to guide my investigation. Tibor provided me an invaluable overview of the structure of the Stalinist party/state; Anna compiled a comprehensive study of the Ministry of Agriculture trial in 1948.

Drawing on his mammoth collection of biographical materials, Árpád Takács generously provided me detailed information on the backgrounds and training of crucial figures in interwar work science and agricultural economics.

My research would not have been possible if not for the generous support I have received from the American Council of Learned Societies, the German Marshall Fund Research Grant, the International Research and Exchanges Board (IREX), and the National Council for Eurasian and East European Research. These agencies are not responsible for the content or findings of my research. I would also like to thank Vera and Ferenc Falus for being such good neighbors and generous friends.

Three influential mentors, Ferenc Szarka, Pál Juhász, and Barney Cohn, encouraged my curiosity about work units early on. Feri *bácsi* patiently answered all my questions about work units and prewar labor contracts when I first visited his home in Sárosd in 1982. I couldn't keep up with the complex calculations he performed to explain the major consequences of apparently minor changes in the proportion of the harvest paid to migrant workers, but I did learn that paying attention to details like this was important. When I happened upon the *Journal of Agricultural Work Science* in the Parliamentary Library in fall of 1983, Pali confirmed my suspicions that socialist wages and interwar work science might be related. After returning from field work in 1984, I happened to mention work units to Barney in a casual conversation. He was immediately intrigued, reassuring me that I was not alone in finding these artifacts interesting, although many would attempt to persuade me otherwise for years afterward.

Gail Kligman and Katherine Verdery have been role models and mentors since I started graduate school. They have nurtured my interest in Eastern European society and culture and shared my commitment to including the less visible but equally significant protagonists of socialist history in their work. The research project they organized about collectivization in Romania, and the beautifully rich scholarship that has issued from it, has been an invaluable resource.

Elizabeth Dunn is a dear friend and invaluable colleague. Her brutal honesty and generosity of spirit rescued this project more than once. I will never be able to repay her for the excellent scholarly advice and emotional support she has given me.

Mark Pittaway, who passed away in 2010 at the tragically early age of 39, was my constant companion when I was doing the first round of archival work for this project in 1996–97. He was working on his dissertation, a marvelous social history of industrial working class conditions in Hungary in the 1950s that was published posthumously in 2012 (*The Workers' State: Industrial

ACKNOWLEDGMENTS xvii

Labor and the Making of Socialist Hungary, 1944–1958, University of Pittsburgh Press). Extremely generous with his time and advice, Mark was the ideal interlocutor to quiz and probe as I waded through the archives. He was not sparing in his criticism of my work. "What about the people?" I assured him that economists and bureaucrats were people too. Following his advice to get closer to the working class, I visited county archives to learn more about village-level politics. I hope I have redeemed myself.

Zsuzsanna Varga's acute appreciation for village life and the trials and tribulations of collectivization have inspired me. Her historical scholarship has enriched our understanding of these events immensely. Agreeing with Mark, Zsuzsa's insistence that I work in county archives to flesh out the picture at the national level has made all the difference to my study, transforming it from a primarily intellectual history to a fully social one.

My dear friend Ákos Róna-Tas has suffered my fascination with work units for years. Time and again when I asked him to help me translate a difficult phrase, he would tell me, "Nobody else in the whole wide world has read this stuff except you." I was quick to reassure him that the authors surely read the text, if no one else had. And then, of course, he did too. For this and much, much more, I thank him.

As a standing member of the Society of People Who Study Boring Things, I realize it is a lot to ask of colleagues to puzzle through the arcane details I sought to understand. For their patience and generosity, I thank Dave Altshuler, Elena Aronova, Michael Burawoy, Éva Darvas, Kriszti Fehérváry, Deborah Fitzgerald, Sheila Fitzpatrick, Michael Geyer, Zsuzsa Gille, Andreas Glaeser, Maya Haber, Paul Hanebrink, Matt Hull, Paul Kenéz, Kati Kovács, György Köver, Ruth Mandel, Diana Mincyte, Maya Nadkarni, Juan Pablo Pardo-Guerra, Patrick Patterson, György Péteri, Alex Stern, László Váradi, Helen Verran, and Lynne Viola. Several colleagues generously read drafts of this book at various stages of development, and patiently explained—in the midst of my tantrums—how to improve the manuscript: Rick Biernacki, Elizabeth Dunn, Jeff Haydu, and Isaac Martin. The research group Leigh Star and I organized in 2001—Geof Bowker, Steve Epstein, Rogers Hall, Jean Lave, Martin Lengwiler, Janice Neri, Ted Porter, L. K. (Mimi) Saunders, and Judith Treas—explained to me what I was looking to find. The Humanities Research Institute at the University of California, Irvine, was our gracious host.

I have had the good fortune of participating in the Science Studies Program at UC San Diego, first as a guest and then as a core faculty member, for more than twenty years. The interdisciplinary community Science Studies fosters has nourished my intellectual appetites and surrounded me with marvelous colleagues. Attending the great talks held every week in the colloquium series fed my curiosity about science studies as a discipline and taught me how to ask questions differently.

Teaching, as we know, is a two-way street, but I have often felt that graduate students, especially those in the Science Studies Program, have been leading the way. Thank you. For assistance directly related to this manuscript I would like to thank Natalie Forssman, Christine Payne, and Kara Wentworth, who read an early draft and provided feedback. Thanks are also due to Joan Donovan, Shannon McMullen, and Jonathan Walton, who shared their precious skills with me.

And then there are my friends and family. Here I really falter. For want of words, let me simply say thank you to Judit Hersko, Adam Hersko-Rona-Tas, Lorna Reading, Sally Rumbaugh, and my sisters Sarah Lampland and Susan L. Woodward. For all the rest of you who remain unnamed know that I am grateful every day for your support and love.

Introduction

How does one quantify the value of an event or put a price on a possibility? Solving novel pricing questions, like figuring out the value of ad space on the web or calculating monetary incentives to mitigate deforestation, requires complex calculations that commensurate the once incommensurate. Reckoning the value of labor also entailed similar procedures: How does one reward skill, what is the cost of productivity, and when is time worthwhile? While markets may be the final arbiter of value, these are problems markets cannot solve. A myriad of factors enter into these calculations, demanding time and expertise to work out, all of this preliminary to releasing products on the market or hiring workers in a firm. How is this done and by whom? In the following account, I analyze how, at a specific moment in Hungarian history, a small coterie of work scientists and government bureaucrats allied under the banner of modernist rationalization to assess and determine the value of agricultural labor scientifically.

In the first volume of *Capital*, Marx walks readers through his argument about value and the fetishism of commodities using simple examples of commensuration by describing goods being exchanged in the town square, an analogy he soon dismantles as he proposes an alternative theory based on labor value.[1] Although 150 years have passed since Marx wrote *Capital*, common sense still assumes that to commodify goods or labor, they must be bought and sold on the marketplace. This may be a useful way of defining commodification in general terms, but it does little to help us understand the process of commodifying labor, that is, the means whereby a wide variety of tasks conducted by disparate groups in many places are rendered commensurate for the purposes of hiring employees and paying them for their services. As a consequence of these complex processes, people come to understand their time as having a monetary value and their efforts to be defined by skill, difficulty, and temporal duration. In other words, labor is conceived of as a unit: of money, time, and activity. We take these procedures for granted, yet they are clearly subject to historical and cultural variation. I will argue that markets are not the only means by which this transformation occurs. The purpose of my intervention is to draw attention to the substantial infrastructure that makes markets possible, usually overlooked by focusing primarily on the reach or efficiency of markets. Modern labor markets rely upon a wide range of capacities and institutions that make markets work, so much so that their combined influence alone is sufficient to commodify labor. The fact that in many instances markets dominate the process of determining the value of labor does not preclude the possibility that other means of achieving the same end are possible. The question then becomes one of discerning how and where crucial moments in the commodification of labor occur and whether they precede, accompany, or exclude marketization.

What does a scientifically calibrated wage system look like? Let me illustrate by citing a few examples from the evaluative matrix designed in Hungary in the 1950s to reward labor contributions at cooperative farms. Each task in cooperative farming was assigned a daily value, the so-called work unit. Harvesting wheat was worth 1 work unit. That is, cutting 1,200 square fathoms[2] of standing wheat with a scythe over the course of 10 hours was worth 1 work unit. In fact, in 1949 cutting 1,200 square fathoms of standing wheat with a scythe over the course of 10 hours at a cooperative farm in Hungary was worth 1 work unit, whereas by 1952 it was worth 2 work units. In 1949 two men each earned one work unit for spreading manure, calibrated according to the following qualifications: the distance travelled, the amount distributed, and whether the team pulling the cart was made up of horses or oxen. By 1952, only horses were listed as draught animals, and artificial fertilizers dominated the manure section. Now only distance and manner of dispersal (mechanized or by hand) were relevant to the classification. For those working with livestock, different variables were taken into consideration. Tending to 15 cattle daily in 1949 was worth 0.4 work units, whereas 2.4 work units were earned for every 100 kilograms of weight increase. In 1952, the number of cattle tended daily had risen to 20, which earned 12 work units on the last day of every month. For every 100 kilogram weight gain, 3 work units were recorded. Furthermore, for every fattened cow delivered to the slaughterhouse in good shape 4 work units were assigned (Földmívelésügyi Minisztérium Termelőszövetkezeti Főosztálya 1949; Földművelésügyi Minisztérium 1952). Designing this matrix was a monumental effort. Every single activity—from hoeing tomatoes to planting tobacco to artificially inseminating livestock—

carried a specific numerical value, the daily level of effort calibrated to accommodate skill and difficulty, as well as local conditions such as the density of the soil or the steepness of the grade. Precision was paramount.

I first encountered the work unit system while conducting field work at a cooperative farm in Hungary during the early 1980s. The least productive member at the farm, the village party secretary's dissolute younger brother, gave me a copy of the two-inch-thick handbook. No longer in use, work units were remembered fondly. Having heard numerous complaints about the poor working habits of farm members, I finally asked a brigade leader why he thought morale was so poor. He replied without hesitation: the work ethic deteriorated when work units were replaced by money. I was intrigued by his response. Why would paying people in money for their time degrade the quality of work performed? What was the difference between working to accumulate work units and working for an hourly wage or piece rate? Most perplexing of all was the question of who designed the work unit system in the first place, and why.

In the book that chronicled my field work, *The Object of Labor* (1995), I essayed the notion that labor could be commodified in the absence of a labor market. Curious about what had been the social and economic consequences of collectivization in Hungary, I worked alongside villagers at the cooperative farm and in their homes to learn what it meant to work and what sorts of work were worthwhile. I was surprised to see that the attitudes and behaviors of younger villagers I observed were far more reminiscent of capitalism than Soviet collectivism: heightened individualism, rampant utilitarianism, economic determinism. The older generation did not share these views, but neither did they embrace socialist ethics or mouth Communist slogans. When I cited the proverb "Group work is better work," people laughed, quickly countering with the proverb warning that property shared is property mishandled. Private enterprise was respected, although more as a zone of individual achievement than as an opportunity to acquire profit by taking risks. Differing attitudes within the community about work and its value were most evident when comparing the views of older and younger generations. Younger folk talked about their labor time in terms of its monetary value, while their elders considered work to be ennobling, valuable for its own sake. Young people talked about keeping a cow as a means to buying a bedroom set while older villagers kept cows for milk. The contrast between an instrumentalist logic and a code of self-sufficiency was glaring. At the time, I attributed these changes to the substantial reorganization and modernization of the economy during the socialist period. Substantial changes had taken place in how and where people worked: factory and farm regimens introduced new managerial hierarchies and shopfloor regulations, bureaucrats designed new wage scales and incentive packages, and schools educated the youth for new skills and trades. Day in and day out radio waves were flooded with the Communist Party's latest economic figures, newspapers spilled ink praising reform packages, and tv commentators boasted about fulfilling the plan. Quantified indicators replaced prices as the coin of the realm. Department stores opened, although shoppers were still forced to scour the shelves for new goods. By the 1970s and 1980s, regulations preventing workers from moving between jobs had been relaxed, making it possible for a labor market to develop. Meanwhile, a thriving second economy subsidized collective farming, expanded the services sector, and raised incomes. All of these had become well-entrenched practices by late socialism, but they were clearly insufficient as explanations for the more comprehensive transformation under way.

I had assumed throughout my previous project that commodification was an inadvertent consequence of socialist policies. The regime was clearly not capitalist, or even state capitalist, since in the Hungarian economy profit was always secondary to politics. A long-standing debate over what in fact distinguishes Marxist-Leninist economies from capitalist ones—such as Trotsky's trenchant critiques and the later convergence debates in the 1970s—suggested strong similarities in form, but beyond broad studies of industrial development, they were quiet on the means by which these similarities were produced (Trotsky 1937; Corrigan, Ramsay, and Sayer 1978; Dimock 1960; Horowitz 1964; Sorokin 1960; Tipps 1973; Weinberg 1969). Furthermore, commodification may have been achieved under socialism, but its genesis and development had surely begun in the capitalist period. More fine-grained studies of economic processes, and in particular historically nuanced analyses, were needed to resolve the question. In the absence of sufficient documentation, however, such projects were impossible. All that changed in 1989. With access to previously secret government documents, I would now be able to trace the connections between policies and techniques devised to commodify labor in agriculture by capitalist economists to those implemented in early socialism.

I undertook the initial archival work and interviews for this project in the mid-1990s, when debates over the character of postsocialist transitions were raging.[3] What had begun as a study of agrarian work science in mid-century Hungary expanded to consider the dynamics of political and economic transition, though in this case from capitalism to socialism. Studying the transition to Stalinism[4] proved to be an excellent comparative lens on the post-1989 era, sharing as it did so many of the same features: radical economic and political restructuring, reshuffling of politicians and bureaucrats, the significant influence of foreign advisors, a ready-to-hand economic model, and the social upheaval all this entailed. More to the point, optimistic views of the ease of the transition to capitalism in the 1990s relied either explicitly or implicitly on a common view that the Stalinist transition had been swift and comprehensive (e.g., Berend 1996, 39; Kovrig 1979, 236; Swain 1992, 38). The dominant notion was that in 1947 Soviet officials arrived with a model for an economic and political overhaul in their back pocket, which they gladly shared with their new comrades in arms. Dictated is the better term, as the Soviets and their Hungarian cronies were willing and able to use extensive brute force to achieve their goals. So, too, Western advisors swept into Eastern Europe and the former Soviet Union in the 1990s, offering a neat package of institutional reforms that would pave the way to capitalism. While the political and economic strains of transition lent urgency to their admonishments, this paled in comparison to Stalinist methods of persuasion and terror. Had the transition to Stalinism been as swift and effective as we had been taught? If not, then why not? What were the social constraints faced by the Communist party/state at its inception, and did they bear any resemblance to the fits and starts of the postsocialist era?

A question about means—how commodification took place—led me to explore agrarian work science from the interwar period, while an interest in the character of change motivated the focus on the transition to Stalinism. Tying these two issues together is a third element in the study, that is, a curiosity about how scientific ideas and formalizing practices move from place to place. In recent years, science studies scholarship has consistently demonstrated that scientific practices take place in particular, circumscribed places (e.g., Henke 2000; Livingstone 2003; Powell 2007; Smith and Agar 1998); claims to general knowledge are fashioned post hoc and constitute additional practices required to secure their warrant (Canguilhem 1991; Casper and Clarke 1998; Fleck 1979; Foucault 1971; Kuhn 2012; Latour 1988). Broad policy innovations travel the world easily in coat pockets and brief cases and festoon the pages of trade publications and academic journals. Actually implementing policies, on the other hand, is a whole different matter, a commonplace in the vast literature on knowledge transfer and institutional change (Braun and Kropp 2010; Breslau 1998; Collins and Evans 2009; DiMaggio and Powell 1991; Drori, Meyer, Ramirez, and Schofer 2003; Glaser 1973; Jasanoff 2004; Maasen and Weingart 2006; Szulanski 2000; Tihanyi and Roath 2002). If doing science and making policy are contingent processes, does this hold true for technical procedures as well? Or, posed a little differently, does it matter if mathematical formulae are understood to be universal signs or contingent projects? Might approaching formalizing practices this way give us greater purchase on understanding the social in social science? Is it possible to disentangle the scientific and the technical from the social, or, to be more precise, ascertain in what ways social sciences and sociocultural, historical practices intertwine, and with what consequences?

The Technicalities of Commodification

The following questions drive my inquiry: Does theorizing and formal modeling in the social sciences have real world effects, and if so, how and why does that occur? Is it possible to trace the role of new fields like agricultural economics and work science in developing specific techniques for commodifying labor? Might one discover empirical evidence to demonstrate how economists and policy makers self-consciously design formal procedures for assessing the value of labor? What sorts of demands—in time, training, and data collection—would realizing such a system make on business and government agencies? And how would it work?

In the introduction to the collection *The Laws of the Markets*, Michel Callon called for histories drawing together the work of economists and their role in crafting policy:

> In order fully to assess the contribution of economics to the constitution of the economy we would need to write a history which has yet to be invented.... it would be fascinating to construct a social history of economics which would show how abstract notions such as that of supply and demand, or those of interconnected markets ... imperfect competition or incentives, have been formulated in constant relation to the practical questions which, in turn, they help reformulate. (Callon 1998, 2)

Since Callon's call for a new history, several path-breaking book-length studies of economists and policy in this vein have been written (Babb 2001; Çalişkan 2010; Fourcade 2009; Mirowski 2002; Mitchell 2002; Roitman 2005).

Theorizing and empirically verifying the notion that the discipline of economics influences economic processes has come to be known as the performativity question. Prompted by Callon's insistence that "economics, in the broad sense of the term, performs, shapes, and formats the economy, rather than observing how it functions" (1998, 2), analysts have proceeded to investigate whether this claim is a useful way of studying the political consequences of economics in relation to specific sectors or economic practices. Economic experiments and public auctions have been a common site for analysis, since the parameters and participants of the economic phenomena under investigation are easily delineated, though markets of all sorts are of interest.[5] In a useful review, Santos and Rodrigues suggest distinguishing between two senses of performativity: weak and strong. They would characterize the following definition by Callon as performativity in the weak sense: "Talking of the performativity of economics means assuming that agency is distributed and that concrete markets constitute collective calculative devices with variable, adjustable configurations" (2005, 3). In contrast, "the strong notion of performativity asserts that the engineering efforts of economists make economics true by construction" (2009, 986).

The concept of performativity has been controversial from the start. As would be expected with any protracted debate, the clarity of Callon's initial formulation has been muddied by a wide range of interpretations of the meaning of economics as performative and how this event is to be identified empirically. The crucial issues at the heart of the disputes are familiar and rehearse debates once played out in economic anthropology on formalism versus substantivism. What do markets do? Is Callon correct to argue that long-standing social relations of exchange are torn asunder by markets to be replaced by new forms of buying and selling, a process he describes as entanglement? Or is Daniel Miller's concept of embeddedness apposite, that is, do predominant cultural and moral practices enfold and entwine social relations as they shift in market society (2002; 2005)? Is there any reason to adopt the notion of *homo economicus* from neoclassical economics? Is the calculative agent a necessary consequence of engaging in market practices or is *homo economicus* a useless ideological fantasy (ibid.)? Who decides the results of experiments? Were game theorists accurate when they claimed to have successfully engineered the Federal Communication Commission's auction of spectrum licences in 1994 (Guala 2001; Muneisa and Callon 2007) or did computerization of the auction play a far more important, though unanticipated, role in the outcomes (Mirowski and Nik-Khah 2007)?

A crucial feature of the argument about performativity concerns the means by which economists exercise influence on markets. As MacKenzie explains, "One important aspect of Callon's work is his insistence that economics itself is a part of the infrastructure of modern markets" (2006, 15). An important element of the debate, this issue is often overlooked or misunderstood, except in work influenced by science studies:

> If we agree that, in order to exist and function, concrete markets require a set of investments and operations to shape calculative agencies, to qualify and singularize goods, and to organize and stabilize the encounters between goods and agencies, the question that arises is: How can this collective work be described and analysed, and who are the actors involved? The notion of an institution (or of a social network), at least in its traditional sense, is too restricted to describe the socio-technical arrangements that markets are. Institutions, as

defined by the neo-institutionalists who emphasize property rights and their enforcement... or more recently by the new economic sociology, leave little room (if any) to material devices, technologies, and more generally, action and distributed cognition. (Callon 2005, 8)

It is precisely this dimension that holds center stage in my analysis: training our attention on a discrete set of material devices fashioned by business economists and work scientists to measure the value of labor, practices that formalize a set of relationships eventually incorporated into wage schemes at work. In short, this is a history of building the infrastructure for labor markets, tarrying in the details of who built it, when, and how.

The attention MacKenzie has brought to material devices and technologies in the debate over performativity (see also Barry and Slater 2002a, 2002b) is also central to recent studies of commensuration and valuation. Puzzling about the ways goods and activities are commensurated—a question at the heart of Marx's analysis of commodity fetishism—has gained renewed interest as problems of pricing intangibles loom in web marketing or among experts in the new field of contingent valuation, in which the "existence value" of nature is calculated (Bateman and Willis 1999; Champ, Boyle, and Brown 2003; Espeland 1998; Fourcade 2011a, 2011b; see also Espeland and Stevens 1998). Discussions deciding environmental policy by figuring out how to offset carbon emissions hinge on similar practices. Careful exegeses of pricing mechanisms are required to figure out the dynamics of more unusual markets, such as those for art works or "singularities" (Karpik 2010; Pardo-Guerra 2013; Velthuis 2005), but are just as useful for discerning mundane features of market formation. I am indebted to those who have engaged in lively debates over performativity and committed themselves to discerning the mysteries of commensuration and pricing, but my theoretical intervention lies elsewhere. I intend to demonstrate that excavating the manner in which formalizing practices themselves are produced and deployed brings us much closer to understanding the social and cultural practices at the heart of markets.

DECIPHERING THE BOOKS

Another crucial intervention in the study of business economics, finance, and markets has been the enormously influential body of work on accounting as a social and cultural practice. A group of researchers clustered around the journal *Accounting, Organizations, and Society*, clearly influenced by Foucault, produced a flurry of work centered on accounting as a practice (Miller,

Hopper, and Laughlin 1991). Peter Miller described the innovative approach as being characterized by three features: (1) insisting on treating accounting as a technology, (2) examining the meanings of numbers and charts, and (3) interrogating how knowing the economy through accounting and its calculative practices alters economic configurations (Miller 1994, 2–4; for a critique of claims to innovation, see Arnold and McCartney 2003). An entirely new world of numbers and charts was revealed. Classic Foucauldian topics of disciplining the subject and governmentality (Hoskin and Macve 1986; Miller and Rose 1990) appeared alongside articles on concerns current in science studies, such as transcription and ontology (Law 1996; Robson 1992). Fascinating historical work freed the study of accounting from its isolation in economic reductionism, proposing that hierarchies of accountability in religious orders—Jesuit pedagogy "an accounting for sins"—could not be explained simply in terms of resource management and capitalist business (Quattrone 2004). For my purposes, examining accounting as a rhetorical practice was particularly helpful (Carruthers and Espeland 1991; Davis, Menon, and Morgan 1982; Thompson 1991). This way of thinking prompted me to ask what a number actually means. Studies of historical crises—inflation and war—dealt with in accounting practice led me to wonder why a number means what it does, but also when (Gallhofer and Haslan 1991; Loft 1986; Thompson 1987). These questions spurred my interrogation into technologies of commodification.

CONSTRUCTING THE INSTITUTIONAL SCAFFOLD

The term infrastructure usually brings to mind the classic forms of communications and transportation networks: interstate highways, transatlantic telegraph cables, train tracks, or telephone lines. In more recent years, more inaccessible components occupy our attention: satellites and the amorphous cloud. Recent work in infrastructure studies demonstrates, however, that a more capacious concept of infrastructure affords greater analytic purchase on the variety of platforms and networks we rely upon than would be the case if we limited ourselves to a specific sector of the economy or type of support service (Barry and Slater 2002a, 2002b; Çalişkan and Callon 2009; Larkin 2013; Star and Bowker 2006; Tassey 1991; Weill and Broadbent 1998). I am using the term infrastructure more expansively to refer to a variety of implements and practices, taking inspiration from Laurent Thévenot's prescient article on investment in form (Thévenot 1984). In his analysis of Taylorism, Thévenot draws attention to the wide range of activities and technologies required to building "a mechanism of scientific management," all of which require effort and time to create and sustain. Discussing the sorts of activities Taylor prescribed for building, Thevenot writes: "Taylor's handbook... contains a large repertoire of form-giving instruments... [among these] are to be found objects used in production on the shop floor, instruments, plans, conventions, scientific formulae, school precepts, ways of giving instructions which are close to military orders, methods of payment to be used in companies, principles, advice and examples to follow when deciding what action to take" (ibid., 8–9). Only in concert do these instruments and activities constitute a working mechanism. One might ask, then, what "form-giving instruments" would be needed to assess the value of labor and facilitate a viable labor market.

In Hungary, field studies in agrarian work science (*mezőgazdasági munkatan*) required the use of slide rules, cameras, respirators, and a stop watch; manorial management (*uradalmi számadás* or *számtartóság*) relied upon new kinds of accounting procedures and managers trained to use these tools; agricultural economics (*mezőgazdasági üzemtan*) required a well-organized national statistical collection agency so that cross-firm comparisons (nationally and internationally) would permit accurate assessments of what farmers were actually doing and with what effects.[6] The university needed a new department of economics to establish a field of agricultural economics, faculty qualified to teach the new field, and students willing to study what until then had been dismissed as irrelevant for running a good farm (not to mention the dismissal among the wealthy of anything that resembled business). Landowners had to be convinced that adopting cost accounting and hiring new staff would be worth the effort, and schools teaching cost accounting for agricultural firms had to be established. Building an infrastructure for labor value and creating the conditions for a market in labor were going to take time and enormous investments by private business and government. As Thévenot remarks, "[that] the establishment of equivalent forms... [is] extremely costly is usually ignored in economic formalizations which, for the purposes of analysis, tend to assume that they have been established from the outset, rather like the currency" (ibid., 3). Currency in fact can be made and unmade, as the history of inflationary spirals and banking problems in Hungary demonstrates all too well (see chapter 4). And once the new scientific wage system was established, cooperative farm members had to be convinced to actually use it and all the accompanying trappings like written records and cost accounting the new system brought with it.

There is a long and venerable history of the role of technology in business and an equally voluminous literature on scientific management and Taylorism (Beissinger 1988; Edwards 1979; Nelson 1980, 1992; Shenhav 1995, 1999;

Slonim 1922). Unfortunately, these histories do not concern themselves with the more mundane technical problems that are of central importance here, that is, the techniques of commensuration required to monetize labor value and the formalizing practices underlying job standardization.[7] It is not difficult to find manuals written by industrial economists and management theorists describing the procedures recommended to design wages scientifically. For example, Alford engages in a lengthy discussion of Merrill Lott's weighted point system to establish wage rate equalization, identifying five factors to consider when rating jobs: mental effort, skill, physical effort, responsibility, and working conditions (Alford 1938, 587–93; see also Lytle 1942).[8] Comparable factors would be used when the Hungarian work unit system was devised. To my knowledge, however, we have no documentation showing how Alford or Lott's designs were actually used in practice by business managers, and with what results. Mathematical formulae for various wage schemes, accompanied by charts comparing ratios of costs saved or productivity achieved for each model, were published in journals and trade manuals, but how they were used to design a specific wage form is a mystery. I hope to rectify this problem by providing an empirical case study of how labor was valued and remunerated. A larger purpose also motivates my analysis, that of discerning how formalizing practices of valuing and commensurating labor were actually deployed by work scientists, agricultural economists, and government bureaucrats.

FORMALIZING PRACTICES

How in fact are work habits dissected into little bits of activity and bursts of time that carry increments of value, facilitating their commensuration with other timed activities or even value-laden goods? The daily tasks of assessing labor value—the actual work of work science—were primarily devoted to establishing standard units of analysis, devising means of measuring the relative value commonly attributed to farm chores, and performing commensuration exercises to render labor values comparable across the agrarian sector. Only after the proper scientific implements had been crafted locally and numerical data compiled would it be possible to consider adopting algorithms advocated by prominent industrial engineers and agricultural economists promising greater efficiency or higher productivity. The intensely local and particular features constituting the system of labor value being designed tend to be obscured, however, because they appear to issue from the application of the algorithm rather than being understood as the very conditions that make it possible to use the algorithm in the first place. Granted, these assumptions are difficult to excavate from the historical record. Moreover, the privileged epistemological status of formalized devices as powerful tools overshadows the work the instruments actually perform. How would we think differently about formalizing practices if we resisted confusing the means with the ends? I offer a different approach by discussing two common misperceptions: (1) what scientific models are used for and the rhetorical practices depicting what social science models are purported to do, and (2) the failure to appreciate the indexicality of mathematical figures.

Scientific Models

Scientific models usually function as provisional hypotheses, working tools with which to specify mechanisms, discover causal relations, and reveal counterintuitive connections. As Naomi Oreskes has argued in the context of ecosystem science, for many purposes modeling is a necessarily iterative process, requiring reformulation as it develops over time. The process of discerning the crucial features over time means these models are necessarily provisional by nature:

> *All* models are open systems. That is to say, their conclusions are not true by virtue of the definition of our terms, like "2" and "+," but only insofar as they encompass the systems that they represent. Alas, no model completely encompasses any natural system. By definition, a model is a simplification—an idealization—of the natural world. We simplify problems to make them tractable, and the same process of idealization that makes problems tractable also makes our models of them open. (Oreskes 2003, 17)

No matter how much models may be tweaked and refined over time—as the exigencies of the material and social world are accommodated through experiment and analysis—models still remain idealizations of empirical conditions. They are never accurate and thorough depictions of physical or social dynamics. In fact, the conditional status of models is a general feature of scientific practice. Thus the process of idealization to facilitate modeling is not a problem in itself. Simplifying conditions and granting unrealistic assumptions are part and parcel of the modeling process. This is as true of physics as it is of economics, as Nancy Cartwright has argued (1983; 2005). A problem arises, however, when one forgets that the idealization is a sharpened instrument to think with, not a true-to-life portrayal. This is a common error, especially if one is outside the community of modelers.

In the social sciences, actually seeing a model applied is difficult to accomplish without downplaying its provisional character. To strengthen their

INTRODUCTION 13

case for implementation, researchers may promote the value of a model in such a way that temporary variables become confused with prescriptive features.[9] Unfortunately, the conviction one conveys rhetorically about the positive outcome of a model is not compatible with the conditional character of the model itself. To make an argument for implementation—e.g., in the case of innovative but experimental policies—one must insist upon the necessity of this model over another, of this policy over others. In other words, to actually implement a policy, and so to see whether it will produce results, one must gloss over the provisional character of its development. Rhetorical strategies to promote the potential value of a model, and to see results, are as much a part of model construction as is its ongoing development, reworking, and refashioning (Lynch 1991). Mistaking provisional indices for normative characterizations can cause difficulties, but it does not lessen the degree to which scientific modeling relies on provisional numbers. This feature of modeling holds in particular for the field of scientific management in the 1920s and 1930s. Advocates touting the miracle of Taylorism were convinced of the potential their innovations offered, downplaying historical contingencies of development and underestimating the complex dynamics of emergent practices. The models looked so good, they had to work. Provisional metrics became necessary measures.[10] This optimism and urgency is reflected again and again in print, even when the immediate benefits of the program were unclear. This misapprehension would haunt attempts to implement various versions of scientific management.

Indexicality

If we are to take material devices and technologies seriously as constitutive of rationalization—as crucial building blocks of the infrastructure of labor value—then specific practices draw our attention, notably the ways that calculations were conducted and standardizing metrics were designed.[11] Work scientists put great store in formal modeling—mathematical formulae and graphs—to guide (and legitimate) their scientific explorations. Formulae used to compare the relative success of new wage systems were widely circulated across the globe in treatises on scientific management, adopted from classic studies of energy and movement in thermodynamics. But how were they used?

One of the seductive features of using mathematical formulae to think through difficult questions is the assumption that numbers can exist outside the context of their construction, that is, that numbers are free-standing signifiers. Rotman explains this view clearly: "By writing its codes in a single tense of the constant present, within which addressees have no physical presence, mathematics dispenses entirely with the linguistic apparatus of deixis" (2000, 15). In this view, then, mathematics has only referential meaning; formulae only refer directly to the numerical quantity or set of relationships they depict graphically. I argue, on the contrary, that the purpose for which the equation was designed—the context in which it was derived—is just as significant as its abstract referential structure. To wit, the indexical meanings of an equation or a measurement are what make it useful. They constitute its pragmatic function.[12] This becomes obvious when reading texts discussing scientific wage schemes, the designs of which were consistently qualified and modified for their application to local conditions. Considering models in practice requires acknowledging that, in this instance, models of labor productivity are designed to target a single relationship (the ratio of energy to output) and as such are too narrowly defined to address working as a complex activity. At the same time, these models are embedded in an array of assumptions about social relationships that hamper their use elsewhere. Therefore, exploring work science as a meaningful practice requires that I examine abstract mathematical formulae as projects embedded in forms of reasoning that are premised on far more than an isolated number or equation would suggest. That is, the contextual features of a formula are as relevant to its application as its concise expression. The necessity of fleshing out the meaning of a formula is particularly important when an innovative policy is being proposed, a situation in which many of the constitutive features of a policy have yet to be established.[13]

Analyzing formalizing practices as tools engineered to tasks shifts our attention away from peripatetic models of scientific managers to the groundwork required to rationalize the workplace itself. Far from the image of scientists adopting a bunch of formulae which they immediately put to work rationalizing the workplace, that is, a top-down approach to the study of innovation, we must in fact work from the bottom up. It is only by plowing through the cumbersome, heavy technical details that we will understand how formal systems are born. The stark numbers we will eventually see in the work unit value matrix are the product of thoroughly social processes, not the crystallization of isolated elements. In short, contrary to their usual depiction, projects of commensuration and quantification are tethered to time and place. They do not float away from history. It follows, then, that questions need to be asked about the cultural specificity and historical contingency of formalizing practices, be they wage schemes in mid-century Hungary or the numbers, formulae, and models used in the social sciences more generally.

Hungary, for Example

Why consider Hungary as a site to think about scientific models and the value of labor? Two specific features of Hungary's history make it a compelling case for this investigation. The first relates to scientific management and economic regime; the second has to do with the availability of research materials. To the first point, devising scientific management practices to improve economic performance fired the imagination of business leaders and government officials across the globe in the mid-twentieth century; it was an international movement. Science promised success; it also bestowed the imprimatur of neutrality, elevating its agenda above the awkward peculiarities of local economies. Studying the ways scientific management and agricultural economics were practiced in two ostensibly different economic systems—Hungary straddled the capitalist/socialist divide in this period—allows us to focus more intensely on the scientific techniques being developed as technocratic tools rather than as crass business strategy.

The other singular feature of Hungarian history that justifies this analysis is the unique source of documentation making it possible to follow the trail of calculating labor value. Scientific engineering of wages and work took place in many countries in the 1930s and 1940s. We may have access to the scientific publications, dissertations, and memoirs of its proponents—as we do in Hungary—but we do not have access to materials providing us minute details of the implementation of new wage systems in factories and businesses. As private enterprises, they were not subject to the kind of oversight and massive recordkeeping socialist enterprises were forced to endure. Once the Communist Party took control of the government in 1948, its mission was to control the economy at all levels. Every single phase of policy design and its various phases of implementation have been carefully documented, from discussions about which draft of the new wage system to adopt to the frustrations of villagers and county bureaucrats about the difficulties they encountered on the ground. (Never underestimate the value of a centralized party/state for conducting historical research.[14]) This makes the Hungarian archive unique; it gives us an unprecedented opportunity to witness a coherent picture of design and implementation over the course of three decades.

Uncommon Times

In the historiography of Eastern and Central Europe, periodization usually follows the contours of wars fought in the region, punctuated by occasional revolutions, new nations forming, and changes in government. The questions I ask in this study prompted an alternative periodization. Were there continuities in the ways labor value was conceived and measured between the interwar period and the Stalinist era? Did bureaucratic practices change? Was expertise redefined? Why was the pursuit of scientific techniques the connecting tissue between capitalist agrarian economists and Communist Party ideologues? And, perhaps most importantly, would adopting an alternative stretch of time to examine these historical dynamics provide insights that might shed new light or challenge long-standing assumptions in the field?

The major strides to modernize the economy after World War II, and to collectivize agriculture in particular, has long been ascribed to the influence of the Soviets. Conventional histories from the Cold War assumed that the Communist Party simply accepted the model presented by the Soviet Union, relying on Russians' advice and support (e.g., Barany 1995) and succumbing when necessary to their not so subtle pressures to conform to the goal of collectivization (e.g., Wädekin and Jacobs 1982). I offer an alternative reading of the historical record. I argue that substantial portions of the new socialist economy we associate with its modernization *in agriculture*, most notably scientific forms of wage calculations, predated the Soviet invasion and Communist Party takeover. Hungarian business economists and work scientists in the 1920s and 1930s laid the groundwork for a new scientific system calculating labor value independently of the ebbs and flows of labor markets. Their inspiration was German agricultural work science and "firm studies" (*Betriebswirtschaft*; e.g., Aereboe 1920, 1928; Laur 1930; Seedorf 1919), not the attempts at scientific management essayed by the Soviets in the 1920 and 1930s (Bailes 1981; Beissinger 1988; Rogger 1981; Siegelbaum 1988). It bears emphasis that although Hungarian work scientists in the interwar period devoted themselves to designing precise measurements of all elements of the laboring process independently of market forces, they did so as committed capitalists. Their goal was to modernize agriculture and so lead Hungary into the ranks of developed capitalist economies. Hence the calculations and commensuration exercises produced were intended to ease the way into a monetary wage system supporting rational business goals and extracting reasonable profits. The fact that they were only implemented in a socialist economy does not matter from a technical point of view. This history tells us as much about the need to rethink periodization in economic history as it does about the shared vision of modernity between interwar capitalist ideologues and Marxist-Leninist policy makers.

Marxism-Leninism rode into Budapest on a tank, bringing back to Central Europe many ideas initially developed by political economists there. Marxist-Leninist practitioners were committed to the scientific retooling of society, sharing with scientific managers of the West many specific techniques and certainly many hopes for higher productivity and greater efficiency in national economies. Yet Marxist-Leninist advocates—Hungarian or Soviet—had to contend with a crowded field of social reformers and political agents on the Hungarian scene. The "interregnum" between small-town fascism and confident Stalinism (1945–48) was an intensely creative period, with lively debates over social welfare and national purpose, fits and starts over land reform and the nationalization of industry, and convoluted strategies of party formation and dismemberment, as well as prosaic tasks like discarding old monies and making new ones. Innovative economic schemata were to be found as much in the offices of Social Democrats and Peasant Parties as in the Central Committee of the Communist Party. The socialist project was built brick by brick, as were the new bridges over the Danube, with found objects—local colleagues, crucial coalitions, and bright ideas. This, after all, is the art of transition.

The new Stalinist state was better suited to implementing the policies of agrarian work scientists than the previous regime would have been, and certainly more effective than a postwar regime in which the interests of private capital would have dominated. The new state's strong commitment to developing a scientifically sound program for increasing labor productivity, the centralization of economic management, and its willingness to use force to achieve its goals constituted a favorable environment in which to institutionalize the wage policies agrarian work scientists and agricultural economists had envisioned. Nonetheless, the state's hopes to impose a uniform wage system for cooperatives were frustrated when county administrators, party personnel, and cooperative members struggled to comprehend, much less implement, the new techniques for calculating wages. Similar problems with the technical infrastructure familiar from the interwar period arose. There were not enough qualified management and accounting personnel—in government or at cooperative farms—to usher in the changes planned. Plans were constantly rewritten, forcing rapid changes in the calculation of targets. Official documentation was unreliable and recordkeeping was neglected or falsified. Power struggles among party leaders and between ministries, and a relentless class war in the countryside, drained resources needed to strengthen cooperative farming. In a few years, economists judged the wage system unworkable. It would take more time to build up an institutional infrastructure to ensure the new practices would take hold.

New Perspectives on Eastern European Socialism

Unprecedented access to archival resources (official materials as well as privately preserved documents) in Eastern Europe over the last twenty-five years has bred a wealth of new scholarship on the postwar recovery and transition to socialism. A far more nuanced picture emerges of the political machinations and social dynamics of postwar societies across the region. Crucial to this flowering of scholarship has also been a generational shift away from Cold War political commitments that loomed large in a literature written to a great extent by economic and political refugees. From this perspective, countries in the Soviet bloc were portrayed as lackeys or reluctant puppets of Soviet politics, ironically instantiating Lenin's path-breaking analysis of imperialism and its consequences. We are now coming to question the validity of this perspective. Few doubt the significance of an occupying army to constraining political debates and narrowing accepted political solutions (a point also relevant to the western zones occupied by English, US, and French armies). Everything else is under renewed scrutiny, most notably Stalin's postwar plans, the "behind the scenes" actions the United States took toward the region,[15] and the relative balance of power between Soviet and Eastern European regimes.

First and foremost, the decisive role of Eastern European regimes in shaping policy in the region has now been recognized. The practices of what had long been called the process of Sovietization following the war have been substantially reexamined. Norman Naimark's study of the Soviet zone in Germany from 1945 to 1949 offers us a far more complex and empirically rich account of the ways a foreign occupying force works with incoherent policy decisions sent by Moscow and adjusts to unanticipated events on the ground (1997). John Connelly's comparative study of the restructuring of higher education in East Germany, Czechoslovakia, and Poland demonstrates clearly how past academic traditions and contemporary political configurations in each country led to different coalitions of faculty and students, their relative politicization, and diverging relationships with the Communist Party (2000). Another valuable comparative study, in this instance between Romania and the Soviet zone of Germany between 1944 and 1948, is Liesbeth van de Grift's analysis of the way the security apparatus in both countries was reconstituted (2012). Recent biographies of crucial Communist leaders in the 1950s—Lakatos and Losonczy in Hungary and Pauker in Romania—provide far more than life histories, contributing to a deeper understanding of the opportunities and constraints East European leaders confronted in the early years of socialism (Kadvany 2001; Kövér 1998; Levy 2001).

Social history has also blossomed. Excellent histories of the class and ethnic conflicts of the immediate postwar period provide new perspectives on the power politics of the transition (Kenney 1997; Frommer 2005). Kate Lebow's study of Nowa Huta, the symbolic center of Polish working class life in the 1950s (2013) and Mark Pittaway's comparative study of four working class communities in Hungary from 1937 to 1958 (2012) open our eyes to communities once shrouded in Communist propaganda. James Mark's studies of the middle classes in Budapest during the 1950s substantially revise our views of the role of non-Communist Party members in building the new state and economy (2005a; 2005b). Studies of the GDR, once hard to find, have also flourished in recent years. Two books focusing on the first decades include Landsman's analysis of the politics of consumerism (2005) and a useful volume edited by Major and Osmond with articles on politics, social dynamics, and culture (2002).

The most valuable work for my project has been Gail Kligman and Katherine Verdery's marvelous book on collectivization in Romania (2011) and the companion volume of case studies edited by Constantin Iordachi and Dorin Dobrincu (2009). We are all indebted to them for ensuring that the views of those who lived through this period have been captured, just in time. Iordachi has gifted us with another wonderful volume on collectivization, coedited with Arnd Bauerkämper (2014), which offers case studies of collectivization in individual states alongside comparative analyses informed by recent interest in cross- and transnational phenomena. Zsuzsanna Varga's enlightening research on agrarian politics and collectivization in Hungary has deepened my understanding immensely (2010; 2013; 2014; Estók et al. 2003).

Excellent ethnographies on village life in Eastern Europe illuminate the enormous changes collectivization wrought, positive and negative alike (e.g., Bukovoy 1998; Creed 1998; Hann 1980; and Nagengast 1991). These works supplement the foundational study of agrarian policies in Eastern Europe by Wädekin and Jacobs (1982).

Sources

I consulted a range of sources in the course of my research, but one significant difference stands out between the materials I examined from the pre- and post-1945 periods. Nearly all of the pre-1945 sources used in this analysis were published documents, whereas the bulk of materials examined post-1945 were previously secret government and party documents. This warrants a short note. For those of us who conducted research in the socialist era, acquiring access to government and party documents in the 1990s was a godsend. The complex machinations of party operatives, government bureaucracies, and state agencies were now open to scrutiny. Long-unanswered questions could be resolved and new debates initiated. Our experiences in the socialist period also taught us, however, that faith in the veracity of party/state documents was unwise. The question of what has been preserved in documents and why is a perennial historiographical problem. In the politically repressive regimes of Eastern Europe, this problem was complicated by the fact that the consequences of documenting evidence or registering an opinion—i.e., taking sides in factional disputes—could be deadly. So, too, exaggerating claims or providing false testimony were common tactics and often directly sanctioned by the party/state. Approaching materials with a high level of skepticism was an absolute must.

This kind of caveat on the reliability of socialist documents is found in virtually every work relying on government and party documents in Eastern Europe during the socialist period. I worry, however, that our acute sense of the difficulty of reading documents produced by a Marxist-Leninist party/state has led us to exoticize the materials unnecessarily. A good number of the documents I reviewed from the 1950s could have easily been produced by the USDA. For example, among the papers I read from the Labor Department of the Ministry of Agriculture were straightforward queries sent by cooperative farm managers into the state of current policies; they were answered in an equally matter-of-fact tone. In terms of the actual materiality of practices, the genre of official forms—the format of letters, bulletins, official accounts—stayed virtually the same across the capitalist-socialist divide. The ways documents were sorted, dated, numbered, and bundled remained unchanged as well, with minor changes in style distinguishing different institutions, not documentation within an agency.[16] Exaggerating the unreliability of party/state materials may have prevented us from considering the similarities between these documents and those produced by any modern bureaucracy: as a partial view and limited source of information written by people whose political and personal agendas may be very difficult to reconstruct with confidence.

The primary difficulty I faced in relation to materials from the pre-1945 era was war damage. A large number of prewar primary materials, such as founding documents from the Agricultural Work Science Institute and the Rationalization Institute, were lost. Accordingly, the work of academic institutes and state policy could only be analyzed through public journals, political pamphlets, newspapers, and a limited number of ministry documents. I reviewed academic journals, university curricula, more widely read magazines on political and social affairs, and weekly newspapers for manorial management personnel from 1920 to 1945 to gather a series of articles and pamphlets that made the case in varying ways for introducing work science and business economics into agriculture. Book reviews introduced the readership to new ideas promoted elsewhere, précis of dissertations described research being done locally to verify results of new theories, and lectures given to informed audiences illustrated the rhetoric of persuasion, while questions raised in the business economics column of a national agrarian newspaper demonstrated how problems with productivity and wages were being identified as such by readers and editors. These debates were not conducted in a vacuum; the academic community and business circles in Hungary kept an eye on events in neighboring or far-flung regions. Magazine and newspaper articles frequently reported on the international politics of agrarian innovation in business practices and educational institutions. Soviet collectivization was a recurring issue in the 1930s, as were Mussolini's agrarian reforms in Italy (Fellner 1933; Heller 1941a; Jánossy 1933; Kovács 1940; Rézler 1940). The German state's developing system of teaching and extracting statistics on agricultural production warranted attention, as did the land-grant system in the United States, a year-long series of newspaper articles on its genesis and workings accompanying debates over the teaching of business economics, agrarian work science, and accounting in universities and technical schools in Hungary (Károly 1924, 1925; Kesztyűs 1929, 1942; Rege 1929; *Köztelek*, 13 Dec. 1919, 22 Jun. 1925, 9 Jul. 1925, 13 Sept. 1925; Soproni 1935).[17] Pamphlets and longer tracts issued by political groups and professional associations articulated positions on land reform, wage levels, and even the value of cooperative production.

To investigate the Stalinist phase of social engineering, I shifted away from an almost exclusive focus on more generally available published materials to include previously secret party/state documents, union materials, and reports from research institutes. Entering the once secret corridors of the party/state archives, I gained access to primary documents on the nitty-gritty processes of policy formation that I had not found for the prewar era. (This is partly also due to the character of state intervention prior to 1946, which was far more limited than would become the case under the Communist Party.) Large numbers of documents chronicling the crafting of policy become available as of 1946. At this stage, I continued to review academic journals and newspapers, but I was also privy to the debates and discussions of a wide range of party officials and state bureaucrats. Between 1945 and 1948, a number of political parties were in play, so I examined their materials to gauge the complex debate over coming economic policy. In particular, I focused on the Social Democrats and on trade union publications. As the Communist Party consolidated power in 1948, the primary focus of my research turned to the two wings of the party/state: party agencies and government offices.[18] At this point, the number of materials I reviewed increased substantially.

When I began my research, I had to resist the insistent urging of colleagues—former Communist Party historians and dissident scholars alike—that the only documents worth reading were those emanating from the highest levels of the party/state. By this logic, reading the minutes of the Central Committee's meetings would be sufficient to know what happened and why, details to be filled in later. I chose to ignore their advice from the start, assuming that the hard work of policy development would be delegated to lower levels of the bureaucracy, where debates over alternatives would shed important light on the issues at stake. I was not mistaken. Final decisions were made at the top, but the work of making policy happened elsewhere. Some archivists voiced concerns that I would not exercise sufficient skepticism when reading the documents, at whatever level of the bureaucracy they were housed. The former head (now retired) of the county archive in Debrecen tried more than once to discourage me from reading any documents from the party/state, since he thought they were all unreliable. I tried to explain to him the difference between believing individual claims on paper and identifying social patterns across the documents, but he could not be swayed. He tried to limit my access to party materials by restricting the days of the week and hours of the day that I could work at the archive. Luckily, other archivists took pity on me and snuck me in behind his back.

Party/state documents varied in tone and style, some wallowing in "shameless forms of boosterism" (Kotkin 1995, 68) and others prickly with criticism. As Kligman and Verdery remark in their invaluable discussion of using documents from this period (2011, 15–24), "Different documents, then, warrant different readings" (ibid., 24). I have worked consistently to survey materials from a variety of institutions so that it would be possible to triangulate information effectively, a crucial task when conducting a historical ethnography of the state. I also found it extremely useful that party writings were so heavily didactic.[19] As Stephen Kotkin observed, Soviet "Stalinism could not stop speaking about itself. It produced an almost endless flow of words about what it was trying to do, why, how, and with what results" (1995, 367).

Moving between central executive agencies and lower-level offices offered a clear view of the way policy was crafted over time. In addition to party/state documents, I reviewed textbooks and syllabi for new courses in socialist business economics and agrarian policy, as well as determining who taught what courses. These provide a fascinating comparison with university texts on business economics published before 1945. Materials from the Academy of Sciences and the Institute of Economic Research in the early 1950s also provided valuable insight into who wielded intellectual power over party policy, since personnel in these two bodies were drawn into (or expelled from) state decision-making processes over the course of the 1948 to 1956 period. All throughout this period, I paid attention to both the content of policy and the personnel who made it, since one of the most interesting elements of this history is the continuity in personnel in ministries and academic institutions from the pre-1948 to the post-1948 period. Reading the personal papers of leading party officials and ministers provided insight into the dynamics of party politics, as well as illustrated how ostensibly congenial offices of the state battled with each other over public and personal priorities.

A final and crucial difference in source materials for this period is represented by interviews I conducted with former party/state officials, economists, and cooperative and state farm personnel. While party/state documents provided an unprecedented look into the workings of the Stalinist state, they also obscure important issues, such as the informal relations among bureaucrats, party stalwarts, and academicians. Official documents from any state at any time must be read carefully and cautiously, but the absence of widely available alternative accounts of the Stalinist period—e.g., memoirs, diaries, polemical histories or autobiographies, and candid interviews—makes it even more difficult to weigh claims made in official correspondence and regulations.

The final phase of research examines the actual process of implementing a new kind of wage system in cooperative farms in the eight years between the full ascension of the Communist Party to power and the revolution of 1956. Policies envisioned in the cushy offices of the centralized bureaus of the party/state were sent down to the less glamorous offices of county agencies and cooperative farms. It bears mentioning that we also move at this point from the offices of relatively positive advocates who have devised these plans almost exclusively on paper to the difficult task of making it work in lukewarm, if not directly hostile, communities. The fairly cozy epistemological community of economists and social policy makers was no longer of primary interest, except when tweaking bits and pieces of the regulations designed. Now the far more complex social communities of local bureaucrats, ignorant party bosses, and skeptical peasants had to figure out just how to put these ideas into practice, even as they grappled with whether they make any sense whatsoever. Extensive review of party/state documents, local newspaper articles, academic reports, and historical analyses form the bulk of my analysis. At this stage, state and party documents at the county and district level were reviewed, to augment the national level materials examined. I was fully cognizant of the limitations of using party/state documents from the 1950s as a proxy for original materials written from the view of villagers and attempted to the best of my ability to read against the grain. I was pleasantly surprised, however, by how much I could learn about local village affairs from county and district reports. Weekly mood reports sent from each district office were extremely valuable sources of information on everyday goings-on and shared with materials I had seen at the national level from the Ministry of Interior a clearheaded, no-nonsense style. Reports written by county and district officials regarding specific policy issues or problems were far more likely to be formulaic, and often ideologically wooden, than mood reports, which included public complaints, disagreements, and fascinating examples of rumors circulating in the countryside.

Studying county- and district-level documents serves an important function, in concert with national-level materials, in opening up the black box of the socialist state. Bimonthly reports on the mood of rural communities compiled by the Ministry of the Interior provide fascinating anecdotes about social unrest, political battles, and even views of international politics.[20] These materials contrast in interesting ways with the reports written up by the agency responsible for overseeing fiscal records, the State Audit Center.[21] Complaints raised against political officials or lawsuits lodged against unfair labor practices or unresolved property claims illustrated levels of frustration and social unrest. Their patience exhausted, villagers would turn to what may have otherwise been considered a volatile and unreliable bureaucracy to adjudicate legal suits, conferring greater legitimacy on the party/state than I would have imagined based on the general tenor of the historiography of rural communities during this period. I was able to examine a few farm-specific materials in each county, among which the reports for the annual meeting were of greatest value. Annual reports described in detail the distribution of work units, for example the proportion of goods in kind and monetary compensation, as well as including information on the distribution to each member of the farm.

Organization of the Book

The book is divided into two major sections, falling roughly into chapters describing events before and after 1946. Following the introduction, part 1 consists of four chapters that portray the field of agrarian work science in Hungarian public life between 1920 and 1945 as an economic imperative, a cultural necessity, a profitable business strategy, and promising science. The arguments made on behalf of agrarian work science mimic those being advocated elsewhere in Europe—Germany in particular—as solutions to economic distress and cultural malaise. Chapter 1 relates the debates over agricultural modernization centering on economies of scale and relative productivity. Just as important to these arguments are disagreements over moral issues, whether it be the sanctity of private property, Catholic charity, appeals for social justice, or nationalist concerns with racial purity. Chapter 2 describes the rise of agrarian work science in relation to the emergence of business economics as a discipline and a profession and the struggles to improve statistical data collection and introduce new accounting procedures nationwide. Dilemmas surrounding the adoption of foreign economic models are a recurring theme. Chapter 3 zeroes in on the techniques being developed to analyze the productivity of labor in agrarian work science: new conceptual tools, practices of commensuration between units of analysis (in particular in relation to the problem of translating in-kind contracts to monetary wages), the role of psychology in management strategies, and the centrality of monetary incentives. Chapter 4 addresses attitudes towards money and its complicated history in Hungary. Long-standing ambivalence about the moral status of money in many communities existed alongside efforts to expand the reach of the monetary economy; confidence in the modernizing pull of money lingered even when national currency was frequently fragile and unreliable between 1918 and 1946.

Part 2 focuses on the transition to socialism and the implementation of a scientifically designed wage system in cooperative farms. Chapter 5 describes crucial changes in state organization, economic policy, and higher education pushed by the Communist Party. The grand aspirations envisioning rapid regime change were impossible to achieve, forcing the party/state into a series of compromises. Staffing government bureaucracies, universities, and research institutes with eager and competent party stalwarts failed, sowing tensions within the party/state over the need for expertise and the distrust of class enemies. In chapter 6, the design of the new wage system is described, clarifying the significant influence of agrarian work scientists from the interwar period. This finding contradicts the long-standing assumption that the new socialist wage system was adopted entirely from Soviet collective farms. Chapter 7 starts by questioning the prominent role attributed to peasant resistance to collectivization, since the difficulties of implementing the new wage system primarily arose from widespread indifference and frequent confusion over the purpose of measuring work in the first place. I then offer an exploration of class warfare the party/state waged against the countryside, complicated by an inept party apparatus and recurring clashes between branches of government and among party leaders over policy and personalities. Confused command structures and internal dissension handicapped the party/state's ability to exercise coercion effectively, suggesting that despite the frequent use of force against its citizens, the all-powerful party/state was incapable of achieving its own goals. Chapter 8 chronicles the three-year period in Hungary following Stalin's death, during which economic policies were revised, the majority of political prisoners released, and voices critical of the party publicly heard. Attempts initiated by party conservatives in 1952 to adopt a more faithful copy of the Soviet wage system were abandoned by 1955 as unworkable, and steps were taken to restructure the entire edifice tying effort to reward. In the meantime, class war erupted as cooperative farms disintegrated, pitting poorer farm members against wealthier peasant families quick to abandon collective production. All the party/state's efforts to insist on new scientific wage calculations foundered until the fate of collective property was at stake. Differing understandings of the basis of property claims and the right to wages ended up being adjudicated in the courts, where the once incomprehensible wage system became a tool for resolving disagreements. Reliance upon written records to solve disputes abetted the party/state's increasingly strident enforcement of regulations on accounting procedures at cooperative farms. The party/state was less successful in forcing cooperative farm members to follow socialist principles of collective labor organization and turned a blind eye to sharecropping contracts with individual members and their families. In the conclusion I consider the lessons one might learn from an approach to regime change that attends closely and rigorously to the social and political dynamics of transition. I also return to the question of how the formalizing practices of social science and the fetishization of labor are mutually implicated.

PART ONE

1

Moral Imperatives, Political Objectives

Agriculture was the backbone of the Hungarian economy, the productivity of labor its Achilles heel. Bemoaned as the "country of three million beggars," Hungary suffered from a surfeit of labor and a paucity of enthusiasm for modern business. Large manorial estates sprawled across the countryside, relying on an indifferent community of impoverished rural families to farm the land, while peasant landowners struggled to make do. Proposals to remedy the problem fell into two broad camps: advocates for land reform anticipated the growth of an intensive peasant agriculture while proponents of scientific engineering imagined a highly rationalized sector of large agrarian enterprises. The debate over these alternatives during the interwar period was framed in terms of economic advantage, both sides citing numbers and graphs to bolster their claims. The divisions, however, were grounded in deep moral convictions about the dignity of labor, the right to property, and the national purpose. Land reform advocates were associated with left wing politics; those endorsing rationalized modernization were strongly committed to conservative, right wing politics. At the end of World War II, the political compass of the debates pointed in new directions. Scientific rationalization became the byword of the Communist Party, making them strange bedfellows with conservative champions of capitalist business whose modernist vision supported collectivization in the late 1940s. Those who received land in the land reform in 1945—most of whom had numbered among the agricultural proletariat—did not vote in the 1947 elections for the Communists but sided with parties on the right guaranteeing property rights. As we shall see, this about-face makes sense once the principles debated in the 1920s and 1930s are outlined. In the following chapter, I will discuss four crucial points of disagreement between the two factions during the interwar period—(1) economies of scale, (2) standard of living, (3) property rights, and (4) morality and the demographic crisis—to show how the debates unfolded. Arguments were legion but shared a common language: a reliance on statistical data and, when that was lacking, a recourse to complex commensuration exercises to produce comparable numbers. After the war certain claims lost legitimacy, such as arguments about property and race; institutions with which claims to land were associated, such as the Catholic Church, were subject to open criticism by social reformers and leftist politicians. Yet the core of the economic argument on behalf of large-scale production continued to be legitimate in the eyes of agrarian economists and work scientists, their position strengthened when thousands of new property owners failed to establish viable farms by the end of the 1940s.

The Lay of the Land

The Austro-Hungarian Empire collapsed at the end of World War I. New nations were formed, such as Czechoslovakia and the Kingdom of Serbs, Croats, and Slovenes, while the territory of Romania vastly increased by absorbing the territories of Transylvania and Bukovina. Hungary shrunk to one-third its prewar size, losing thousands of acres of plowland, meadows, and forests. In the decade following the war land reform was enacted by virtually all countries in the region, but Hungary's conservative forces forestalled any significant change, leaving it with the most skewed distribution of property in Central Europe. Close to 75 percent (74.7) of the farms in Hungary had less than 5 kh (2.85 ha).[1] These small holdings covered only one-tenth of the land under cultivation. At the other extreme, we find that more than a third of the land (37.7 percent) was distributed among 0.15 percent of the total number of farms in the country. Farms between 50 and 100 kh (28.5 and 57 ha)—i.e., those considered large enough to be self-supporting—were scarce; they constituted only 1 percent of the farms and covered only 6.4 percent of arable land (Kerék 1939, 300; see table 1.1).

Manorial estates in Hungary resembled latifundia and colonial plantations familiar in the New World and Far East: an expanse of land, dotted with small outposts for resident workers, and a wealthy owner or renter rarely to be seen in the estate's vicinity. Among its residents we would find a steward, ten coachmen, three farmhands assigned to the coachmen, a farm boss, seven general-use farmhands, someone in charge of feed for the animals, two dairymen, three cowboys, two pigherders, and one shepherd. Craftsmen lived on

TABLE 1.1. The size and number of landed properties in Hungary

Size of properties		No. of farms	%	Acreage		%
kat. hold	hectare			kat. hold	hectare	
below 1	below .57	628,431	38.5	236,417	134,758	1.5
1–5 kh	.57–2.85	556,352	34.2	1,394,829	795,053	8.7
5–10 kh	2.85–5.7	204,471	12.5	1,477,376	842,104	9.2
10–20 kh	5.7–11.4	144,186	8.7	2,025,946	1,154,789	12.6
20–50	11.4–28.5	73,663	4.5	2,172,300	1,238,211	13.6
50–100	28.5–57	15,240	1.0	1,036,162	590,612	6.4
100–200	57–114	5,792	0.3	805,164	458,943	5.0
200–300	114–171	2,126	0.1	516,875	294,619	3.2
300–500	171–285	1,714	0.1	663,676	378,295	4.1
500–1,000	285–570	1,362	0.1	944,250	538,223	5.9
1,000–2,000	570–1,140	581	—	798,490	455,139	5.0
2,000–3,000	1,140–1,710	187	—	452,109	257,702	2.8
3,000–5,000	1,710–2,850	117	—	451,376	257,284	2.8
5,000–10,000	2,850–5,700	101	—	680,084	387,648	4.2
10,000–20,000	5,700–11,400	48	—	690,953	393,843	4.2
20,000–50,000	11,400–28,500	25	—	855,106	487,410	5.3
50,000–100,000	28,500–57,000	10	—	671,475	382,741	4.2
above 200,000	above 114,000	1	—	209,256	119,276	1.3

Source: Kerék 1939, 298

the estate as well. There would be several machinists, a blacksmith, an assistant to the blacksmith, a cartwright with his apprentices, a stonemason, and someone in charge of the warehouse. Two gardeners and two viticulturists would be counted among the specialists employed on the estate. The families of estate workers (*cseléd*) were housed in isolated clumps of buildings scattered across the land. In the summer months approximately fifty migrant workers would come to do all the fieldwork, and all additional tasks throughout the year would require hiring an additional 3,715 days of adult male labor. In contrast to the workers living on the estate, day laborers lived in nearby villages, whereas migrant workers came from farther away, sometimes far-flung counties in poorer regions.[2] Landowners who lived in villages adjacent to or in the vicinity of manorial estates treated estate workers poorly—viewing them as little more than animals—and considered them to be as foreign to the region as migrant workers were, even though their families may have lived on the estate for generations.

Workers residing on the estate were paid on a yearly contract, the majority of which was comprised of goods in kind: a fixed amount of grain (wheat, rye, barley), firewood, and in some cases a kilogram of bacon and a measure of salt. A small amount of cash was allotted to each household. Housing was provided, if one could call it that: four large families lived in one room measuring approximately seven square meters. Families were given the right to a small garden plot and often had the opportunity to raise piglets until the age of one. In exceptional circumstances, manorial residents were allowed to keep a cow for milk. Migrant workers were paid with grains and room and board; only day laborers were paid in cash. What workers earned in yearly contracts or by day labor varied from county to county and sometimes from manorial estate to manorial estate. While the means of recruiting migrant workers resembled a labor market—a labor boss recruited a band of workers every spring—the contractual relationship was usually based on long-term ties between specific villages and particular manorial estates that siphoned off better workers, leaving only a motley crew of obstinate and lazy folk for everyone else. Under these conditions, labor productivity was extremely poor, difficult to improve, and even more difficult to measure. Modernizing production—rationalizing farms to compete internationally—was only a a pipe dream. Undaunted, agrarian work scientists and economic engineers sought to make this dream a reality.

Disagreements Economic and Moral

The conservative government often depicted Hungarian village communities as populated by the self-supporting, proud, and deeply nationalistic peasant landowner, wrapping itself in populist clothing while downplaying the need for land reform. Numerous essays and monographs strongly contested the image of a comfortable peasant lifestyle promoted by the conservative government, penned by leftist social critics known under the collective name of "sociographs" (e.g., Illyés 1936; Kovács 1937; Szabó 1937; see also Tóth 1984 and Esbenshade 2006). The land question was rehearsed interminably in publications of every sort (weekly newspapers, journals, pamphlets, dissertations, and book-length monographs). Academics, businessmen, and political activists of every stripe participated. Those who advocated on behalf of manorial farms included professional economists, religious leaders, conservative nationalists, and intellectuals associated with rightist causes, though their motivations differed sharply. Business economists and work scientists, for example, were primarily motivated by their interests in modernization and improved productivity in agriculture, while the Catholic Church emphasized its moral obligations to society. These claims rang hollow, since "the Catholic Church was one of the largest landowners in Hungary, deriving a good part of its wealth from vast estates in the south and west. Given this vested material interest, the episcopate could hardly support any serious effort at land reform that might improve the lot of Hungary's ethnic Magyar peasantry and so remained skeptical of anything that smacked of populism" (Hanebrink 2006, 130). Advocates for land reform clustered on the left of the political spectrum, though some conservative religious associations numbered among their ranks. Beyond simple arguments in terms of the equitable distribution of land, many in this camp envisioned a rejuvenated agricultural economy based on small farms specializing in intensified production as an alternative to manorial estates dominating the landscape.

PRODUCTIVITY AND ECONOMIES OF SCALE

The relative advantages of small versus large farms preoccupied those on both sides of the land reform debate. Economic arguments advanced by pro-manorial advocates claimed that large farms were more productive and offered more stable, secure employment. It was a matter of gospel in this camp that large estates were more able to modernize than small farms, in part because manorial properties were better equipped than their poorer neighbors. To assuage their leftist colleagues, conservative writers were fond of citing the debate Kautsky led over proper economies for agriculture as evidence for their position. Land reform advocates were fully cognizant of the distinct disadvantage small farms confronted in relation to economies of scale. Family landownings—accumulated through inheritance and marriage—were frequently divided up into small plots scattered across the outskirts of the village boundaries. In the absence of any substantial land reform, peasant advocates argued that consolidating holdings would significantly improve the viability of small farms. A popular alternative was a vision of "Garden Hungary,"[3] intensifying agricultural production to replace the obsolete "wheat factory" (*gabonagyár*) model of large estates inherited from the previous century (Kerék 1942, 92). This approach was considered a far more competitive and sustainable model for farming, better suited to the nation's climate and geographical position within Europe. Moreover, small-scale but intensive farming had the definite advantage of requiring greater labor inputs, a must in a country burdened with an unemployed agrarian proletariat.

The economy of scale argument did not necessarily support the idea that manorial estates should be kept intact, since large sections of land on manorial estates were underexploited. This was particularly true of the larger estates. Approximately one-third of the acreage of manorial estates larger than 500 kh (285 ha) was not cultivated (Kerék 1939, 316); the percentage of land devoted to forests at large estates varied from 12.9 to 42.8 percent (ibid.,

302). Smaller properties, on the other hand, devoted a larger proportion of lands to cultivation: "The proportion of land under agricultural cultivation (plowlands, gardens, vineyeards, meadows and pastures) is largest at small farms (87.3–95.8 percent), somewhat smaller at medium-sized farms (78.4–80.9 percent), and much smaller at large estates (50.6–77.6 percent)" (ibid.). The discrepancies in land use were seen as evidence that large estates crowded smaller farms, preventing them from growing into more viable economies of scale, challenging the need to keeping estates intact.[4]

Policies designed to resettle peasants within the country—to distribute the population more evenly and landownings more fairly—were a recurring feature of debates over agricultural modernization, reaching back into the nineteenth century following the abolition of feudalism in 1848. During the debate over the 1935 settlement plan in parliament, Samuel Mándy rejected the view that large estates crowded smaller farms: "In general we don't have overcrowded small farms. . . . Even though the price of land is more than half as cheap as before, buyers are as rare as a white raven. Small landowners in particular are hard to find" (*Köztelek*, 20 Jan. 1935). Mándy overlooked the possibility that lands would not be sold off but simply parcelled out to neighboring communities.

To some minds it was folly to spend huge amounts of money dismantling manorial farms when the money could be better spent improving smaller farms (e.g., *Köztelek*, 24 Nov. 1935). A smarter way to remove the barriers to modernization and specialization at small farms would be to improve access to decent sources of credit. And of course, this would eliminate the need for radical land reform (Juhász 1983, 73):

> As a helpful tool one could take advantage of hereditary leases, yearly rental properties [*járadékbirtok*], partial rental or a long-term tenancy, but more important than all of these is the re-creation of a land credit service in disarray from the war and inflation. This would make property credit with long-term, cheap mortgage deeds, as well as firm and investment credit cooperative bonds, accessible to the "acquisitive" person who has no money. (Czettler 1995:52)

Greater attention to the variety of resources available would assist the transformation of production and property relations.

The interminable debate over farm size was also debated in relation to how many people were supported by various types of farms. In 1935 the government proposed legislation allowing the resettlement of the poor on underused or abandoned land. The position defending manorial estates was articulated in Baron Gyula Károlyi's intervention based on statistics he had compiled demonstrating the greater economic viability of manorial estates over small farms. According to his calculations, 1,000 kh (570 ha) of manorial production would sustain 156.9 families (784 souls), whereas if this land were to be broken up into farms of 20 kh (11.4 ha) in size, only 49 families (539 souls) could be provided for (*Köztelek*, 13 Jan. 1935). In response to criticism that his numbers were not generalizable, as they were based on only twenty-nine estates, Károlyi argued that his selection of these farms was not intended to be representative, since in his mind only well run estates should be the basis for discussions of social policy:

> Every large estate only fulfills its calling and is desirable to maintain in the interest of society if it provides a secure living to more people than were the land distributed one by one to small, viable farms. Breaking these [estates] up is not desirable from the point of view of society, in fact it is dangerous because it deprives people the possibility of a living that the state is unable to provide by other means. In the case of those estates that, for some reason, cannot provide more people a living than a small farm . . . then the state is justified in making arrangements for a better distribution. (ibid.)

Kesztyűs, speaking on behalf of the Committee on Firm Statistics of OMGE (Országos Magyar Gazdasági Egyesülete, or the National Hungarian Agricultural Association), agreed with Károlyi that that the state could play a role, though his attention was focused on small and middle-sized farms. Based on his 1934 study of small farms in various regions of the country, Kesztyűs believed that these farms could transform the value of their land and family labor into profitable income. The conditions for making that possible required solving a series of problems: "*adjusting farm tax, acquisition of agricultural capital, consolidation, resettlement, tax reform, producer and marketing cooperatives, foreign trade contracts, industrial customs protection, railroad tariffs, agricultural vocational training, and the question of cartels*" (*Köztelek*, 11 Aug. 1935; italics in original). If need be, the state would have to bear some of the costs of the agrarian poor, since as another participant in the debate explained, "wages are not a charitable institution" (*Köztelek*, 11 Feb. 1940). In the final analysis, agrarian economists argued that the question of appropriate size for a farm depended on a number of factors, such as climate, quality of land, degree of intensification, and crop profile (Heller 1941b, 705–6).

STANDARD OF LIVING ON MANORIAL ESTATES

The promanorial faction was firmly convinced that manorial servants had a better standard of living than their village compatriots whose income was cobbled together from migrant work in the summer or occasional day-labor contracts. Landowners claimed privileged knowledge of village life and so were confident of their views (e.g. *Köztelek*, 11 Feb. 1940).[5] Estate owners believed their farms were stable, secure sources of employment for poor agricultural workers who would otherwise be left to the mercies of a fickle labor market and poorly managed family farms. They backed up their claims by showing that unemployment was less common in regions dominated by large estates (Szeibert 1939, 20). Heller firmly believed, like Károlyi, that if manorial estates were dismantled, then a large number of manorial servants would become homeless (1937, 21).

The public was unaware of these true conditions on contemporary farms, they argued, blinded as they were by prejudice and ignorance. They laid responsibility for this misrepresentation on the sociographs and other land reform advocates:

> Popular village research and demographic reports have cast the saddest jumble of data in front of the thoughtful readers who have been indifferent 'til now. There are those who consider the most important cause of people's abject poverty to be the unhealthy distribution of property relations. Among these there are many who are inclined to paint the situation of *cseléd-s* at manorial estates as far darker than is the case in order to realize their conception of land reform. . . . On average the situation of the manorial *cseléd* community is better than that of small farmers with 10–15 kh (5.7–8.55 ha), but in any case it is much higher than the social and economic condition of dwarf farmers, the landless, or our brethren who barely subsist on day labor. (Heller 1936, 455)

The better income manorial servants enjoyed was easily demonstrated by figures compiled by OMGE: "While in a large firm the costs of one man's work day was 2.04–2.75 pengő in 1935, the income of working family members of a small farm could be expressed as 1.52–1.84 pengő. The goal of agricultural social policy should be to produce a well-to-do working class rather than the troubling pipe dream of an aggressive policy promoting small farms" (Hartstein 1937, 10). Nonetheless, the author did argue that the state had an important role in looking after its citizens, justifying the state's responsibility for bearing some of the costs of the agrarian poor.

Housing was another element in the argument for manorial estates and against land reform. Shining examples of modernized housing stock were trumpeted, such as the estates of the Bishop of Kalocsa or Gróf Gyula Zichy. These estates had built housing providing families with an apartment (consisting of a main room and kitchen), as well as an adjacent kitchen garden: "Every family lives separately, can live its life independently, like any urban citizen" ("Az Egyházi Vagyon és a Rerum Novarum" 1941, 56). Scherer mentions properties owned by Duke Pál Eszterházy and Gróf Gyula Károlyi that provided electricity to manorial housing. Improvements in the availability of clean fresh water had also been made (ibid., 51–53), a crucial consideration for residents often forced to draw water from contaminated wells. Indeed, in the interest of public health, the Ministry of Agriculture issued regulations in 1930 that privies be placed fifteen to twenty meters from the well and below the "height" of the well's incline. The isolation and treatment of excrement was also stipulated in the regulation (Heller 1937, 70).

Heller did not paint as rosy a picture of manorial housing, conceding that these regulations were rarely implemented (ibid.). Heller was also more forthcoming in describing the wide disparity in the quality of housing on manorial estates across the country. Though policies designed to improve manorial housing had been in place since 1907, those implementing these innovations were stymied first by fiscal demands of the First World War and then by the economic downturn of the 1930s (ibid., 67–68). In fact, the Ministry of Agriculture took nine years to pass into law the resolution reached at the meeting of the International Labor Board in 1921 stipulating the minimum requirements for workers' housing. Substantial changes were also hampered, unfortunately, by the stipulation the Minister of Agriculture added to the law in 1930, that improvements to housing would only have to be in effect if the costs of providing the enhancements did not increase the production costs of the manorial estate (ibid., 71). In short, the glowing descriptions that Scherer offers us of comfortable accommodations were the exception, rather than the rule.

Scherer devoted an entire article to listing the numerous services a select group of manorial estates offered.[6] Included in his list of benefits were old-age pensions; schools on site; access to doctor and medical services, including hospital stays; midwives; clothing allowance; sport facilities; a movie theater (with talking movies); and courses on popular topics taught by the managers of the estate (1941, 51–53). Who would choose to live in a village when such luxurious accommodations were waiting at the local manorial estate? Ernő Éber performed a series of careful calculations to translate the complex bundle of manorial servants' income—e.g., payment in kind (grains mostly), rights to keep a few animals, use of a small plot of land to garden, firewood—into a monetary value, providing solid empirical evidence for his claim that manorial servants were decently compensated. His results convinced him that manorial workers' needs were well met, a position strengthened by comparing the less favorable figures on the working conditions of industrial workers. In the end, Éber did qualify his statements: "We don't want to embellish the situation of manorial servants. . . . Not for all the world do we want to say that manorial servants abound in wealth and live off the fat of the land" (1941, 59–60).

Addressing the question of living standards from another angle, promanorial advocates cited figures concerning infant mortality and access to medical care. "As calculated by the demographic department of the statistical office, the infant mortality rate among small farmers and renters was 75 percent, that of manorial servants 42 percent. In the category of small farmers and renters, of those who had died before their seventh birthday, 80 percent had received medical attention; over the age of 7 it was 68 percent. In the group of manorial servants, 88 percent of those who died before their seventh birthday received medical aid, whereas 75 percent of those were seen above the age of 7" (Éber 1941, 60). With regard to nutrition, Heller compiled a comparison of nutritional habits among villagers (1936, 456–57), using data from the students enrolled in one elementary school. He quoted the nutritional standard advocated by Professor Károly Waltner of the Pediatric Clinic in Szeged. "[W]e only rated the nutritional habits as satisfactory if they still provided the possibility of ingesting the minimally necessary caloric amount, so that children fed in this way are not yet starving in the strict sense of the word" (Heller 1936, 457). Not the most generous standard, surely, but for Heller's purposes, sufficient, as the following passage makes clear:

> Dividing the units counted by the number of children, we get the most favorable picture of the nutrition of the children of *cseléd-s* serving at the estates of the magnates, because there were 5.85 units for a child. The children of those hired at renters' estates followed, with 5.7 units, and the postwar property owners 5.33. The units for servants at other long established properties and so-called peasants (5.24 and 5.25) stood alongside each other, which—in contrast to the new property owners—can be explained by the indebtedness of a large segment of landowners in this category. Among herdsmen working for the village community the number drops to 4.93 and those serving at farms of less than 100 kh (285 ha) was 3.92, which strongly approaches the nutritional scale of the children of unemployed parents, 2.5. (ibid., 458–59).[7]

The excitement that promanorial advocates exuded about the comfortable conditions of life on manorial estates was greeted skeptically by many observers of rural life, whose descriptions were bleaker. The sociograph Szabó's famous study of life in Tard, a poor village in the northeast, also addressed the issue of nutrition, using information from school children. He came to a very different conclusion than Heller: "The researcher examines the lists with alarm and doesn't understand that with such nourishment how they can do difficult work and remain healthy" (1937, 46–47). He also relied on students' self-reporting, examples of which are included in the book. What is striking is that even families owning five to ten *hold* (2.8–5.7 ha) relied on bread and grains for 40 percent of their diet and on meats for 20 percent. The percentage of bread and grains over meat products increased substantially as the size of family farms dwindled. The diet among those with one *hold* or less [0.57 ha]—which would include manorial servants—was composed of 60 percent bread and grains, whereas only 3 percent came from meats (Szabó 1937, 229). The consumption of fruits and vegetables was minimal, as were dairy products and eggs. Heller did not understand why the diets of manorial residents were so meagre, lambasting manorial families for their poor culinary traditions: "[T]he reason that children's nourishment doesn't reach the desired level of calories, doesn't include the primary essential nutrients, and doesn't deliver vitamins to the system is not just the bad economic situation, but also the nutritional customs of our people" (1936, 460). Heller clearly was ignorant of the common practice among the poor in rural communities to sell produce for cash, to augment the family's meagre income, in order to buy salt and matches at the store in town and pay their taxes (Szabó 1935, 133).[8] "Cheap food, cheap meat, and eggs are only cheap in the city" (ibid.).

Imre Kovács, a well-known sociograph, contributed to the debate over nutrition by conducting his own analysis of calorie intake among manorial servants (1935). Although at odds with his usual essayist style, this foray into work physiology was a valuable way of engaging in the debate with promanorial advocates. He proposed to use two units of energy—the caloric requirements of physical exertion and its kilowatt hours—to figure out how hard manorial servants were working and the amount of calories needed to perform their jobs (1935, 219). He walked the reader through his calculations, calibrating the physical exertion unit—the kilogrammeter—according to the gender and age of the worker, the kind of work done in various jobs on the manor, and the seasonal character of work performed.[9] Having arrived at the yearly caloric requirements for an entire family of manorial servants (parents and three children between the ages of six and twelve), he proceeded to investigate just how many calories they actually consumed. Heller and Kovács's description of manorial servants' diets were similar: bread and meat (poultry and pork), milk (if owning cows was permitted on the estate), vegetables, and some eggs, with bread and starches dominating (Kovács 1935, 221; Heller 1936, 460). Like their compatriots living in the village, they were more apt to sell or barter with eggs, poultry, and piglets than to consume them:

> Though the calorie value swings between certain limits, on average they have 4–4.5–5 million calories versus the caloric requirement of six million. Therefore manorial servants are deficient in calories, which they can supplement

slightly with goods taken to market, but then there would be nothing for clothes, shoes, etc. . . . Today approximately 5 million people [half of the nation's population] do not eat adequately. Their standard of living is below that of general expectations. In practice and in numbers this means that if everyone would have enough to eat, then nothing would be exported. . . . What we transport abroad for currency and other reasons we deprive from ourselves. (Kovács 1935, 222)

Szabó made the point more simply: "According to these examples there is no way we can call the menu of the wealthier [peasant children] lavish, only more plentiful than that of the poor; there is much more meat but the foods are hardly more varied. Rather their wealth is demonstrated by the fact that they don't run out of bacon [*szalonna*], the child eats when he wants to, meat is not an unattainable dream and the children don't write, as the poorer ones do, that '*I would really like to be fat*'" (1937, 53; italics in the original).

PROPERTY RIGHTS

Promanorial advocates did have to confront the inconvenient truth that manorial servants found their standard of living less than ideal and dreamed of moving into the village. The ardent manorial apologist Ernő Éber did concede that finding themselves at the bottom of the agrarian status hierarchy could have motivated manorial servants to leave the estate (1941, 60), but Éber could not reconcile this attitude with his own view of manorial life. Why abandon security of employment and housing for insecurity? The strange desire to acquire land was misguided, Éber argued, based on the unfortunately mistaken view that owning land promised a life of comfort and independence. "Manorial servants also dream about acquiring property. But this desire for property—a village house, economic support—would serve to free them from worry for their old-age and to offset the danger of being left without work. It is not driven by the ideal of production on an independent farm" (1937, 164). Heller cited numbers from Fejér County that showed that few manorial servants actually chose to purchase land in town during various phases of the limited land reform in the 1920s and 1930s; he considered this evidence that manorial servants did not actually expect much from the redistribution of land (1937, 165–66).

Heller's calculations about nutritional patterns in rural communities proved that property itself was no guarantee of good nutrition. Why, he asked, "is property envied" when the children of manorial servants were better fed than those of villagers owning small farms (Heller 1937, 39)? This discrepancy called into question the entire enterprise of "property redistribution eating up considerable costs," which would increase the number of small farms but not the quality of life among villagers (ibid.). Éber's explanation was less sympathetic. The only reason for this irrational view was the inappropriate draw of bygone days, not the conditions of manorial farming in the present. "Above all we have here the atavistic and the indestructibly powerful idea in agrarian society that wealth and respect are connected to landed property.... It is nearly impossible to completely dispel these motives ingrained in their spirit by improving the material conditions of farm servants" (1941, 60–61).

It is curious to read Éber's claim that desiring property in land (starving for land; *földéhség*) was old-fashioned and out of date in light of the very strong arguments promanorial advocates provided for respecting the rights of landowners to their property. Balás argued forcefully that the nation's prosperity depended on the institution of private property, citing a correlation between citizens' well-being and relations of private property. Moreover, he argued, the benefits of private property didn't just accrue to the owner but in fact to the public good (1941, 430). The possibility of owning property was as significant to the growth of the nation as the distribution of property at any one time, since the freedom to pursue business opportunities itself bred prosperity and well-being. Balás firmly believed that in the absence of private property economic activity, even diligence would wither on the vine. This, he claimed, had a basis in biological and psychological factors:

> Therefore, the diligence and energy of the individual, and the *will* and *inclination* that promote these, can be the only true source of creating wealth. And we may add that biological and physical, that is individual and human reality, is the true source of creating wealth.... Not only is it worthwhile to invest in the most work directly, but also the extent to which there is investment in the most appetite, confidence, optimism, and judgment. And at this point we see most clearly the social, psychological, and historical superiority of a system of private property over a system of collective property. (ibid., 428–29)

Efforts to enforce collective property, as in the Soviet Union, were not the only danger to economic development. More radical plans to redistribute land by simply taking land from the propertied and giving it away to the landless would, in Count Teleki's terms, be demoralizing: "No one can receive a national gift without recompense and at the detriment of others" (*Köztelek*, 4 Nov. 1935).

Threats to private property among advocates for land reform were viewed among social conservatives as attacks on the integrity of the Catholic Church (e.g., Czettler 1995 [1923]). Jenő Czettler, a prominent conservative who wrote frequently on social policy, waged a campaign throughout the 1930s on behalf of the dignity of the Church and its invaluable social role in Hungarian society. The economic order, Czettler argued, could not be built on the basis of individual self interest or the comparable notion of class self-interest, found at the heart of socialism. Brotherly love was the true foundation of the economic order, best illustrated by the generosity of the Catholic Church (1995, 159). The Church protected the poor and the weak, not least against the ravages of capitalism (ibid., 158). In his discussions of landed property and social justice, Czettler tacked back and forth between the Middle Ages and the twentieth century. The current order was built over centuries and provided a model for solving contemporary problems. "And whatever fables the scribblers of the now declining economic system relate about the backwardness and darkness of the Middle Ages, all solutions to today's big agrarian questions were obtained in the economic system of the Middle Ages. No democratic regime has ever distributed as much land, created as many new villages and houses as the Catholic Church" (ibid., 159–60).

Few outside the community of social conservatives found the social tasks of the Catholic Church a convincing argument against land reform:

> Hungarian public opinion and those in Hungarian politics rarely discussed the character, role, and function of this type of real estate without passion. Bourgeois criticism of a Protestant provenance and occasionally from liberals rarely rose above the tangle of prejudices motivated by denominational sensibilities. The Catholic Church and the Catholic spokespersons in public life—insofar as they were forced to touch on this issue—defended the historical necessity and legality of the moral, political, and social status of large landowners. (Gergely 1999, 270)

Even among Catholics there was disagreement over the sanctity of Church properties. Elements of the middle classes and intelligentsia were dissatisfied with the conservative politics of the Catholic Church. Catholic youth groups peopled by eager activists spoke out in favor of land reform, drawing inspiration from the platform advocated by the conservative Bishop Prohászka during World War I (Gergely 1989, 68). The relatively moderate position taken by KALOT (The National Body of Catholic Agrarian Youth Men's Groups, Katolikus Agrárifjúsági Legényegyletek Országos Testület) favoring provisioning of house plots and resettlement in the 1920s and 1930s was abandoned in 1938 when it allied with EMSZO (Parish Workers' Departments, Egyházközségi Munkásszakosztályok) to call for all-out land reform (Balogh 1998, 44–45). This was a brave move on the part of KALOT. "And so a strange contradiction arose: while trying to win the peasant masses for the Church, KALOT could have ended up at odds with the Catholic hierarchy" (ibid., 46).[10]

MORALITY AND THE DEMOGRAPHIC CRISIS

Defenders of manorial estates were not satisfied with a purely economic argument. This would have been a crassly materialist position, roundly condemned by social conservatives (Hartstein 1937, 3). Crucial social questions were also worthy of consideration: the moral life of the peasantry, their ethnic origins, and their role in forestalling the demographic collapse of the Hungarian ethnic community. The demographic collapse of the Hungarian nation preoccupied a broad swath of Hungarian political and religious elite. Hungary was not alone in experiencing a demographic crisis; all across Europe politicians and concerned citizens pointed to the substantial drop in the birthrate (Horn 1994). This was a complex debate, touching on issues of ethnic nationalism, religious persecution, and impoverishment. Writers from the left and the right denounced the growing practice among the peasantry of restricting the size of their family to one child (the "one-child" or *egyke* problem). Explanations for the falling birthrate differed according to one's political views. The left tended to emphasize economic factors, specifically fears amongst poor peasants of endangering the viability of their farms through partible inheritance practices. The right saw the drop in the birthrate as a sign of moral decay: "The paganization of marriage, free love, in other words the slackening of religiosity plays a crucial role" (Heller 1937, 25). For embittered nationalists, the problem of a dropping birthrate in Hungary was exacerbated by the loss of former territories after World War I, leaving the country surrounded by hostile and fertile neighbors (Szeibert 1938, 402; see also Heller 1937, 24). Yet not all foreigners lived outside the boundaries of the new nation, making it all the more important that social policy encourage Hungarian families within Hungary to increase in size. Conservatives consistently claimed that the truly Hungarian (i.e., racially pure) stock was to be found primarily among manorial servants, hence protecting these families—keeping them safe and secure on manorial estates—was of the highest priority (e.g., Heller 1937, 26), so that "It is undeniable that the strata of those employed in agriculture constitute the most significant section of the nation's labor stock, and that the most valuable factors in the regeneration of that nation is the greater fertility of agriculturally employed families" (Szeibert 1938, 401). Promanorial advocates were firmly convinced that manorial servants were primarily of Hungarian ethnic origin. In their eyes, this meant that manorial servants were worthy of greater support *as Hungarian employees.*

Promanorial advocates proudly declared that manorial workers consistently had larger families than landowners (Szeibert 1938, 406), a trend that in itself was worthy of support for all true nationalists. Heller went so far as to say that fecundity constituted manorial servants' true calling (1937, 24).[11] Lajos Thirring, the well-respected statistician, ascertained that the demographic profile of class strata demonstrated that larger families tended to be found among the agrarian population (Szeibert 1938, 406): "According to the table, mothers with many children do indeed figure only in agrarian circles, and primarily among those in this stratum who have need of support due to their circumstances, life prospects, and income" (ibid.). Teleki relied on a similar argument to explain the higher birthrate among manorial servants, that is, that children were able to contribute to the family purse by taking on jobs at the estate (*Köztelek*, 24 Nov. 1935).[12] Éber cited figures from the demographic department of the Central Statistical Office: "The number of *births* in 1932 for the entire population is 2.6 percent. At the same time the number of births of manorial servants numbering roughly 216,000 souls and their 382,000 dependents was 19,801, which equals 3 percent. At the same time the number of births for 1,186,462 earning industrial workers and their dependents was 27,561, which only comes to 2.3 percent" (1941, 60). To refute the claim that manorial estates were responsible for the one-child policy—that they prevented peasant families from acquiring land—Heller cited evidence that if one mapped the incidence of *egyke*, the majority of those areas were not in regions of large estates (1937, 26).[13] Indeed, he saw the problem in exactly the opposite terms: "In this area the sad French conditions are worthy of note, where the years in which the large estate system was annihilated coincides with the decline of the population" (ibid.).

Heller also argued vehemently that manorial servants were protected from the evils of village life: pubs, pride, and idleness (1937, 129). Observers were wrong to believe that "the morass of spiritual topor from villages reaches us here, holding the spirit of the *puszta* in the thrall of indifference to religion born in the last century" (128). Life on the *puszta*[14] was superior, as the families of manorial workers did not suffer the indignities of depravity, nor were they exposed to corrosive political disaffection among the poor or the politics of leftist rabble-rousers (129–30, 135). Sexual mores were better at manorial estates than in villages; far fewer children were born out of wedlock at manorial estates, and marriages were more stable (130–31). Manorial estate workers were less inclined to frequent the barns housing migrant workers, whose women were notorious for their loose lifestyle. Sexual diseases were extremely rare, as was abortion, in contrast to the practices of *egyke* families (130). Heller's depiction of the *puszta* as a world apart was common among promanorial advocates; any sort of unrest or dissatisfaction expressed by manorial servants had to have been sown by outsiders (for example, *Köztelek*, 24 Mar. 1940). The isolation of *puszta* life could certainly account for some of these differences, but not all. Heller argued that, in stark contrast to the industrial worker, manorial workers shared the interests of their employer, so old-fashioned patriarchal authority was not falling by the by (18). In these comments, Heller echoes views commonly articulated in the nineteenth century criticizing the intrusion of the market into labor relations at manorial estates. The market was inhuman, whereas wealthy landowners felt responsible for their employees. As one wealthy landowner explained, "Manorial estates [noblesse] *oblige*" ("A nagybirtok kötelez," *Köztelek*, 13 Aug. 1944). This sentiment was reciprocated by manorial servants, Heller argued. They did not share villagers' rampant anticlericalism and hatred of the wealthy (133–34). "The soul of the estate servant is a pliable raw material easiest to mould, the best soil for moral and religious ideas" (128–129); they were not haunted by "the devil of the Hungarian peasant's soul: pride" (129).

Radicalization and Jewish Property

In the mid-1930s, concerns increased among the political elite about the consequences of landlessness, that is, the radicalization of the poor. Election results in May of 1939 showed a large swing to the right—votes among the agrarian proletariat moved away from the Smallholders and toward the less radical of the two extreme right parties, the Hungarian Life Party (Magyar Élet Párt) (Juhász 1983, 68). For many land reform became a national issue in their thinking, but not in the sense that in populist circles it had been been conceived, that the social elevation and distribution of land to the most Hungarian strata of several million are national issues. Rather, it was because the feeling had arisen that land reform would be the prerequisite for preserving independence and averting the danger of domestic right-wing extremists allying with foreigners. On the other hand, there were those who for the very same reason spoke out against the radical solution, even though they were believers in reform. In this way demanding land reform as "social national defense" increasingly became a slogan, which of course those on the extreme right also manipulated (ibid., 69).

Questions of national security only became more urgent as Europe headed into war. Fears were raised by pro–land reform advocates that a disenfranchised populace would refuse to fight for the nation; others in this camp worried that the existence of large estates made them easy prey to foreign invaders. Kodolányi reported that disillusioned Polish peasants were surrendering to the Soviet army, which quickly took over the abandoned estates of wealthy landowners (Juhász 1983, 73–77). In the eyes of promanorial advocates, such as Scherer, the presence of manorial estates in border regions constituted a guarantee that estate employees would protect the land against foreign invaders, notably the Germans (ibid., 73). György Parragi, a journalist with the right-leaning but anti-German newspaper *The Hungarian Nation* (*Magyar Nemzet*), argued in favor of the status quo, despite his clear recognition that Hungarian society was rift with social injustice. Land reform, he believed, was precipitous and untimely: "To begin the radical reconstruction of the social structure in times such as these would be just like the suicidal endeavor 'of beginning a largescale renovation of our house when a growing storm is over our heads: taking everything out to the yard underneath the strawberry tree, dismantling the old roof, and replacing the old foundation'" (ibid., 75).

Ethnicity and race also figured prominently in the debates. As discussed earlier, promanorial advocates were firmly convinced that manorial servants were primarily of Hungarian ethnic origin. Depriving citizens of questionable racial identity—most notably Jews—of their rights to property was already well underway by 1942.[15] Heller acknowledged this fact when he claimed that the government's agrarian policies had already carved out a significant chunk of land from manorial estates: "The resettlement and Jewish laws plan to requisition one and a half million *hold* of land, from which by the end of 1940 they have distributed 178,000 *kat. hold* to about 35,000 claimants" (1941b, 706).

In some quarters, racial purity was less important than the integrity of agricultural production. Among reasons for dismantling large estates, Heller saw several considerations as "honorable" (*tiszteletre méltó*): Bismarck's treatment of ethnic minorities in Eastern Prussia, spreading the population more evenly across the country, as a means to a healthier distribution of income, and "transferring landed property into the hands of the racially pure" (1941b, 710). But this could not excuse poor farm management or the loss of productive capacity. One could not justify taking over Jewish estates without ensuring that farms producing comparable value be established. By 1942, the Hungarian state had anticipated these concerns. Section 12 of the law depriving Jews of their property stipulated that "the person . . . or usufructuary . . . or renter is required to maintain the property in the existing condition, with the same meticulous care as a proper farmer, at least from the day the present law came into effect, as well as to perform the customary field and forest tasks . . . with the meticulous care of a proper farmer and according to the order in which they had been farmed until the state, i.e. the credit institute or buyer appointed by the Minister of Agriculture, takes possession of the property" (Vértes 1997, 194).

Kerék, a vocal advocate for land reform, also addressed the status of Jewish estates in his 1942 book *The Road to Land Reform* (*A földreform útja*), though with far less enthusiasm than others at the time. "These must end up in the hands of agriculturalists, even if the appropriation is not exclusively motivated by the interests of policies for landed property. We are pleased that we can expand the 'Lebensraum' of our agricultural people with thousands of *hold*, the title of land that we would never have acquired solely for reasons of land policy" (1942, 84). In contrast to Heller, he felt strongly that turning the lands over to small farms would not endanger the provisioning of supplies to the nation (ibid.). Unfortunately for Kerék and his allies promoting land reform, the law prohibited any move to dismantle estates. Recognizing the resistance parcellization faced, Kerék bluntly asked, "Yet where are all those reliable Christian (large estate) renters who would be needed to lease the substantial numbers of Jewish estates. After all, those estate owners who—sooner or later whether they want to or not—will be required to give notice to their Jewish renters are already having problems with the question of replacement" (ibid., 87). As an alternative, Kerék proposed a system where public welfare cooperatives (*közjóléti szövetkezetek*) could become the entity renting an estate, keeping the workers employed and the farm going. While considering this a transitional form to be replaced eventually by private ownership, he also saw his plan as a way of training workers to become farmers on their own (ibid., 86–89). Needless to say, this did not come to pass, at least as long as the rightist governments were in power.

Conclusion

Moral imperatives fed the debates about how to improve farming in Hungary. Potentially simple questions like economies of scale were mired in age-old disagreements about rights to land and fair use. Though opponents readily sought quantitative evidence to support their positions, the numbers could not resolve their disagreements because they started from different assumptions. Wealthy estate owners considered the writings of leftist sociographers to be the irresponsible scribblings of outsiders who lacked any connection to the communities that had been entrusted to the manor's care for generations. Advocates for the poor bristled when shown clean calculations of caloric intake; they had heard village schoolchildren dream of being fat one day. Agrarian economists had an alternative vision: wipe the slate clean of musty old manors and backward peasant farms. Peasants unable to compete in the modern marketplace would join their poor brethren as wage laborers in the truly modern farm where efficiency, productivity, and profit would be guaranteed by enlightened management professionals. Science would save the day.

In the 1920s and 1930s, scientific management served many masters. As Rabinbach explains, "the vision of a society in which social conflict was eliminated in favor of technological and scientific imperatives could embrace liberal, socialist, authoritarian, and even communist and fascist solutions. Productivism, in short, was politically promiscuous" (Rabinbach 1990, 272). And as we shall see, the Communist Party was quick to adopt scientific techniques that were designed for a very different politics. The undiscriminating character of scientific management should not blind us, however, to the political commitments agrarian economists and work scientists embraced in the interwar years. Theirs was a "reactionary modernism" (Herf 1984), a social imaginary rife with steep social hierarchies and a conservative moral agenda. Scientifically calibrated wage systems would preserve privileges of the elite by improving their economic fortunes, or at least, this was the argument being made by work scientists to win over those skeptical of innovation. It is to these arguments we now turn.

2

Rationalizing the Economic Infrastructure

"Scientific labor organization." I know. It sounds rather strange at first glance. But we'll get accustomed to it!

KÖZTELEK, 18 Sept. 1927

Scientifically engineering agriculture was an audacious undertaking. Conceptually foreign to estate managers and workers alike, work science arrived in the Hungarian countryside unbidden, ushered in by a small coterie of agrarian economists. Increased labor output, improved efficiencies, and higher incomes were promised. To embark on this road, however, required a progressive transformation of minds, bodies and communities few would willingly embrace.

Following on the heels of German innovations in business management and cost accounting, Hungarians committed to modernizing agricultural production mounted a sustained campaign to establish fields of study and institutions of learning to make these changes possible. Efforts to create university departments and develop college curricula to hone the expertise needed to retool and rationalize agriculture bore fruit by the 1930s, but the broader campaign promoting economics as legitimate science and business management as honorable engineering was far less successful. While quick to blame intransigence among farmers, experts underestimated the substantial commitments in time and energy, much less capital, that these innovations would entail. Divisions within the agrarian business community over what knowledge mattered—played out in classic boundary squabbles over the integrity and status of various disciplines—slowed progress, as did the conflicting interests of those with expertise and those with property. The chapter begins with a discussion of building infrastructures for business management, followed by a brief history of agrarian scientific management in Europe, and proceeds to a discussion of the scientific aspirations of work scientists and agrarian economists and their efforts to build an academic and government infrastructure to support their work. The use of numbers in studies and reports on agrarian modernization is next, and the chapter ends by describing the specific problems posed by accounting and bookkeeping. Chapter 3 will be devoted to the practices of agrarian work scientists in estimating the value of labor.

Building an Infrastructure

A central claim of this book is that the role of markets in setting the value of labor in twentieth-century capitalism has been exaggerated. The reason for this, I argue, is that this approach completely ignores the tremendous amount of time, energy, and money invested in a range of entities and practices that must be created and sustained to ascertain a reasonable approximation of the value of goods, labor, and services in an economy. Just what sorts of practices and what kinds of entities were required at any historical moment is an empirical question. In the following chapter, I identify four domains agrarian work scientists and agricultural economists saw as requiring immediate attention in Hungary: (1) recognizing economics as a scientific discipline; (2) creating institutions to train economists and estate managers, as well as advance research; (3) establishing a network of state agencies devoted to collecting data on the economy, and agricultural production in particular; and (4) promoting the use of bookkeeping in business.

Pleas directed to state authorities and private businessmen in Hungary for improved business practices often pointed to the success of laudable policies in Germany, Switzerland, and the United States as evidence that changes needed to be made to remain competitive in the world economy. It is important to note, however, that despite the widespread fascination with scientific management and business economics worldwide, a point I will discuss in the next section, whether and how policies were actually implemented remained an open question. As Mary Nolan points out in her excellent book *Visions of Modernity: American Business and the Modernization of Germany* (1994), the hype surrounding Fordism in the press and professional journals in 1920s Germany was not matched by similar enthusiasm in management offices or union halls when it came to implementing changes in the organization of labor or management. Nolan discusses a variety of debates waged in the mid- to late 1920s over the feasibility of adopting American business practices, for example, the absence of sufficiently large markets to grow consumption that Fordism relied upon. A more important consideration was the widespread conviction that German workers were craftsmen of refined sensibility, whereas Americans were ignorant and unskilled (Nolan 1994, 74, 84–85). But what actually was done in the name of "rationalization," as it came to be called in Germany, was another matter. Nolan quotes a disappointed doctoral candidate who bemoaned the state of affairs in the 1930s: "'Let's think about all the exaggerated hopes and illusions that the slogan of rationalization stirred up! And what remains on close inspection? Basically, precious little . . .'" (1994, 227). In the worsening economic climate of the 1930s, efforts to rationalize factory production slowed dramatically, while institutions established by the German state, such as the National Productivity Board and the German Institute for Technical Labor Training, hobbled along as best they could. Nolan argues, in fact, that the failure of rationalization to live up to its promises of rapid modernization, improved living standards, and a congenial accommodation between owners and workers set the stage for alternative solutions to be considered:

> The Nazis were the dual beneficiaries of the rationalization movement. . . . [t]hey were aided by the paralysis and demoralization of the workers' movement, to which rationalization in no small measure contributed. On the other hand, the Nazis adopted many of the ideas and policies of the rationalization movement, especially in the area of company management. Their antimodern rhetoric masked distinctly modern and rationalized forms of work organization, intensification, and wage policies. (ibid., 232)

It is important, therefore, to consider carefully whether and how rationalized, scientifically engineered business arose, since the conditions making this possible—such as the Nazis' banning of organized labor—did not and could not hold in many other contexts.

Another consideration is the role of various agencies in advocating for change, for example, professional associations, government bodies, or labor organizations. As we shall see, Hungarian "boosters" of agricultural economics called on the state to intervene on their behalf, believing that proper policies would trump the indifference shown by wealthy landowners to changes in managing their farms. Demanding state resources be deployed to their ends was also the strategy among Hungarian economists rallying allies to found an economics faculty at the university. Relying on the state to solve economic problems was common among the conservative elites in Hungary. This was not the case elsewhere. As Tomlinson explains, in a fascinating article on the post–World War II policies of the Labor government in Britain, private businesses balked when called upon to alter business practices in order to serve the state's need for data on productivity. They did so even when the government had consulted employers' associations in the course of developing policy. "[E]mployers were largely able to channel and resist the government's initiatives, especially insofar as they threatened managerial autonomy.

The reasons for such opposition were partly political—a desire to resist what was widely seen as the excessive ambitions of a meddling government" (1994, 175). Another reason for their reluctance to comply was the flourishing private sector in the late 1940s, giving them little incentive to invest in changes when profit was there for the taking (ibid.). In Hungary, resisting change was also the case when economic fortunes were endangered in the midst of the depression; scarce resources would not be expended on unproven management innovations.

Modernizing the infrastructure for business made great demands on the material and social conditions of office work itself, a less obvious but equally significant factor for business owners. Yates argues that as American businesses increased in scale and complexity in the latter half of the nineteenth century—epitomized by the growth of railroad companies—the control of information and personnel within the enterprise became a crucial consideration.

> [T]he informal, incidental, and primarily oral communication within the traditional small firm could satisfy neither the requirements of systematic management nor those of later attempts to humanize the workplace. More formal and systematic modes of communication, primarily written but also including documented oral communication, were essential. Regular flows of upward, downward, and lateral communication as well as detailed record-keeping procedures played a critical role in the new "systems." (1989, 2)

A short list of new machines and procedures introduced over the course of fifty years suffices: telegraph, telephone, typewriter, mimeograph, adding machines, dating stamps, paper clips, filing cabinets, manila folders, and photocopying machines, not to mention other, less familiar innovations such as the hectograph and Schapirograph designed to duplicate massive numbers of documents in house (ibid., 21–64). These new gadgets brought about serious changes in the way business was conducted, introducing efficiencies but also raising questions about legitimate concerns of dating and authenticating documents. In the United States debates raged over adopting carbon paper in place of the press book, a means of creating copies of correspondence that had been widely used in the nineteenth century. Opinions differed on what constituted a legal copy of a document: one with all corrections and signature or the copy made while the original itself was being typed. The Taft Commission on Economy and Efficiency eventually requested legal authorities decide the matter (ibid., 48–49).[1] In Germany, a debate ensued when businesses attempted to alter the way in which accounting books were being kept. Traditionally bound accounting volumes were being replaced with a system of free-standing pages. In his treatise on the German science of business management, András Schranz notes, "After heated debates lawyers and experts in accounting resigned themselves to the use of free-standing pages; the practice is barely tolerated in the legal community, while German commercial law has yet to recognize it" (1930, 122).[2] In other words, demands to change business practices required huge investments for business owners: the need to recruit new personnel, buy new equipment, and, most importantly, spend the time it took for new procedures to be incorporated into the daily routine. Solomons (1968) relates similar problems faced when cost accounting was being promoted in the United States during the 1920s and 1930s. One might say that, in the view of business owners, adopting cost accounting was not cost effective. In the minds of business scientists, however, it was considered an absolute necessity.

International Business and Agrarian Scientific Management

> The rationalization movement is spreading across the world like a new gospel. To rationalize is the slogan of today's economic life, which conquers and forces itself into the most isolated corners of the world.
> RAITH 1930, 248

The history of scientific management in European industry is documented in numerous studies (Maier 1975; Nolan 1994; Guillén 1994; Rabinbach 1992). Contemporary bibliographies are also available to augment these synthetic accounts: "There is no country in Europe in which studies have not been published on problems of industrial organization; even in the Tartar language such studies are found" (Devinat 1927, 10). Devinat discusses "original contributions to the [scientific management] movement" in "the advanced industrial countries of Europe," a list that includes fourteen countries (ibid., 74).[3] What is less well known is the move toward scientific management in agrarian businesses. A brief report, "Scientific Management in Agriculture," was published by the International Labor Review to acquaint its readers with what it calls "the first really international discussion on [the application of the principles of scientific management] on the subject of agriculture" held at the Thirteenth International Congress of Agriculture in 1927. Devinat specifically mentions agriculture as a focus of scientific management in Czechoslovkia (Institute for the Rationalisation of Agricultural Labor), Finland (the Finnish Society for Scientific Management in Agriculture), France (Commission d'étude de la Confédération nationale des Associations agricoles), Germany (the Union for the Improvement of Technical Methods in Agriculture), Great Britain (Agricultural Economics Institute, Oxford), and Russia (Station of

Normalisation Experiments in Agricultural Labor) (Devinat 1927, 57, 74–90, 216–22). Other institutes conducting research in agriculture include the Czechoslovak Academy of Agriculture (ILO 1930, 855), the National Institute of Industrial Psychology in England that devoted attention to studies of picking bush fruit (ILO 1927b, 703), and the Central Office of Farm Accounting in Poland (ILO 1926, 56).

Government officials, private businessmen, and engaged scholars across the continent went in search of new ideas and better practices.[4] This was just as true of specialists in agricultural production as of industrial magnates and business elites. Germans and Soviets traveled to the United States to study farming methods, while American farmers went to the Soviet Union to observe collectivization efforts (Aereboe 1930; Bailes 1981; Brinkman 1925; Fitzgerald 1996, 2003). The International Labor Office in Geneva kept track of innovations in the rationalization of farm labor (see ILO 1927, 1930, 1932), while the International Institute of Agriculture based in Rome promoted scientific management strategies (Macara 1926, 73–188).

The kinds of farms studied by business specialists varied in size. Analysis of labor costs in agriculture were conducted on smaller farms (from 50 to 450 acres), as were found in Cornwall and Devon in England and Illinois, US (ILO 1928). Devinat suggests that, at least in Central and Eastern Europe, postwar land reforms were a powerful stimulus for implementing scientific management: "[A]s a result of the Agrarian Reform involving the breaking up of large estates, the necessity of introducing scientific farming methods for the purpose of avoiding a reduction in agricultural output has been very widely realised" (Devinat 1927, 77). While no doubt true, this estimation overlooks other sites for scientific management in agriculture, neglecting other economies of scale with different labor needs. This is demonstrated in examples of labor studies targeted specifically at large farms. Within Europe, Germany was touted as the only place where "detailed studies . . . have been undertaken. . . . Germany is one of the countries having the best chance of obtaining good results in this direction; the idea of scientific management will always be most applicable on large-scale farms and in such farms Germany is especially rich" (ILO 1927a, 380). Hungary also fell into this category. Beyond Europe, a similar picture emerges. An article appeared in the *International Labor Review* describing "An experiment in the Management of Indian Labor," conducted by Albert Howard, the Imperial Economic Botanist who was both the Director of the Institute of Plant Industry in Indore and Agricultural Adviser to States in Central India and Rajputana (1931). As the editors explained, "[o]ne of the outstanding problems of the present phase of colonial development in Asia and Africa is that of the best and most scientific methods for the organization of work in large-scale agricultural undertakings" (ibid., 636).[5]

The primary inspiration for Hungarian agrarian modernizers was the German academic community and initiatives taken by the German state and agrarian interests to reform business practices and develop reliable data collection methods for statistical compilations. Discussions of German policies recur frequently in the literature, and as models of success and emulation, German sources dominate (Aereboe 1920, 1930; Laur 1930; Ries 1924, 1930). Experiment stations at Pommritz and Bornim, where new techniques of rationalization were tested, were frequently mentioned. It is not surprising that the work of German academics would have such a firm hold on Hungarian economists and work scientists, since the ties between German and Hungarian academic and artistic communities were deep and long-lived. Hungarian agricultural economists and work scientists were also well informed about other innovations in agricultural modernization, such as Mussolini's Agrarian Programme (Heller 1941a) and Soviet collectivization (Fellner 1933; Jánossy 1933; Kovács 1940; Rézler 1940). In the midst of debates over the character of agrarian education and state policies toward the agrarian sector, Siegescu penned a series in 1925 in Köztelek (Commons)[6] on the establishment of the Department of Agriculture and the workings of land grant universities in the United States (22 Jun. 1925; 9 Jul. 1925; 13 Sept. 1925; see also Ferleger and Lazonick 1993, 1994 on land grant colleges and the managerial revolution in the United States).[7]

The Needs of Infrastructure

Agrarian business economists and work scientists were well aware of the enormous barriers to entry their plans for modernizing production created. Their campaign to improve the fortunes of agricultural big business had to be fought on many fronts: appealing to officials in the ministry and politicians in other branches of government; conducting research experiments and training young scholars to take the lead; and promoting their views in public lectures and conferences. Issue after issue of the Commons (Köztelek), the weekly newspaper published by OMGE for wealthy landowners, contained editorials admonishing short-sighted politicians and apathetic landowners; weekly columns were devoted to the promise of business science and the value of bookkeeping. Restructuring production would require great demands on public resources, but those pushing for modernization saw few other options.

CHAPTER TWO

CREATING A DISCIPLINE: BUSINESS AS SCIENCE

The impact of engineering and scientific advances on industrial production in the late nineteenth century can hardly be overestimated: harnessing electricity, improvements in mechanization, innovations in telecommunications and transportation. Innovations in agricultural tools and machinery shared in these developments; just as significant for the agrarian sector were crucial advances in chemistry (soil science, fertilizers, irrigation). Age-old practices of cultivating new strains of crops and breeding animals were accelerated by advances in agronomy and physiology. In contrast, studies of the workings of the firm, commercial relations, accounting, and management studies remained a small subset of studies at the most influential schools for manorial estate managers well into the 1930s. This was no longer acceptable to the growing community of business scientists and agrarian economists. "It is odd that, in this country, the study of questions in agriculture relating to economics have scarcely received any attention, in contrast to questions of the natural sciences" (Károly 1925, 2). Business scientists, and agrarian economists more generally, constantly fought against the strong prejudice among manorial estate managers that the success of an enterprise lay solely in the skilled use of agronomy and its related physical sciences. Agricultural management had to be recast as an industry subject to the same economic laws as other branches of the economy: "The fact is that in the field of agrarian politics and especially agricultural economists the current situation of economic research is not satisfactory. . . . We as managers must constantly insist that the *economic part* (strictly speaking) of agriculture is also science, and it deserves and demands every possible means and instrument for it to thrive in the interests of agrarian Hungary"(Kesztyűs 1929b, 49; italics in the original). The success of agricultural production hinged on changing attitudes about the relative significance of technical versus economic considerations. "The most important question of agriculture production is undoubtedly not technical, but one of business management, by which we mean *the modern study of firm development,* i.e. the organizational issues, sales, revenue, questions of production, in conjunction with questions of taxation, credit and customs policy, to wit, agrarian policy" (Kesztyűs 1929, 15; italics in original). So important was the science of business that Kesztyűs—following the well-respected Aereboe's lead in Berlin—advocated reducing the number of courses in the natural sciences required for a degree at agrarian high schools and technical colleges, in favor of economics and firm organization (Kesztyűs 1932, 128).

The indifference of Hungarian farmers to innovation and improvement was a recurring theme in *Köztelek*. Writing on the fortieth anniversary of its publication, the editor, Barna Buda, cited the pervasive indifference of the agrarian community to innovation and change as one of the central battles waged since the inception of the periodical. Whether it be the self-satisfied aristocrat living comfortably off the manor's proceeds or the ignorant farmer set in his ways, the result was the same: intransigence. Not mincing his words, László Tokaji Nagy blamed false pride and age-old prejudice: "[Hungarians] consider business activity and management unworthy of them. Neither deep contemplation nor much logic is required to figure out that the aversion to a business career we have long fostered is primarily responsible for the impossible material situation we find ourselves in now" (*Köztelek*, 20 Nov. 1930).

A crucial element of the conceptual development of business management was the shift in perspective concerning the dynamic properties characterizing the unit of analysis: the firm or business (*üzem*). This analytic innovation was frequently illustrated by organic metaphors, although in and of themselves organic metaphors were not new. Árpád Hensch, the preeminent business economist in agriculture of an earlier generation, used this trope in his 1901 treatise on "the study of management and equipment": "the normal operation of the human organism is premised on the operation of various organs and their proper relationship. The smallest deviation from this affects the organism unfavorably. So too in the organization of a business firm the process of production can only proceed unhindered if the productive requirements are guaranteed balanced collaboration" (1901, 171). While Hensch recognized the interconnectedness of various elements of the system, the emphasis in post-World War I writings increasingly became the dynamic qualities of the system. "The statistics of individual firms must be dynamic, that is, they must reveal the movement and influence of economic forces. The statistics comparing several similar firms, however, must be biological, so that the internal life of the firm and its developmental capacities be demonstrated" (Kesztyűs 1943, 4). Analogies between the moving elements of body and the firm abounded: "Working capital is the circulatory system of the economy" (Világhy 1930, 13); "The more regularities we succeed in recognizing, the closer we come to a complete understanding of the nature of the firm" (Lautenberg 1933, 29). Indeed, the firm itself was fetishized. "An agricultural firm is a living organism so it reacts to everything" (Reichenbach, *Köztelek*, 29 Sept. 1940). Or, in the words of Sagawe, an agrarian management studies specialist in Germany: "the firm must be understood as a whole, as a living organism, before the work of detail advising is begun" (Kesztyűs 1942, 5).

Recourse to organic tropes served more than simply allying the study of business with the authority of the physical sciences. It also gave the business scientist as scientist an important role in public affairs. Bearing the mantle of science—having a privileged relationship to crucial knowledge—carried important obligations: treat, and perhaps cure, the economy.

> Therefore with calculation once again we tie the branches of the firm together; the strict accounting break down of firm branches does not interfere with the full picture of the organic interconnections of the firm.... When the lungs of a human body are sick, the doctor won't cut out the lungs, because he knows that then the entire body will expire. On the contrary, he will heal the sick lungs, and so heal the entire human body.... For experts, however, it is necessary to attend to branches of the firm, just as the doctor examines the sick part or even the individual organs of the healthy man in the scope of the entire body. (Kesztyűs 1932, 146)[8]

While one could consider innovations in business management to be a simple consequence of the sustained analysis of the systematic features of the factory or farm, this would underestimate the degree to which rationalization of business methods was allied with broader conceptions of planning and collective organization in the period. Frequent allusions to the varieties of state intervention essayed across the globe make clear that the relative success of different approaches drew attention but, of equal importance, also heightened concerns among agrarian economists and wealthy landowners about the viability of a less than effective agrarian policy at home. "For forty years I have dealt with agricultural economic questions in theory and in practice, and I declare that life has never laid more difficult tasks before domestic agriculture to solve than now . . ." (Ormándy, *Köztelek*, 22 Dec. 1940). His confidence in the scientific development of his field held firm. "In its current state of development, the economic branch of agricultural science is suited to solving the questions arising from managed agriculture expertly and without disappointment" (ibid.).

BUILDING A UNIVERSITY

In the first decades of the twentieth century, increasing pressure was being exerted on the government by a wide range of interest groups to establish a university devoted solely to economics. Within the agrarian community, these demands were part of broader debates waged among elites—wealthy landowners, manorial personnel, and school teachers—concerning the proper distribution of advanced theoretical training and practical instruction for farming at different levels of the existing educational hierarchy: middle school, high school, and technical schools. The founding of the School of Economics (within the Technical University) in 1920 did not satisfy all par-

RATIONALIZING THE ECONOMIC INFRASTRUCTURE 59

ties, but it went a long way in promising improved training for the next generation of leaders. "Up to this point, higher education in agriculture, commerce, and industry was placed on separate altars, but now we have been given a common temple, with common priests. Let's hope that the students will also unite in the common goal of building the economy" (Buday, *Köztelek*, 3 Jan. 1920). Elevating business and economics to serious fields of study also entailed convincing large numbers of well situated families that these were respectable occupations. "... [T]he middle class must break away from the notion that holds commerce to be an inferior career and that the sciences of commerce are only 'artfulness'" (*Köztelek*, 14 June 1925). A united front of economic specialists was required to break the monopoly of the legal profession on positions in government administration. "The general mission of the Economic Faculty is, in the face of the by now unjustified dominance of pure legal training, to introduce an economic mentality based on the study of the laws of practical life and for leading men to make it their own" (Buday, *Köztelek*, 6 Nov. 1920). But clearly more was at stake in educating governing elites about economic processes than simple professional rivalries. "The conditions compel us now more than ever to place our economic life on a more scientific basis, more accurately, to employ sciences in the service of economic life" (ibid.).

STATE RESOURCES FOR TRAINING AND RESEARCH

In tandem with attempts to establish economics as an academic discipline, advocates of business management demanded further resources be devoted at the national and regional level to build a research and training infrastructure, new institutes of agricultural economics and research stations. Agricultural economists claimed that mastering the theoretical tools of agricultural economists required far greater investment than in other agricultural sciences, where long experience in farming was often sufficient (Kesztyűs 1929b, 52). Valuable strides in research also required investment in proper tools. "Germans refer to estate management as the *microscope* of the economy, and consider the estate management center to be the laboratory of economics or business management, that is, the site where many such microscopes are collected" (Kesztyűs 1943, 3; italics in the original). In 1920 the Hungarian Ministry of Agriculture established the National Agricultural Business Institute (Országos Mezőgazdasági Üzemi Intézet), modeled on comparable national institutes in Austria, Denmark, Germany, Italy, Norway, and the United States (Károly 1925, 2–3).[9] These were lean years for founding a new research institute: "at the worst time, during the most adverse financial conditions"

(Károly 1923, 3). Nonetheless, the new institute was cause for celebration. OMGE had been lobbying for it for more than ten years. In *Köztelek*, the new institute was acclaimed as "the new home for scientific exploration, a newer springboard for the progress of Hungarian agriculture" (ibid.). Dr. Károly Rezső, ministerial counsellor and proponent for introducing Taylorism into agriculture, was named its director. The committee overseeing the institute's activities was composed of representatives from the Ministry of Agriculture, OMGE, the Hungarian Farmers' Association (Magyar Gazdaszövetség), and the National Chamber of Agriculture (Országos Mezőgazdasági Kamara) (*Mezőgazdasági Üzemtani Közlemények* I.évf., 1. szám 51). Károly envisioned a broad mandate: to deal with all questions that straddled the boundary between economics and agriculture, that is, issues relating to farm equipment, handling stock, appraisal, accounting, leases, profitability, and firm statistics (ibid., 2). Dedicated to gathering information, conducting research, and widely distributing their findings, Rezső promised to take an impartial and unbiased stand on all issues the institute addressed (ibid., 3). The institute published a journal from 1923 to 1926 entitled *Proceedings of Agricultural Business Studies* (*Mezőgazdasági Üzemtani Közlemények*) with a number of articles on Taylorism in agriculture and manorial accounting (V. Fodor 1925; Károly 1924, 1925, 1926). Another journal, published between 1929 and 1934, *Agricultural Business and Estate Management* (*Mezőgazdasági Üzem és Számtartás*) continued in this vein, focusing much attention on the German rationalization movement (Balogh 1930; Kesztyűs 1929; Rege 1929; Schranz 1930; Vutskits 1934). The National Association of Hungarian Estate Managers (Magyar Gazdatisztek Országos Egyesülete, or MGOE) regularly held seminars for the benefit of their members wishing to keep up with the most current research in agronomy, animal husbandry, and business science. In the 1940s, another research center was established by the Ministry of Agriculture, the National Institute for the Study of Agricultural Organization and Production Costs (Országos Mezőgazdasági Üzemi és Termelési Költségvizsgáló Intézet). The new focus on cost analysis and accounting procedures demonstrates the increasing centrality of these concerns to state agents and researchers. Finally, in 1943–44, the Agricultural Work Science Institute (Mezőgazdasági Munkatudományi Intézet) was founded at the Horticultural University in Budapest; a journal of the same name was also published.

Agrarian business economists published regularly in the house journals of research institutes. A number of prominent figures in this field also published regularly in journals read by estate managers, such as the *Hungarian Farmers' Review* (*Magyar Gazdák Szemléje*), *Agricultural Bulletin* (*Mezőgazdasági Közlöny*), and *Köztelek*. Heartfelt editorial comments, discussions of current events, and book reviews all provide valuable insights into their agenda. A regular business economics column (*Gazdasági üzemtani rovat*) appeared in every issue of *Köztelek*. Agrarian economists also published in journals read by a broader section of the Hungarian intellectual elite, such as the *Hungarian Review* (*Magyar Szemle*) and *Economic Review* (*Közgazdasági Szemle*). Articles in the *Magyar Szemle* addressed questions of agrarian labor problems, social research in the countryside, and the future of peasant society; articles in the *Közgazdasági Szemle* were more consistently professional in tone, illustrated with tables and graphs, as was common at the time.

Experiment stations were established, usually at technical schools, such as Magyaróvár. "Working up bookkeeping results and data, setting up agricultural bookkeeping offices, and establishing agricultural economics institutes represent the kind of turning point in agriculture as the use of a scale did in chemistry, which only reached its high level by this means" (Károly 1925, 3). Scientific advances and the growth of experimental methodologies required the expansion of technical school training (Faber, *Köztelek*, 10 Nov. 1927). The requirements of various branches of business management varied, according to the manner in which data had to be collected and systematized. Careful studies of workers' movements or time trials with new tools could be conducted at school farms; that was not true for the more encompassing field of agricultural economics. "For agricultural economics the entire country is the area of examination and research. *Experimental farms that are absolutely indispensable for the curriculum of productive technical sciences* are suitable for *research on work*, but not from the point of view of research in the study of agricultural economics, organization, bookkeeping (calculation of revenues), appraisal, and marketing" (Kesztyűs 1932, 127; italics in the original). With sufficient data available, collected at centers throughout the country, the health of agriculture could be determined. "The operation of a 'number center' (*számközpont*) is perhaps best compared to the activities of a doctor. They prepare an accurate, reliable diagnosis of the current condition, difficulties, and problems of agriculture, that is, they search for the means and tools to determine and resolve the obstructive and helpful factors of agricultural production" (Juhos, *Köztelek*, 21 Oct. 1934).

Within the broader agrarian community, calls for investments in scientific research and training were treated skeptically. Battles over the relative value of abstract knowledge and practical experience were fought consistently throughout this period. Even the minister of agriculture, Mihály Teleki, found abstract knowledge inadequate. In a speech he gave in 1930 on practical advice concerning surplus production, he argued that "in the field of agriculture we don't need grand science, but rather the known tools of proper and intensive agriculture need to be deployed with diligence and perseverence" (Teleki, *Köztelek*, 23 Mar. 1930). Business interests had little patience for abstract theories; applied knowledge was far more valuable. Among estate managers actually running farms we find some with university degrees and forward-looking managers with technical certifications. But we also find the fairly skilled but less ambitious manager, as well as those far better described with the old-fashioned term farm steward or, more colorfully phrased, the "steward armed with a stick (*botos ispán*)," the belligerent and authoritarian boss whose sole concern was discipline, rather than improving productivity or introducing new crops. Simon Polich, a licensed estate manager, wrote an essay in the agricultural economics column in *Köztelek* targeted at the faculty of agricultural colleges. He was intent on reminding the faculty that 99 percent of those who enrolled in their classes had no plans to become a scientist but did so to earn an income.[10] They would do so only if properly trained in accounting, a skill many of them would have to pursue as individual farmers. With the exception of large estates, where the division of labor was highly developed, no separate accounting office would be kept at most farms. And even there, "above all the manager of an estate office must first and foremost promote income generation by means of the data he has worked up, and only secondarily should he be an accountant, which means auditing money and materials" (Polich 1930, 115).

OCCASIONAL ALLIES

Wealthy landowners tended to regard estate managers as their allies in promoting large farms. For example, Baron Miklós Vay, president of OMGE and member of the upper house of parliament, congratulated the farm stewards' organization in 1940 for not "los[ing] heart, when the Zeitgeist shook the financial basis of large farming and with this the fate of the staff of farm stewards. Instead [the community] advocated the principle that we can replace the decreasing size of large estates with greater knowledge and more work" (*Köztelek*, 11 Feb. 1940). Estate managers did not shy away from representing themselves as loyal servants, when it served their purpose. In the dark days of social unrest following World War I, for example, estate managers protected estates from total destruction, as the association was quick to point out. "The profession of stewards often warded off the passions of the October Revolution bodily and so acted as a sea break in the face of the flood that was unleashed against castles and country houses by a deliberately provoked crowd ... no doubt the behavior of our profession of stewards has been impressed upon the minds of our farm owners" (*Köztelek*, 8 May 1920). As agricultural policies were being formulated by bureaucrats after the Soviet Republic was defeated by conservative forces, the association wished to remind members of the government and parliament of the role of managers as "protectors of the nation" (ibid.).

Yet the interests of professional managers and wealthy landowners did not always coincide. Estate managers had long insisted on their own association representing their interests, despite the strong antipathy wealthy landowners held for anything resembling a union (Faber 1932, 85). The association pledged to assure landowners that an independent association of stewards would strengthen the ties of landowners and managers, who were, after all, dependent on each other (ibid.). Faber, a leading light in MGOE, was not shy in expressing his displeasure with the structure of agricultural administration, notably the fact that the regional agricultural chambers were designed to represent the interests of landowners alone. "It's undeniable that a portion of stewards feel a certain animosity toward the agricultural chambers... because they are offended by the fact that they do not sit in the chambers either in proportion to their numbers or to the importance of their careers.... *Forming the Agricultural Chambers without stewards is incomplete and an enormous mistake, in which the most important and knowledgeable productive stratum is not represented*" (ibid., 54–55; italics in the original). Occasionally the state stepped in to side with the professional interests of estate managers, as when it attempted to widen the purview of agricultural expertise in a proposal put forth by the Ministry of Agriculture in 1927 to require all farms to hire a licensed manager. "The minister wishes no less than that in a decade every patch of Hungarian property be in the hands of qualified men" (*Köztelek*, 18 Sept. 1927). This measure did not pass, leaving Law XXVII from 1900, which stipulated that only estates publicly owned, entailed estates, and estates rented by banks were required to maintain a licensed manager. Indeed, the decades-old national network of consumer cooperatives did not employ one farm manager, and the Ministry of Agriculture continued to prefer to hire those with legal training rather than with professional training (Walleshausen 1993, 142). According to Faber's calculations, this meant that in 1932, more than half (56 percent) the estates over 1,000 cadastral holds (570 ha) were not required to employ a licensed manager, and even those that were required to do so frequently ignored the law (1932, 30). "Our biggest problem is 'the aversion of our property owners to intelligent estate managers [gazdatiszt] with proper training'" (Walleshausen 1993, 126). "It's not the swineherd who's been promoted to farm manager who's at fault—he would be crazy not to grab the chance. It's the farmowner's fault for making an uneducated person a farm manager at his own expense and that of public production" (*Köztelek*,

26 May 1935). Beyond the personal affront Faber felt when his professional integrity was questioned, there were also other considerations, namely the viability of farming. Without trained personnel, "the landowner [was] working against his own interests" (ibid.). In 1941 Faber was able to determine, on the basis of the national census, that approximately 50 percent of estate managers were licensed (Faber 1941, 11).[11]

The divergent interests of estate managers and landowners were most evident in relationship to the ways in which farms should and could be organized. As managerial staff, the interests of estate managers were far less tied to the institution of private property per se, at least in the form of privately owned manorial estates. Their concern was to continue being employed in their profession. Many alternative places of employment were identified: "in various types of cooperatives, model farms, experimental zones, organized with positions in farm management, [for] itinerant teachers and expert advisors" (Walleshausen 1993, 142). In every instance, estate managers would be able to advise small farmers, effectively deploying their knowledge to improve production and profitability. When the curriculum at Mosonmagyaróvár, the preeminent agrarian technical school, was revised in 1935, cooperative accounting was added to the list of requirements (ibid., 149). As a means of providing practical experience in running a cooperative, the school established the Avaria cooperative in 1932. Students began by starting a poultry department and a department devoted to protecting and caring for fruit trees (ibid., 143). In the following year, an entire conference was devoted to cooperative farming by the Farm Association at the college.[12] In his speech opening the academic year 1931–1932, the principal of Magyaróvár, Gábor Groffits, encouraged the student body to pursue these skills. "New knowledge prepares students, when they end up in jobs as farm advisors . . . that they will become the expert leaders of the village. That is why it is important that students here at the academy acquire the ability to communicate with people, and learn how to present their ideas in a way that is understandable by all" (ibid., 138). As the specter of the land reform loomed ever larger during World War II, estate stewards became even stronger advocates for cooperative farming, concerned—as were their colleagues in agrarian economics—that the land reform would result in extreme parcellization, and so endanger the viability of agricultural production overall.

GATHERING DATA

The fields of work science and agrarian firm studies increasingly relied on the use of numbers to represent processes and facilitate comparisons over the course of the early twentieth century. Conveying complex processes in farming did not require numbers; after all, there was a long history of publishing specialized agrarian treatises dating back to the early seventeenth century. János Lippay published a "perpetual calendar" in 1662, Márton Szentiványi's *Curiosiora et selectiora variorum scientiarum miscellanea, 1689–1709* was published posthumously in the mid-eighteenth century, while *De re rustica*, a treatise specifically on Hungarian agriculture by Mátyás Bél, saw the light of day in 1730 (Kosáry 1983, 185). In the latter half of the nineteenth century, however, offices and associations devoted to statistical compilations started to appear in Hungary, in scientific and government communities. The Central Statistical Office came into being with the founding of the semi-independent Hungarian state in 1867. In the latter half of the nineteenth century voices were raised in support of gathering solid empirical materials. Calls were made for the state to promote a broad decentralized network of farmers clubs and circles (the farming community, *gazdaközönség*) to ally with experimental stations to gather and distribute statistics quickly, in the interest of keeping farmers informed about the conditions of supply and demand at markets. "The farming community must get used to facing the facts (*a tényekkel számoljon*)" (Gaal 1885, 105). Yet numbers alone were insufficient. "Official statistics cannot provide adequate information on many, many things. On the other hand, a broader range of studies can discover the true cause of problems, bring to the surface useful ideas which by realizing them in the future in a suitable manner can forestall greater troubles" (ibid.).

Practices encouraging the collection of statistics existed in the nineteenth century, but they were in no way the dominant means of representing economic knowledge. Indeed, in the field of agricultural economics this continued to be the case well into the twentieth century. "In recent times, and even today, business science does not take numerical data as the basis or starting point of its investigations in every branch, even though *Thünen* had already shown the way clearly" (Kesztyűs 1929a, 16; italics in the original).[13] By the 1920s, it was no longer possible to consider statistics solely an option among many methods of analysis and formal representation. As Witthen reminded the readers of *Köztelek* in 1930, the most current literature on agrarian policy published abroad had joined the trend in economics to use statistical data to determine cause and effect and establish regularities (6 Jul. 1930).

A serious impediment to the use of numbers in science, and in business, was the absence of reliable and consistent collection procedures. Agrarian modernizers bemoaned the fact that basic calculations required to guide economic rationalization could not be tallied, as records were simply not kept. Rationalizing production and sales depended, after all, on much finer-grained business records. "Reckoning constitutes the weak side of Hungarian farmers, which may explain the impossible situation that many of them do not even know the values hidden in their farms and so they cannot even provide an account of the earnings on their invested assets" (*Köztelek*, 18 Sept. 1920, 700). These problems were found just as commonly at large manorial estates as on small family farms. Agrarian work scientists and business economists devoted much effort to demonstrating how estate managers and farmers alike could benefit from the knowledge they would gain from proper accounting procedures. Indeed, one of the highest priorities of agrarian business economists was establishing advisory offices on bookkeeping, making it possible for farmers to seek aid in working through their records. "Business science is the compendium, leveler, and pinnacle of all agricultural sciences and yet it still lacks an independent workshop and an established, dedicated workforce. And yet nothing is more necessary, because Hungarian farmers have demonstrated very little appreciation for questions of business, commerce, and bookkeeping" (*Köztelek*, 14 Aug. 1920).

OMGE regularly pressured the government, and the Ministry of Agriculture in particular, to improve the collection of statistics on the agricultural sector. But money was always tight. On one occasion, when denied funding once more, the minister of agriculture explained the lamentable conditions of budgetary constraints by informing OMGE representatives that only in the 1930–1931 budget had he been able to secure funding to *print* the statistical survey conducted in 1927 (*Köztelek*, 16 Nov. 1930). Indeed, as Rege recounts in his newspaper article "Statistics and Agriculture," the collection of firm statistics on Hungarian agriculture had been put on hold since 1895, despite a promise made in 1923 to the International Agricultural Institute in Rome that Hungary would conduct this survey (ibid.), a condition he regarded as shameful and sure to guarantee poor decision making by uninformed officials. Thriftiness demanded that government monies be spent on gathering firm statistics, Rege argued, since this was the only way to design policies leading out of the crisis, using the example of foreign countries to strengthen his case. In industrial countries, agricultural lobbies had been able to fight for customs legislation, armed with firm statistics "like a gun in their hands," policies they would not otherwise have achieved due to their status as minorities in their legislatures (ibid.). It was ironic, Rege noted, that despite Hungary being a primarily agrarian economy, Hungarians were better informed of conditions of profitability in agriculture in other countries than at home, since these data were unknown. In fact, having data on Hungary to provide international committees was extremely important: "For us, especially in our current isolation, if we had internationally recognized data for negotiations of international affairs, this would be a resource of inestimable value" (Witthen, *Köztelek,* 9 Feb. 1930).

Hungarians were particularly impressed with the dual purpose of German advising stations. Not only did they provide farmers with valuable assistance, but staff also extracted data from farm records to be fed into a national agency. This permitted the German state to craft policy on the basis of reliable information to ensure sufficient foodstuffs for the populace, as well as to improve agricultural production. An additional advantage to a national network of advising and information collecting agencies was the data it made available to those in the field of agricultural economics to conduct rigorous comparative studies of agrarian firms. Agrarian economists and interest groups knew they faced serious obstacles to developing a comparable system in Hungary and so looked to the government for leadership.

> The reason so many private accounting and consulting stations were established in Germany was so that the taxing of farmers based on their accounts would be more just and proportionate than here. Since in our country the possibility for accounting is pretty much unknown and not used, there is no basis for establishing private accounting and firm consulting organs. To spread acounting better than this therefore can only be done by the state, so that accounting be standardized as well as uniformly institutionalized. (Kesztyűs 1943, 10)

Unbeknownst to Kesztyűs, his vision of establishing a centralized authority supervising accounting practices would be taken up with a vengeance by the socialist bureaucracy in a few short years.

Crucial to the use of numbers in agricultural economics was their ability to make visible the internal workings of the firm. Indeed, as Schranz argued, a clear distinction was made between economics and the studies of business economics by German specialists. Economics, in this view, was dedicated to the analysis of society at large. "On the other hand, the business economist starts from the operation of the firm, the enterprise, from the various [divisions of the firm] . . . and in the course of these detail studies he arrives at a characterization of economic life" (Schranz 1936, 78). Recording the results of individual branches of a firm was simply insufficient for understanding the complex interrelationships characterizing the workings of a firm. Precisely since firms were dynamic entities, their features demanded representation in quantitative variables. In the words of Sagawe, German accounting expert, "how is it possible to recognize the mutual and profitable interconnections between firm branches other than from bookkeeping?" (Kesztyűs 1942, 6), or, more accurately, from the accounting statistics (*számtartásstatisztika*) compiled from accounting (ibid.). Scherer echoed these sentiments when he argued that the only way to track the connections between profitability, efficiency, and labor inputs would be to "pin them down" in numbers (1943, 442). The fact that the industrial sector took it for granted that all elements of the firm would be recorded in numbers made it even more surprising that the agriculture sector neglected this so shamefully (ibid.).

The role of government or association offices was crucial to the expansion of bookkeeping knowledge and skills. But the very act of compiling management statistics was a powerful force, creating more developed organizations and inculcating a business spirit (Kesztyűs 1929b, 55). Gathering data would transform the individual. Béla Kenéz devoted his opening lecture at the 1935 congress of statisticians to the psychology of the science of statistics, praising the town of Kecskemét, where the congress was held as an example to the entire agrarian population of the country for its "sensibility for the wisdom of progress" (*Köztelek*, 9 Jun. 1935). The power of numbers was grounded in their association with strong moral qualities. Good habits of gathering statistics, nationally and at the firm level, were associated with "higher cultured countries" (*magasabb kultúrájú államok*), such as the United States, Germany, or Denmark (Scherer 1943, 442). Embracing the use of numbers had significant historical consequences. "Let the example of England stand before us. Besides favorable geographic location, its ancient constitution, and the national pride so characteristic of the English people, England's domination rests on proper counting" (Szalay, *Köztelek*, 14 Apr. 1927). In every case, Szalay insisted, it was possible to identify the moment when wise men recognized the value of numbers. Hence one could encourage aspirations within the nation for these achievements. The most commonly mentioned model of effective data gathering, Germany, also had a history of neglecting bookkeeping. Only with a sustained campaign to encourage bookkeeping in the latter half of the nineteenth century and, as of the 1890s, the opening of many private accounting firms, was this level of development achieved (Károly 1926, 50).

KEEPING BOOKS

If the modern farmer wishes to get by, he will be forced to farm with a pencil.
KÖZTELEK, 3 Nov. 1940

Business economists were convinced that the importance of mastering the skills of bookkeeping could not be underestimated. Therefore, it was incumbent upon them to dispel fears while instilling the proper respect for its power. In an article discussing the significance of accounting for small farms, Kulin painted a pleasant and reassuring picture. "Farmers have many enemies in the animal and plant world, and even among men; he has far fewer friends and helpers.... accounting is a reliable and loyal helpmate, worthy of more love and respect...." (*Köztelek*, 3 Mar. 1940). But there were conditions for the friendship: conscientious effort to record changes in the value of the farm accurately and in a timely manner. If a farmer neglected these conditions, accounting could mislead and deceive him. "The discussion above doesn't mean that numbers completely dominate farmers, just as it would also be a mistake if a farmer would take measures haphazardly or allow himself to be led by his feelings. It is not right for numbers to control the farmer, but it is necessary for them to lead him" (ibid., 176)

The campaign to widen the use of bookkeeping had two parallel strains: (1) reassuring efforts to dispel trepidations about accounting, and (2) strong arguments for the specificity and complexity of accounting in an agrarian enterprise. These conveyed the tension between the need to expand the use of bookkeeping among practicing farmers to encourage greater familiarity with the farm as a business and the move to professionalize accounting and estate management so strongly advocated by the research arm of the Ministry of Agriculture and interest groups like OMGE and MGOE. "In itself experience and numbers do not reach the goal. The true solution lies in the fortunate accordance of numbers, practical observation and expertise" (ibid.). Bookkeeping was required to determine the farm's profit and level of efficiency (Kesztyűs 1942, 5–6; Károly 1925, 1). This demanded a clear understanding of how the various branches were internally structured and how different components of the enterprise interacted (*Köztelek*, 25 Sept. 1927). "Accounting is not an end in itself, nor a separate branch of knowledge. Rather, it is the bond between various productive factors, by virtue of which—as air surrounds bodies—itself becomes a productive factor" (Polich 1930, 115). Armed with an effective means of revealing the workings of their business, farmers could use bookkeeping as a way to determine the relative profitability of one branch over others and make investments accordingly. Well aware that farmers were not very conversant with ways to increase production effectively, Kesztyűs described two strategies that had to be avoided: spending too much money to increase efficiency and developing one branch of production at the expense of other branches, thereby reducing overall profitability by neglecting productive branches already in place (1942, 6). As a farmer became used to scrutinizing the workings of his farm, he would learn new skills. "The more that numerical data support the personal observations of the firm manager, the more he is able to anticipate future *possibilities* on the basis of past outcomes and their development. *Regular bookkeeping is the compass and primary lever of profitable increased production"* (Károly 1925, 1; italics in the original).[14]

There were a number of motivations for using bookkeeping to improve production, the most obvious being the necessity of altering Hungarian agricultural practice to survive in the intensified competitive environment of interwar economies. A modern economy demanded modern techniques of management and accounting. Learning to keep careful records also had other positive results. Károly explained that the best defense against improperly estimated taxes was careful accounting (ibid.). By this logic, bookkeeping assisted the transformation of an unprofitable farm into a profitable one, all the while protecting farm owners from unnecessary state burdens. This was all the more important when the agrarian community thought that their tax burden exceeded that of other sectors. Bookkeeping would solve this problem, ensuring "that Hungarian agriculture also attain national economic advantages, and no longer be pushed into the background by industry and commerce, the advanced organization and business mentality of which grew out of accounting" (Kesztyűs 1929b, 55).

Because of the vast indifference to modern recordkeeping practices and firm analysis, there were extensive misunderstandings about what double entry bookkeeping actually meant. Kleindin bemoaned the fact that farmers didn't understand that double-entry bookkeeping was not simply the practice of writing numbers down in two columns instead of one. This confusion led to serious doubts about the value of double-entry bookkeeping. Stein recalls a case in which the integrity of an estate owner running for parliament was questioned, resulting in a suit against slander. On the campaign trail his opponent frequently raised the issue of "double books" (*kétszeres könyvek*, or *zweierlei Bücher*) being kept at the manorial estate, suggesting that the low taxes paid by the estate owner clearly were due to his keeping two sets of books, one of which was designed to mislead authorities. The estate owner fought back—on the campaign trail and in court—asserting that he followed double-entry bookkeeping (*kettős könyvelés*, or *Doppelte Buchführung*) and was not in the habit of avoiding his financial obligations to the nation (Stein 1930, 221). His opponent, a well-established farmer in the county, was effective in his smear campaign because he and all those in his audience on the campaign trail didn't understand this new form of recordkeeping. Stein went so far as to describe the negative attitudes toward agricultural bookkeeping as superstitious and full of fantasy (1930, 221). Even as late as 1944, writers complained about the glaring absence of bookkeeping skills and willingness to adopt new practices (*Köztelek*, 7 May 1944).

Farming required its own forms of bookkeeping and specialized knowledge to make that knowledge effective. Yet even in the pages of *Köztelek*, ads for "chartered accountant (in double bookkeeping)" stipulated that estate managers must have a diploma from a commercial high school. Elemér Papp, a licensed farmer and estate manager, found this very strange. "Their fundamentals are naturally the same, but just as one cannot start a steam engine with benzin, so you can't apply commercial double-entry bookkeeping in a hackneyed manner to farm [double-entry] bookkeeping" (*Köztelek*, 1 Jan. 1925). In the words of another estate manager, more was required of a farm accountant than just the basic training of a commercial accountant. Professional competence also demanded "farm knowledge directly affected by bookkeeping, because—above and beyond general commercial bookkeeping—there will be hundreds of kinds of work performed, materials used, and stock evaluated that will be judged differently" (Naschitz, *Köztelek*, 29 Mar. 1925). Indeed, the reverse was also true, that bank and commercial accountants found themselves in unknown waters when reviewing estate books. Stein quotes a well known accounting luminary to this effect: "He said that he feels secure and on firm ground in the face of the balance [sheets] of industrial large enterprises and their evaluation, [whereas] he is overpowered by a feeling of uncertainty, dizziness as it were, when he must deal with the accounts of large estates" (Stein 1930, 222). Pretensions to expertise and specialized knowledge were ridiculed by Scherer, who saw the move to a more refined science of business management and accounting as superfluous.

> The task of agricultural accounting is not so difficult or complicated if we understand the accounting tasks in agriculture in their own natural *logical simplicity*, and if we don't spice up our textbooks with mathematical theories. Because there is no sense in using "formulas" and theories difficult to understand in order to arrive at the essence of the question in roundabout and puzzling ways. It only frightens and alarms people; it doesn't produce results. . . . Agricultural bookkeeping is not a diabolical science; it is easy to understand . . . if we explain it naturally in a *graceful* style. There are many among us who like to present everything, to teach in the most circuitous fashion, let's say *more scientifically*. They don't consider, however, that farmers *in their own natural way of thinking* and simplemindedness always understand and follow *the short and clear* way. And anyway, practice can never conform to exaggerated theory. In the world of accounting this causes confusion and fear, and isn't even suited to the nature of agriculture. (Scherer 1943, 443; italics in the original)

Stein expressed similar outrage at the ability of specialists to muddy the waters of accounting, to lead people away from the simple observation that accounting results are simply a combination of the changes in profit and the changes in the value of shares distributed and capital invested in the firm. "Indeed it is amazing that a few experts—with their spectacle of audacious mental acrobatics—can obscure in the lay and semilay audience the infinitely simple, let's say primitive recognition by which the clarity of accounting could never have been clouded over; so too, the fear of losing one's way that overcomes most people while wandering in the forest of numbers would not be able to develop" (Stein 1930, 225–26). It is worth pointing out that both Scherer and Stein held doctorates and were practicing estate managers; Stein was even the head of the bookkeeping department of the Agricultural Society of Baranya County. This tells us that the rift between those calling for increased specialization and those dismissing "scientificity" in accounting did not fall simply along the lines of professionalizing elites and the lay community.

In journal articles or newspaper columns, authors took great pains to explain the advantages of bookkeeping. Since opinions differed as to what was the most effective way of keeping track of the costs of running a farm, it was necessary to spell out what the advantages and disadvantages of each method entailed. Kleindin summarized the three strategies agrarian enterprises pursued (1927, 692). The simplest and least time consuming method was to figure one number to express the entirety of a farm's results, with no attention paid to the various branches of production that constituted the farm's activities. In Kleindin's opinion, the motivating interest for those using this strategy was to increase profitability; a single number at the end of the year was all that was needed to determine if the farm's profits had grown. The second method was to differentiate those branches of the farm that were explicitly dedicated to cash cropping or income generation, for example the dairy, pig farm, or seed growing, and to treat them differently in recording information.

> If, for example, only the dairy and fattening operations are the sole business objective, then the rest of the branches—for example, the breeding stock, field crops, feed cultivation, mill, etc., as well as tools, manorial servants, and draft power—exist to serve the aforementioned major objectives. Then the expenses devoted to these branches, under the slogan "kost was kost, je weniger desto besser" [cost what it may, the less the better], are reckoned at cost.... *at the end of the year accounting will calculate the auxiliary branches as having no profit or loss, since the numerical results of the estate will appear as a result of the branches singled out as the major objective.* (ibid.; italics in the original)

The final and most thorough strategy Kleindin discussed was to treat each branch of production separately, keeping a record of the production costs of each, shown in market prices or current local prices, not in production costs. These figures had to be kept independently of commercial profit and loss, since there often could be a time lag between harvest and selling. The crucial issue was one of relative balance between costs and results. "It's not enough to focus one-sidedly on growing crops and increasing gross yield; one must also find a way to not have the costs of production consume earnings" (Károly 1924, 75).

The major problem agrarian business owners identified with a new accounting strategy was the near impossibility of differentiating between the various branches of production in a farm, making the kind of bookkeeping inherited from commerce and credit agencies impossible to implement. Cattle and pigs were fed from crops grown by the farm, corn stalks were processed into sugar, the manure of animals rejuvenated the soil, and so forth. So the question was not simply, is the dairy profitable, but rather is the farm profitable with a dairy or without? Moreover, what were the consequences of changing the dimensions of or the policies regarding any particular branch of production? These issues had to be foremost in the mind of the estate manager or farmer. As Juhos cautioned, "one must always question the current way of farming" (1927, 1310). Juhos was quick to point out that double-entry bookkeeping was superior in his mind, and much better suited to modern farming, but he understood the amount of energy and time required to transform one's records to a new system. While Kruspach was of the opinion that a good farmer would have neither the time nor the energy to see to the books (*Köztelek*, 22 Nov. 1925), Juhos was pragmatic enough to recognize that it was better to build on existing practices of single-entry bookkeeping than to demand everyone switch over immediately. Double bookkeeping would come in its own time, he believed. Unfortunately, "there is no systematic collection of data that keeps economic interests in view. What exists is old, incomplete, or inaccurate and so cannot be used well either generally or specifically" (Juhos 1927, 1310). So in the final analysis, a greater investment of time and energy would produce much better results. Bookkeeping provided a wider view of the farm, a broader understanding of the role of tasks and their value. "Without adequate information the landowner many times only sees what appears in front of his eyes and perceives only those things and the work of those who are near at hand" (Scherer 1943, 445). The reigning principle of the firm or farm as a complex system of relationships—relationships reflecting divergent activities within the branches of production—clearly had not been widely accepted. Questions like where capital should be invested and to what extent within the farm's branching structure were difficult to pose without any idea of what was productive, what was not, and why.

These arguments did not quiet the naysayers. Keeping careful records of farm expenses, assets, and income was in principle a good idea, but it seemed to collide with the fact that numbers—prices—were constantly changing. The market was not a fixed bulletin board with numbers posted for the season, much less the year. Doubters questioned the point of writing down all this information if it rapidly became obsolete. This perception was particularly relevant in the early 1930s, when there had been several years of intense price fluctuation. Kulin acknowledged that changing prices did make it harder to keep track of the firm's activities. The crucial point was simpler.

> Cost accounting is always only looking for probability; it will never completely cover reality. Only accounting can determine reality (naturally on the basis of past data). With the calculative procedure indicated, however, we can attain—in the case of price displacements—a guarantee of the most stable and reliable income for our agriculture.... *The purpose of calculation is to discover the relatively best.* To belittle calculation, because many times it is very removed from reality, is improper. (*Köztelek*, 18 Aug. 1935; italics in the original)

So, contrary to the claims made by business economists, adopting bookkeeping required two major changes in the way that a farmer or estate manager conceptualized his business. First it required that he learn to look upon his farm as a complex set of relationships of productivity and profitability. The interdependence of branches of production on the estate—dairy, pig farm, crops, sugar processing plant, etc.—was no longer best conceived in terms of the actual products that moved between the branches: manure to crops, sugar beets to processing, grains to dairy. These interactions had to be conceptualized as flows of value, of capital, of money invested in the farm's various activities. Bookkeeping was meant to record the movement within and between branches, figures that themselves were constantly in flux. Shifting one's perspective away from physical objects—pork, milk, wheat, and rye—also extended to the way wages were understood as an element in the flow of capital. The common practice of remunerating field hands and other farm workers in kind rather than in cash necessitated translating the goods and services distributed to the workforce into a monetary figure. A further complication arose from the on-going debate over whether farm costs or market prices should be the preferred method of recording the costs of labor (Scherer 1943, 445). These sorts of calculations were daunting for economists writing doctoral dissertations; without guidance, figuring these numbers would be far beyond the skill of the average farmer or estate manager.

The second, perhaps more difficult conceptual leap was to understand that the numbers recorded were necessary to assist a process of calculation where the numbers no longer were important. The search was not for stashes of money hidden in the haystack or barn but for abstract probabilities, for speculations on the potentially most profitable business policy. This was a far more difficult notion to grasp than the simple idea that we can only figure out the year's income if we write down all the steps it took to get there. Kesztyűs was not exaggerating, then, when he claimed that "business mentality" [*üzleti szellem*] would be created by accounting (1929, 55).

In light of the emotional and conceptual difficulties the adoption of bookkeeping required, how did business economists imagine these barriers to be removed? This too would demand time and energy from the agrarian community. They pointed to the fact that other advanced nations had conquered these problems; they simply started earlier. The impressive system the Germans demonstrated in the war—"the perfection of the excellently planned and manageable war economy" (Scherer 1943, 442)—was the result of a sustained campaign over many years. "Learning is never demeaning, nor does it offend national dignity, especially if life vindicates the proper economic attempts and system" (ibid.). With the help of OMGE and state agencies, Hungary could also ensure a firm grounding in solid bookkeeping and centralized statistical gathering practices. This was a recurring theme in *Köztelek* and, in particular, a crucial theme in the column on agricultural economics. Regularly offered courses, subsidized by the state, could be established to teach bookkeeping. Equally important, Kruspach argued, was the need to standardize procedures "in recognition of the special conditions" throughout the nation (*Köztelek*, 22 Nov. 1925). Kesztyűs was very explicit about what was needed. Writing in 1929, he said

> *A law must be created* for establishing accounting departments, about their scope of activity and competence. Accounting departments must conduct standardized accounting, which the center will develop in detail and have passed by the agricultural and financial authorities, as in the Swiss and German situation. Preparation of the bill must be preceded by substantial propaganda, the result of which surely will be sufficient numbers of accounting statements of manorial estates for management and reconsideration. (1929b, 51; italics in the original)

Going further, Kesztyűs argued that the country needed to establish a National Bookkeeping Institute (Országos Számtartási Intézet), the costs of which should be borne jointly by the Ministry of Agriculture and the Finance Ministry (ibid., 55).

Conclusion

The audacity of agricultural modernizers knew no bounds. Presumptuous demands were made on state resources to create a new university, fund research in fields of dubious value, and institutionalize entirely new protocols for business records. Hungarians were not alone in calling for improved methods of recordkeeping, data gathering, and training personnel. Experts in the fields of business economics and scientific management throughout Europe, and in the United States, mounted unrelenting campaigns for their cause in this period. They sought to convince their colleagues in business and politicians in government that the only means by which a truly modernized, efficient, and productive economy could be built would be to invest in an infrastructure capable of sustaining their efforts. In Hungary, however, their demands fell on deaf ears; the project to modernize the agrarian economy foundered on skepticism and apathy. The national purse had little to spare, and farmers even less. Journals were established and university dissertations were written, but the more encompassing reforms mandating new accounting standards and encouraging innovative labor practices went nowhere. Warnings that economic distress would only worsen if modernization were to be postponed sparked little interest. None of this dissuaded advocates from devoting their own labors to developing a conceptual and methodological repertoire should times change. It is to this project I now turn.

3

Formalizing Practices

The task of agrarian work scientists was twofold: develop a new branch of agrarian firm studies and promote innovations in organizing labor at large farms. This double commitment to scientific rigor and practical application had long characterized research in agronomy, stockbreeding, and related fields in the natural sciences. Now it was time to do the same for agrarian economics. First and foremost work scientists needed to build an analytic framework and methodological toolkit to conduct studies worthy of a novel disciplinary configuration. Work science began by drawing much of its conceptual repertoire from a selection of existing manorial practices and borrowing the tried and true methods of German experts in business economics. This constituted the base. Then it was necessary to standardize metrics, collect data, and adopt formalizing procedures to represent and analyze processes. Just as important was to discern the features of the agrarian work force, qualities that were seen to vary by class, gender, age, ethnicity, and region. In light of the diversity of the working community, work scientists were keen to discover effective motivational tools and managerial strategies to induce higher productivity for every possible configuration of workers and managers. With formal tools at hand, researchers could proceed to examine the complex dynamics of firm organization, perform crucial comparative studies of economies of scale, figure out the relative value of differing labor regimes, and calculate varying levels of efficiency and profitability. Armed with convincing analyses, agrarian work scientists would be equipped to provide valuable advice to individual farmers and to bureaucrats crafting national policy. In this chapter I describe the work of work science: establishing standard metrics for labor and labor power and determining the proper means of assessing the cost of labor, notably complex calculations rendering goods and services stipulated in labor contracts into commensurate units of analysis. It is here where we will see how formalizing practices take shape bit by bit.

The Science of Work

As we learn from Rabinbach's history *The Human Motor: Energy, Fatigue, and the Origins of Modernity* (1990), the field of work science grew out of the fascination and fixation of European thinkers in the nineteenth century with energy, power, labor, and production. Far more than the preoccupation of political economists like Ricardo or Marx, the concern with labor was as much a product of new ideas about thermodynamics as the object of social reform. Refiguring labor as energy and the body as a machine had consequences that far exceeded simple materialist conceptualizations.

> [U]topian social and political ideologies . . . conceived of the body both as a productive force and as a political instrument whose energies could be subjected to scientifically designed systems of organization. Thus, the classical traditions of nineteenth-century social thought, as well as the radical ideologies of the early twentieth century, shared the belief that human society is ultimately predicated on the unlimited capacity to produce and that this "social imperative" mirrored nature's own unlimited capacity for production. The laboring body was thus interpreted as the site of conversion, or exchange, between nature and society—the medium through which the forces of nature are transformed into the forces that propel society. (Rabinbach 1990, 2–3)

The purposes and political allegiances of work scientists changed over time; well-intentioned social reformers of the nineteenth century hoping to rid society of fatigue were replaced by technocratic managers hoping to quash worker unrest in the twentieth. Fellow travelers along the way included pioneers in the fields of psychotechnique or occupational psychology and specialists in the physiology of work, as well as less scientifically grounded projects like Taylor's innovations in the United States. The study of work was a growth industry, giving birth to a range of new academic specialties and strengthening the capacity of industry to grow in new directions.

An important dimension of the work of assessing labor power and productivity was related to the increasing significance of the values of precision and accuracy among scientists and bureaucrats in European modernity. Just what constituted proper measurement and accurate assessment was a vexing problem in many domains of scientific activity in the nineteenth and twentieth centuries, but the concern was not isolated to scientific scholarship. As Norton Wise and others point out in the excellent collection *The Values of*

Precision (1995), the call for greater precision and accuracy in measurements, and the standardization that issued from these demands, was sparked by complex historical changes beyond the strictures of scientific projects, such as the rise of a centrally organized nation state and technological innovations in industry. Wise emphasizes the enormous efforts at achieving agreement among relevant bodies and unifying practices across communities that were entailed in what may have appeared to be a straightforward task. The work was labor intensive, time consuming, and often contentious, whether it took place in eighteenth-century French chemistry, Victorian engineering, nineteenth-century German physiology, or insurance calculations (see Golinski, Holmes and Olesko, Gooday, and Porter in Wise 1995). In the burgeoning field of economics, the problem of measurement, statistics, and quantification was of particular concern, touching as it did central questions of public policy and private business (see Breslau 1998; Klein and Morgan 2001; Stapleford 2009). As Mirowski (1989) has argued, economists' quest for scientific legitimacy led them to privilege quantitative models adopted from other sciences that were substantially misunderstood and ill suited to their purposes, to the detriment of the field as a whole. For our purposes here, it is noteworthy that Hungarian business economists and work scientists rehearsed similar debates about measurement and assessment while mounting a campaign for scientific legitimacy.

In the chapter to follow, I describe the work that was involved in creating a science of work. To lay the groundwork for the new science, appropriate methods had to be selected for assessing productivity. Great effort was expended in designing abstract concepts: What should be the temporal standard for the work day? How would the standard worker be defined? And, most importantly, what would be the unit of analysis that most effectively captured the process of laboring in a simple and parsimonious fashion? With these tools in hand, work scientists and firm specialists proceeded to figure out the value of wage packages at manorial estates; it was impossible to gauge the level of productivity without first knowing the cost of labor. This would prove difficult in the absence of a market price for labor. Complex techniques of commensuration had to be devised to reckon the value of goods and services in the manorial worker's yearly contract into a common unit of comparison. To successfully implement a wage system designed to increase productivity relied on yet another branch of work science: the psychology of work. Managers had to learn how to handle workers whose temperament and abilities varied widely. The most effective means of motivating workers, however, was out of reach. Experts agreed that paying workers in money would alter everything; they also agreed that convincing the owners of manorial estates to pay wages in cash would be an uphill battle.

The Work of Work Science

METHODS AND MEASURES

To transform the manorial estate from an old-fashioned, hidebound farm to a modern enterprise entailed introducing techniques and tools that were unknown at manorial estates but were in common use among work scientists and scientific managers in other parts of Europe. Reichenbach, author of the definitive study of agrarian work science in interwar Hungary, listed the following aids to the study of agricultural work: stopwatch, slide rule, dynamometer, psychographs, respiratory devices, and moving film (1925, 222).[1] Ries, an influential German specialist, encouraged estate managers to carry a slide ruler at all times to perform the calculations necessary for time studies; he confessed feeling more uncomfortable on the job without his slide ruler than without his watch (1930, 234). Long-established analytic categories, such as labor power, also figured prominently in their studies. In the nineteenth century, the metonymic term "worker's hands" (*munkáskéz*)—comparable to the term farmhands in English—was the common designation in print and continued to appear in a number of work science texts in the twentieth century. By the turn of the century, however, labor power—the energy and activity of working—became the focal point of analysis. The category of manual labor power (*kézimunkaerő*) or, simply, labor power (*munkaerő*) occurred more and more, as Hungarian work scientists sought to bring greater definitional accuracy to their analyses.[2] It is interesting that at the turn of the century, the term "manual labor power" was often paired with discussions of draught animals (*igaerő*) and machines. Hensch, however, specifically distinguished human labor power from other resources.

> In public life it is common to understand under labor power, with the exception of human power, to be the work of draught animals and machines (motors), yet neither animals nor machines are capable of regularly performing productive work. . . . Human work differs substantially from the work of animals and machines in that due to his mental capacities, man acts consciously in performing work whereas machines and animals fit only for mechanical exertion are incapable of this. Thus when speaking of labor power, only human labor power should be understood, while as sources of power animals and machines belong in the rank of capital, namely live and dead inventory. (1901, 140)

Humans only constituted a form of capital, Hensch explained, if they were property, that is, if they were slaves (ibid.). At the time, poor peasants might as well have been slaves. Legislation passed in 1898 stipulating the contractual obligations of agrarian workers was harsh and unforgiving, earning it the epithet the Slave Law. All that mattered to agrarian work scientists was harnessing workers' creative and productive capacities.

The purpose of labor studies differed. Some studies were conducted to ascertain the value of innovations in tool design, for example, Hamar's physiological comparison of the use of the traditional Hungarian hoe versus the new Thomka variety. (The customary Hungarian hoe proved to be better, surprising the observers (ibid., 211)). Fischer conducted a study of tool use to figure the relative costs involved, as well as the resulting quality of the produce. His analysis comparing gathering potatoes by machine or by hand confirmed suspicions; mechanized harvesting proved to be cheaper (1936, 43). Physiological issues of the working body also commanded attention, as in Hamar's article "The Effect of Physical Labor on Kidney Function" (1944) or Ujlaki Nagy's analysis of the thermodynamic and biological exchange values of agricultural work (1943).

Not all experiments in labor rationalization were actually conducted under the supervision of trained work scientists. In some cases, farm managers took it upon themselves to initiate new labor methods and systems of payment. Although the motivations for innovating came from practical problems they encountered at their own manorial estates, managers' hopes extended beyond the boundary of their farm, since they published their results to encourage others to follow in their footsteps. Kemptner's success in altering sharecropping contracts for sugar beets to ensure both the quantity and quality of crops grown—primarily by evaluating and recording the work completed at each phase of production—demonstrated that rationalizing elements of the labor process could redound to the benefit of both employers and employees (*Köztelek*, 20 Mar. 1938).[3] Ivanics's experience with introducing hourly wages for day laborers made him a particularly credible voice in debates being waged over the wisdom of hourly wages in agriculture (*Köztelek*, 10 Mar. 1940). Moreover, in both cases mentioned the authors could testify to years of experience—nine years in one instance, fifteen in the other—that could disarm skeptical readers.

STATUS QUO IN AGRICULTURAL WORK

A variety of labor contracts peopled the agrarian landscape when work scientists came onto the scene. These needed to be analyzed in the terms set by the agrarian work science literature. While some attention was devoted to the role of agricultural workers employed to supplement family labor, nearly all of the discussion focused on the labor force at manorial estates. Labor contracts on the manor differed along the axes of temporal duration, form of remuneration, and, for some, completion of specific task. For nearly all employed, their productivity (the results of their daily activities) was not a consideration in contemporary contracts, a problem work scientists and business economists wished to change. Manorial servants, that is, those living on the manor, signed on for a year's labor for which they were remunerated with a basic package of goods and services, called the *kommenció*. The exact content of the *kommenció* varied from manor to manor, and by region, but as a rule the contract stipulated manorial workers receive a fixed amount of grains (wheat, rye, barley), distributed four times over the course of the year; a measure of salt; the use of a small plot of land to farm; the right to keep a limited number of animals (a cow, a young calf, a few pigs, and poultry); firewood; housing (either the corner of a big room with other families, or a separate room for the family); and a small amount of cash. Occasionally a pair of boots was also distributed every other year.

Wheat harvesters were contracted for the length of the wheat harvest, were fed, and were paid a set portion of the wheat harvested. They tended to be poor peasants from nearby villagers, unlike the migrant labor force that traditionally came from areas dominated by small peasant holdings and less than desirable land. Migrant contracts—for two-month or six-month spans—were provided payment in grains, temporary housing, and meals. There were also costs involved in paying for their travel expenses and for the agent who negotiated the contract (Reichenbach 1930, 210).[4] Sharecroppers specializing in particular crops, like tobacco, arrived in the spring and stayed until the the crop was harvested. They saw to their own daily needs and were paid a portion of the crop they tended. Day laborers, also poor people from local communities, were hired for cash to augment the work of other staff on the estate (Szeibert 1939, 822–28). Only sharecroppers had any pecuniary interest in their output.[5] In all other cases, the pace and intensity of work were at the whim of the labor force. This had to change, if productivity were to improve. Before all else, business economists and work scientists had to come to agreement on a standard set of analytic categories.

Many of the methods and metrics work scientists employed did not constitute a radical departure from everyday farming, such as the use of time to distinguish categories of workers and customary measures of output, for example, sheaves of wheat, quintals of grain, and acres of ploughland. These standard measurements, such as how long it took a team of oxen to plough an acre of soil over the course of a day (in contrast to a horse-drawn plough), were published in handy pocketbooks by OMGE, along with a wealth of data arrayed in tables and charts designed to assist estate managers in the field. (As a rule, manorial workers and peasants alike were skilled in estimating volume, weight, and area from years of practice, which may not have been the case for some estate managers fresh from technical school.)

Three techniques were necessary for the work of work science: techniques to standardize, commensurate, and motivate.

Standardization

TEMPORAL UNITS

At the turn of the century, Hensch stipulated in his treatise on managing farms that for the purposes of calculating costs of production, the length of a day's labor should be understood on the average as ten hours. Of course, the length of the work day varied seasonally, as Hensch acknowledged. To approximate the differences across seasons, Hensch defined a winter day of labor as eight hours and a summer's day as twelve; the ten-hour day held for spring and fall work (Hensch 1901, 105).[6] This standard measure performed two condensations simultaneously. It set the unit of measurement on a temporal scale, and it collapsed a series of tasks into a generic unit of time. For the purposes of measurement, it erased the distinct tasks of harvesting wheat, hoeing and weeding row crops, and tending to animals, categories that set the terms of social hierarchy at the manor and within village communities (see Lampland 1995). This was an important innovation, as this condensation allowed management to use accounting strategies for calculating the labor needs of the estate without reference to the great variety of agricultural tasks conducted at a large estate. Providing a base unit of analysis was particularly important as it addressed one of the primary concerns voiced by observers and even strong advocates of scientific management, that is, the far greater diversity of tasks found in agriculture than in industry (e.g., Farkas 1941, 1; Halács 1928–29, 44; Károly 1924, 81; Szeibert 1939, 820; V. Fodor 1925, 44). Removing this stumbling block paved the way for implementing scientific management techniques into agricultural management. Using a single temporal unit of analysis did create other accounting problems, though. Occasional day labor was the only job in agrarian employment for which a set monetary payment was stipulated. All other jobs entailed complex forms of remuneration in kind. Techniques of commensuration were designed to deal with this question, a point to which I shall return.

The ten-hour labor day also figured in publications designed for estate owners and managers, which isn't surprising considering the influence of manorial practice on work science. In the 1915 issue of the *Köztelek Pocketbook*, for example, the table listing "the working capacity of manual labor" had been designed on the basis of a ten-hour day for one worker (*Köztelek Zsebnaptár*, 239). The degree of standardization is illustrated in the fact that the 1915 table was reproduced nearly word for word, number for number in the fiftieth volume of *Köztelek Zsebnaptár* published thirty years later (1944, II. kötet, 192). On the other hand, the practice of stipulating ten hours of work as the unit of labor analysis diverged from customary village descriptions of the work day as lasting from sunrise to sunset. It also contravened legal practice. It bears mentioning that the practice of defining the work day as lasting from sunrise to sunset wasn't unknown to agrarian business scientists. Reichenbach himself makes mention of it (1930, 196–97).

STANDARD WORKER

In all cases, a work day was calculated assuming the worker was male. If women or children were hired, the value of a work day was reduced according to customary percentages: 80 percent for women and 50 percent for children (under the age of twelve). Kesztyűs makes the gendering of this category explicit when defining the temporal unit as "wage income expressed in terms of the wage for a man's work day" ("egy férfimunkanapra redukált munkabérkereset"; *Köztelek*, 11 Aug. 1935). Szalay gives a more precise description of how he transforms women's and children's labor contribution into men's units: "Men's day labor constitutes half of the 600 composite days of labor; approximately 300 days are women and children's day labor. Taking the latter as 0.6 of a man's day labor, the 600 day labor days are equal to 480 male day labor days" (1931, 15). Reichenbach followed the example set by Sedlmayer, a faculty member of the Management Studies Department at the Agricultural Technical School in Vienna. This entailed combining seasons and gender in his calculations, for example, one summer day of man's work day is equal to only two-thirds of a man's work day in winter; a woman's summer workday is only two-thirds of a man's and, in winter, only one-third of male workers (Reichenbach 1930, 219). These calculations were subject to additional manipulation if the workforce was composed of ethnic minorities or foreigners, whose work habits differed from those of the standard Hungarian male.

UNIT OF ACTIVITY, OR NORMAL OUTPUT

Dissecting the labor process also entailed creating a general rubric for task or activity as such. In Károly's essay describing the use of Taylor's system in agriculture, he describes a task-wage (*pensum*) but also makes reference to a task system, using the German term *Pensumarbeit* (1924, 78, 83). After creating a unit for activity per se, the next step was to set its parameters. For everyday calculations, farm managers could consult the charts published in pocketbooks on daily output. Yet these constituted only a loose approximation. This may not have bothered practicing farmers, but Reichenbach needed to underscore their provisional status when introducing the columns he wrote for the *Köztelek* pocketbooks on daily labor capacity with the following caveat: "Normal work performance varies greatly according to the workers, quality of the soil, the weather, etc. Therefore the average data referred hereto are not sufficient in practice" (*Köztelelek zsebtár* 1927, 362). As approximations, they provided a useful gauge for managerial personnel. But they did not accord with what work scientists had come to call, following Taylor, normal output (Világhy 1930, 18), that is, they were not based on carefully conducted studies that would measure an average index of performance. Although this kind of information was sorely lacking, Reichenbach nonetheless cautioned amateurs against tackling the difficulties of time and motion studies, suggesting instead that these difficult procedures could only effectively be performed at experiment stations and teaching farms with properly qualified personnel and equipment (Reichenbach 1930, 228). "Indeed, it isn't even right at experimental farms to start complicated movement and tool studies, i.e. to start with the most difficult work, since the difficulties at the beginning may ruin the inexperienced researchers' mood for further experimentation. Only after long, tiring, and persistent experimental work is it possible to achieve results in this field" (ibid.).

Determining reliable estimations of average or typical results constituted a cornerstone of labor analysis. The precise procedures for estimating normal performance differed, however, depending on one's intellectual lineage or practical experience.[7] Citing the path-breaking work of Lüders, the estate manager of the Halle-Trothai research farm in Germany, Károly insisted that the simple tasks of observing work performance and time study were necessary to determine "standardized work performance" (1924, 83).[8] Farkas's study of digging included recording the amount of time devoted to the act of digging, supplemental actions needed to dig (cleaning and fixing the shovel), and time otherwise lost to the task. These calculations allowed him to establish the daily normal output, proper tool use, and an approximation of proper wages for digging (1941). The method favored by German experimentalists at the Pommritz estate in Saxony, on the other hand, entailed keeping a work diary recording all the information for future analysis, for example, time, weather, soil, the shape of the field, and the strength of workers and draft power (Kölber 1932, 3). This was a relatively easy approach and accorded to a degree with bookkeeping practices already in place at some manorial estates. Another method was the "production index card" kept by the estate where Fischer had conducted labor studies. "The farm steward keeps an index file based on the daily labor reports. Each productive branch has its own card file, in which the time units for animal teams and by manual labor [*gyalogos*] have been accurately determined on the basis of the daily report" (1936, 11). The German specialist Ries, on the other hand, promoted a much more sustained and fine-grained analysis, requiring observers to break down all tasks into units of action (work elements, *munkaelemek*), which were then calibrated by their temporal duration (time units, *időegység*; Kölber 1932). Ries's careful dissection of work enabled observers to assess the typical level of performance for each task. (This would be the technique preferred in the early years of socialist wage calculations.)

Other researchers combined calculations of time and output with additional concerns. Chronicling the "output problem" in horticulture, Liszka insisted on pairing quantitative with qualitative improvements, citing in particular the focus on quality in the German and Italian literature (1943). Szakáll's analysis of the impact of high temperatures on performance was intended to provide a more rigorous estimation of physiological features of agricultural work. He measured energy circulation during "normal hoeing" with a respiration device. This required additional calculations to accommodate the impact on tool use of wearing a cumbersome device (1943b, 423). Szakáll called earlier researchers to task for considering only the rate of cooling in various climatic conditions in their models. "Working in high temperatures overburdens other functions of our system: circulation and water and salt reserves due to increased sweat production. Overburdening these functions could have a much larger role in inducing the drop in productive capacity than the accumulation of surplus heat" (1943a, 249). Since the wheat harvest—one of the most physically arduous agricultural tasks of the year—occurred at the height of the summer months, it was crucial for work physiologists to examine performance in conditions that reproduced those workers experienced (ibid., 249–56). Observers not committed to the scientific imperative of work science found these activities bizarre; Leopold's comment about the absurdity of measuring sweat captures this sentiment.

> Not long ago the *Deutsche Landwirtschaftliche Presse* published the results of an experiment, which kept track of the sweat of German workers . . . feeding the threshing machine for several hours throughout the day. There have been similar German-type attempts here as well. We will not bother with such un-

fruitful, arduous measuring. Unlike the Germans, we consider it God's command that He didn't want us to measure our sweat and our tears by grams. (1934, 462)

Apparently work scientists were in the minority.

Commensuration

Farm managers and business economists knew full well that labor costs constituted a substantial portion of business expense. Less well known was the relative cost and value of the agrarian labor force: differences among different segments of the laboring population, the rate of productivity of various types of labor, and differences evident when engaged in intensive versus extensive farming.[9] Eventually money would be the primary means of commensurating tasks and people. Like many of their contemporaries, agrarian business economists assumed that the shift to money was a general, and in fact important, accompaniment to modernizing production. "Generally in more developed economic and agricultural conditions payment in kind is replaced by monetary payment; this trend is already evident in individual places here" (Szeibert 1939, 842). It was also assumed that paying workers in cash constituted a better means of encouraging more and better work than remunerating them in kind. (I will return to this issue in the discussion of motive forces below.) For all these reasons, then—managing, rationalizing, and modernizing agriculture—money was a crucial element in work science and is worthy of lengthier treatment than other projects of commensuration.

The question of wages occupied an important place in management studies, since the process of evaluating performance in relation to particular wage systems posed very specific problems. It was difficult in Hungary to determine the costs of labor when a large portion of workers' income came from payments in kind: grains, housing, use of a garden, right to raise farm animals, and so on. To solve the commensuration problem, a number of experts attempted to devise formulae or means to reach a reasonable approximation of labor costs. The range of solutions ran from pure deduction to hands-on figuring of prices and value.

The German economist J. H. von Thünen advocated a deductive approach to determining "the natural wage" in his influential treatise the *Isolated City*, published in the 1820s (Reichenbach 1925, 66). Unfortunately, this formula did not work when Thünen tried it at his own estate in Mecklenburg. Reichenbach notes that other German theoreticians—Waterstadt, Schuchmacher—were not dissuaded by Thünen's failure from still believing that this problem could be solved deductively (ibid.). The more common approach in Hungary was to build analyses from empirical data. Two examples are offered to illustrate different means by which indices were created to figure out labor costs. In both examples, the calculations performed required distinguishing between various modes of payment for agricultural workers: money wage, in kind, and a combination of the two. The first example is József Badics's doctoral thesis, "The Influence of Changes in Prices on Profitability in Agriculture during the Years 1913–1925" (1929), in which he considered productivity in light of changing prices in goods and labor over about a ten-year period. The second was István Szalay's book *Requirements and Costs of Work by Hand at Agricultural Firms Based on a Study of 121 Hungarian Medium and Large Estates*, published in 1931.

In figuring the cost of labor, Badics recognized that both monetary wages and payment in kind had to be included, and categories of workers whose combination of goods and money differed had to be calculated separately. Badics was also keen to incorporate metrics that conveyed the perspective of both workers and managers, that were able to chronicle the consequences of economic changes for the livelihood of workers as well as the profitability of enterprises. Accordingly, he designed three metrics to represent changes in wages: (1) a figure in 1913 gold crowns; (2) in wheat; and (3) in a unit he called the "wheat production area" (*búzatermőterület*) (1929, 14).[10] The first metric—the wage value expressed in gold crowns—would be useful during periods of "normal price development and stable currency" (ibid.). Unfortunately, the decade he studied was anything but, marked by both fluctuations caused by the war and the constraints imposed by the government's compulsory economic policies (ibid., 6–7). Hence the numbers in Badics's analysis of the decade 1913–1925 reflected neither the actual value they represented to workers nor the costs imposed on estate owners at the time. Once normal conditions returned, he explained, this index would prove helpful. In relation to the second metric (the wheat scale), he explained that it was useful to gauge quantitative changes in workers' compensation, a dimension poorly calibrated by changes in price. After all, he reasoned, the price of wheat was irrelevant to manorial servants: "it makes absolutely no difference how the price of natural products consumed in the household is quoted" (1929, 14). Finally, his third metric represented a correlation between the cost of wheat and the area required to grow this wheat in terms of average yield. This provided farm owners an index with "approximate accuracy" of the burdens workers' wages represented for the business, since these were the two factors that could affect the cost of payment in kind: changes in average yields and the amount of in-kind goods distributed (ibid., 15).

Szalay offered a different approach to figuring a "labor cost index," though it too was built inductively from a large data set. A credentialed farmer (*okleveles mezőgazda*), Szalay conducted an extensive study of manual labor requirements and use at 121 manors across the country (1931). The data he used were provided by OMGE in its 1927 survey. He both participated in the compilation of the statistical data and also conducted a computation of manual labor costs (ibid., 61). Szalay's analysis was designed to identify the factors that influenced—favorably or unfavorably—the profitability of agricultural firms (ibid., 60). In all steps of his analysis, he provided careful records of the means by which he reached his "unit of manual labor days" (*kézimunkanapegység*) and his "index of labor costs" (*munkaköltség index*).

Calculating the average labor use data from the survey, he developed hypothetical units, which were based on the assumption that management was able to take full advantage of workers' time, using the standard metric of ten hours (ibid., 16). He then provided a breakdown of labor needs—both practical and hypothetical—for five separate manorial estates, which represented different distributions of acreage and crop profiles and were found in three different regions in the country (ibid., 22–30). The second section of his analysis addressed the cost of manual labor, expressed in money (*pengő*). He made a further distinction in his calculations between farm price and market price, to accommodate situations in which farms provided laborers with in-kind payments from among their own stores, rather than purchasing them from elsewhere. This was commonplace; Badics worked up his calculations using statistical studies conducted by the US Department of Agriculture on the relation between market and farm prices (Badics 1929, 5n). Nearly all the goods agricultural workers were allocated could be given a monetary price, with the exception of housing, firewood, the use of garden plot, and keeping animals. Szalay walked the reader through the calculations he made to transform goods and services into prices. The value of having access to a garden was determined by figuring the average yield of corn/acre, minus labor costs, then multiplied by the monetary value of an acre's yield times the numbers of acres a manorial servant was provided (1931, 41). In the case of firewood, Szalay faced a more complicated calculation. When farms used wood or coal for heating, the calculation was simply based on the respective market price. However, many manorial estates had a policy of open access to fuel that permitted manorial workers to forage in nearby woods for firewood. "At these kinds of estates firewood is usually composed of straw or brushwood. If some farms allow open access to firewood, it is because firewood is cheaper at estates far from the market that farm extensively. Exactly evaluating [what is required to] heat a room cannot be worked out, but it should not be estimated at more than the value of 6 cubic meters of wood" (ibid.). He had to settle for an approximation.

Szalay's careful and extensive calculations illustrate the processes of condensation and commensuration necessary for business studies analysis. This collapsing of different metrics was crucial to the broader goals of figuring labor costs and their influence on profits. So, for example, in discussing the task of determining the cost of various types of labor to a firm, he explains: "The cost of workers paid according to labor time and labor output must be reduced to a common denominator, that is, we treat the workers paid according to labor time as if they had been paid according to output. This is so because, as I already mentioned, from the point of view of the firm, it is not important how many days the workers worked, but how much they accomplished" (ibid., 53). The relationship between labor cost and gross earnings—the labor cost index—condensed two calculations as well. The premier German specialist in agrarian business economics, F. Aereboe, adopted a method that figured out how many acres of plough land cover the cost of a manorial worker, and a calculation for the percentage of the gross earnings (*brutto hozam*) spent on manual labor. The labor cost index is represented, then, in the following equation (Szalay 1931, 56):

$$x = \frac{100 \times K}{(H-k)}$$

in which H represents the gross earnings/acre and k represents the cost of manual labor (56). In the final analysis, all these calculations made it possible to provide more abstract formulations of costing practices, as well as assisting manorial personnel to estimate the relative costs of the business enterprise. Introducing this equation also propelled Szalay into the company of those committed to the parsimonious depictions of value calculations common in the industrial sector (e.g., see Schlesinger 1949 [1920]).

Motivation

The power of labor could be calculated using thermodynamic models of energy and output, as we have seen. Harnessing that power was a more complicated affair, dependent on identifying the motive forces that drove workers to pursue their labors. These were crucial elements in the equations of profitability and productivity—significant units of analysis deployed by work scientists—yet far less amenable to quantification than other metrics like time and task. They nonetheless were central to business studies, since all other aspects of firm organization depended on the effective manipulation of workers' skills and capacities to produce outcomes.

Workers' participation in production could be understood in various ways: in terms of inherent qualities of the person (gender, age, willingness, mood, desire for material gain), as characteristics of groups (class, ethnicity), and as products of dynamic interactions among workers (competition). Managers were another element in the human equation, evaluated in terms of their personal qualities and for their ability to properly balance social and personal elements of the workforce. Some of these skills could be taught, others were specific to individuals. In this respect, management was just as constrained as workers were by their personal temperament and abilities. "In agriculture . . . almost everything depends on the manager. The administration can be the most well organized, all of the conditions may exist necessary to achieve high labour performance, nonetheless there will be no result, if the appropriate qualities are missing from the manager of the firm. Where the master is sleepy, everyone sleeps" (V. Fodor 1925, 46). In short, agrarian work science and business studies were truly *social* sciences. It bears emphasis, however, that all the features described below—temperament, social position, and moral stance—were assumed to be innate characteristics, some shared by all, such as a motive force like instinct, and some others, like industriousness, inherited through one's family. Techniques were developed to exact particular actions from workers. The most powerful technique, though fraught with its own difficulties, was the monetary wage. (I will return to this point later in the chapter.)

EMBODIED QUALITIES

Workers' mood or inclination to work was a recurring issue in discussions of restructuring agricultural production. "Selecting proper tools and proper manpower, as well as establishing proper work norms, have the purpose of regulating personal questions of rationalization, yet perhaps of these the most important is the question of work mood" (Raith 1930, 256). Qualities like a desire to work, responsibility, and discipline were described as "psychological incentives (*lelki rúgók*)," neglect of which could endanger continued production and cooperation between employers and workers (Dorschung, *Köztelek*, 22 Dec. 1940). Indeed, a whole series of qualities could be described as "psychological motives," all of which must be attended to in the management of labor (ibid., 18).

> Creative instinct is like that. It lives in everyone and encourages us to use our labor power. The creative love of work, however, only truly prevails if we give our workers a certain degree of independence and direction. In many cases

> this makes them more inventive and results in the improvement of work. . . .
> Ambition can also be an important stimulus for a disposition to work. . . .
> Knowledge can also increase love of work. Everyone works more willingly in
> a job where he knows what he's doing rather than in situations where he does
> not. (ibid.)

These qualities were not considered simply important *aspects* of work but as constitutive properties of production. In another text, Reichenbach refers to them as psychological inducements ("*lelki indítóokok*"; 1930, 248). It was crucial for management to stimulate a desire to work and strengthen it regularly (Reichenbach 1925, 17). Reichenbach bemoaned the view among many farm owners (*gazda*) that success was dependent on supervision alone (1930, 261).

Writing in the 1920s and 1930s, all of the specialists assumed that psychological motives were grounded in physical qualities of the person, that is, that they were embodied. This assumption was not unusual at the time. The relative significance of various factors in human behavior—evolution, heredity, genetics, culture, personality—were at the center of debate within psychology and related disciplines worldwide (e.g., Morris-Suzuki 2000; Roediger and Esch 2012; Stern 2005; Turda and Weindling 2007). In the very different world of psychoanalysis, the concept of instinct suffered from being underspecified analytically. I quote here from a Hungarian in the psychoanalytic tradition: "There are primary repressed energies in the psyche, which various psychological schools call instincts, needs, wants, etc. All that we can really determine is that repressed energies exert an influence at the depth of the psyche" (Révay 1946, 86). Distinguishing the dynamics of different social groupings also entailed founding several branches of psychological investigation: folk psychology, social psychology, and mass psychology (*néplélektan, társas-lélektan, tömeglélektan*; Rézler 1944).[11] "[O]ne must lead a people inclined to a life in the barracks entirely differently than the Hungarian worker, even the most enterprising of whom has a difficult time enduring rigid military training. Therefore one must establish work psychology separately for each country and each people, and so without preliminary studies from abroad we must fulfill significant tasks for ourselves" (ibid., 14; see also Thurzó 1934).

References to natural abilities and the language of instinct pervade the literature (cf. Lampland 2009). Reichenbach introduced his discussion of agricultural workers by quoting the German author Krafft: "'The value of agricultural workers depends on their naturally given abilities, their dexterity or clumsiness, stronger or weaker build, their way of eating, their intellectual and moral level'" (1925, 14). Some qualities, on the other hand, were more specific to classes or national groups. Manorial estate managers and work scientists shared the views of their Hungarian compatriots nationwide that workers differed in their skills and abilities, according to their regional identification or racial/ethnic identity.

> With regard to working capacity and suitability as a manorial servant, day laborer or contractual laborer, the pure Hungarian is worth the most. Among these the people from the Great Plain and Dunántúl are especially so. Even though we pay these workers well, they nonetheless work the cheapest because of their substantial performance. Following these are the Slovak workers from these regions, where they haven't become too americanized [lazy].[12] Their working capacity, however, is less. German workers are excellent, if they are working for themselves, but they don't gladly engage in wage labor, especially for a master of another ethnicity. Moreover, they are expensive; together with the Serb they are the most expensive worker. The undemanding Ruthenian and Romanian are fairly obedient, but they are very weak workers and although they make do with less, they still perform very expensive and slipshod work. (Reichenbach 1930, 247–48)[13]

It bears emphasis that Reichenbach's description of ethnic workers in agriculture living within the country's borders could have been drawn from any number of social science texts in interwar Hungary. The disciplines of Hungarian ethnography and rural sociology were firmly grounded in identifying and explaining differences between classes, races, ethnic groups, and religious communities within the nation, as we saw in chapter 1 (e.g., Erdei 1941; Györffy 1942; Illyés 1936; Kovács 1937; Szabó 1937). The experimental and observational work of occupational psychologists and work scientists confirmed these views. "I put a Hungarian from Szekszárd with a German [Schwabian] from the Völgység in a harvest row to little purpose, or a Slovak from Csongrád with an end harvester (*végarató*), the German and the Slovak will fall behind the Hungarian. And the better the wheat lies, the more they fall behind" (Leopold 1934, 455).

Just as common in these texts are differentiations of ability according to class position. The social hierarchy of rural communities between the wars built upon the rock of property ownership, which became the guarantor not just of one's livelihood but also of one's moral character. The poorer one was, the less property one owned, the more disreputable one became. Having to leave home to find work, that is, being relatively mobile in the labor market, was a sign of lack of control and so loss of integrity (see Lampland 1995). We see these sentiments clearly expressed in Reichenbach's differentiation among workers according to property ownership.

> A portion of day laborers are dwarfholders [i.e., own very small properties]. They are more class-conscious and so in general more difficult to manage, but they are more diligent, reliable and skilled workers. The day laborer without property either lives in his own small house, or usually is just a tenant; he often doesn't even have his own domestic animals. The latter are the most mobile, and the least reliable workers politically and morally. (Reichenbach 1930, 196)

There were other considerations as well. Harmath strongly argued against hiring sharecroppers who owned land or had relatives in the vicinity of the estate, claiming that they would spirit off manure to use on their own land or plow their land with the farm's draft animals rather than use these resources solely on the crops they were hired to grow (1937, 50).

Poorer workers were not necessarily unreliable. A family history of faithful service at the manor—a respectable form of diligence—bode well. "Good traits are inherited over generations. That is why we attempt to obtain as many decent, capable, and diligent families as possible working permanently for our estate. At our large estates we often find workers' families, which are descended from many generations of excellent manorial servants and workers" (Reichenbach 1930, 247). On the other hand, not being primarily employed in agriculture robbed one of the ability to work effectively, as the following passage indicates: "Among migrant workers one may also find village craftsmen and industrial workers, who take on migrant labor in the summer to ensure a better living. These latter are usually the least fit for use in the work gang, so it is best if we don't allow industrial workers to be accepted into the crew" (ibid., 209).

Finally, gender and age were central organizing principles for labor studies. As explained earlier, day labor had long been differentiated according to these two axes, women and children earning only a portion of adult men's wages. Particular tasks were coded by gender, such as the division of labor within a wheat harvesting crew: adult men invariably cut the wheat, followed by two women gathering and bundling it after him. Convinced that these differences would affect the results of his experiments, Kölber made a point of stipulating which kind of workers should participate in every phase of his research project (1932, 7–39). In one instance, he explicitly stated that "it is not rational to use adult male workers to thin out corn planted in seed holes, as it entails a lot of stooping and troublesome work," implying that the task was better suited to women and children (ibid., 29; see Jain 2006 for a discussion of comparable arguments about Mexican migrant laborers in California agriculture in the 1970s). Reichenbach devoted several paragraphs in his 1925 treatise to citing Aereboe's careful delineation of gender-specific traits for agricultural workers (1925, 52).

COMMUNITY PRESSURES

While workers were understood to have inherent qualities, the character of their work was also strongly influenced by their fellow workers. "It does happen that the diligent and industrious workers carry the lazier and slower worker with them. However, this is less common. As a rule the latter tend to set the labor tempo" (Károly 1924, 85). Managers were advised to structure the work crew in such a way that the best workers set the pace, for example, placing them at the end of rows and a couple in the middle to ensure that laggards not win out (Reichenbach 1930, 253). The power of the group to bond while working was portrayed by Reichenbach with an image of workers being motivated to work by their own rhythm and, barring that, then by singing (1930, 249). On the other hand, mixing ill-fed and poorly behaved migrant workers with hardworking local laborers could be "demoralizing" (ibid., 210). So too a poorly managed farm demoralized the work force (ibid., 263). Bad management, poor supervision, and inappropriate expectations— e.g., ruining workers' health by setting the pace of work too high ("*kényszer-ütem*")—all defeated attempts to increase output (ibid., 254).

Drawing a community together through work constituted an important means of achieving increased results. Another means of stimulating higher productivity was to use ambition to stir "noble competition" (*nemes vetél-kedés*) within groups (Reichenbach 1930, 249), or what Kemptner called a "marvellous, healthy contest" (*bámulatos egészséges verseny*; Köztelek, 20 Mar. 1938). Halács described the results of an "agricultural work competition" favorably, highlighting advantages to both workers and the national economy, including not only increased output but as an impetus to acquiring new tools and developing innovations in the organization of labor (1928–29, 45–47). Rewards had to be set, which would inspire the more able. Világhy cautioned that when hard workers earned as much as lazy ones, their attitude worsened (1930, 16). Competition was the heart of capitalism, after all. "Records [in work, sport, or aviatics] are the most beautiful documents of human abilities. Entire generations follow those attempts to break the record, the act of which displays their abilities to the fullest. *Vast numbers of goods are produced by the masses doing their utmost in the course of economic competition. Everyone will get his share who works for it*" (Köztelek, 25 Sept. 1927; italics in the original). Both the government and private agencies considered the power of competition an effective tool for increasing production and improving the quality of produce. In 1927, the Ministry of Agriculture announced a "program of competitive production" (*versenytermelési programm*), citing the successes Italy and Austria had acheived with comparable policies (*Köztelek*, 1 Dec. 1927).

In 1935, the National Milk Propaganda Committee mounted a cleanliness campaign (*tejtisztasági verseny*), awarding points for the conditions of milk barns and the health record of the herd (*Köztelek*, 22 Sept. 1935). Reichenbach suggested policies making it an honor to commission someone for particular jobs, as well as holding public ceremonies to recognize workers' accomplishments (1930).

Work scientists spent much time pondering how to use personal aptitudes and social dynamics in their favor. The fields of psychotechnique, occupational psychology, or what Erdélyi called "economic psychology" (1936, 14) were crucial fields for work science. Analysts wished to ascertain the psychological features—skills and temperament—needed in specific jobs and then to develop tests to identify these qualities in job applicants. "The field of psychology includes the investigation of the intellectual and psychological [spiritual] world of man and an examination of whether the directions of thoughts, moods, and feelings, as well as abilities, lend one to be suited for some kind of special work" (Reichenbach 1930, 222). While work scientists and business economists were keenly aware of the intertwining of motivation and aptitude, specific testing instruments for agricultural workers were not developed during this period. "Eventually a separate branch of pedagogy will deal with the training of agricultural workers. Until now, however, in the absence of material tools and equipment, investigations of agricultural work took place by practical observation and experimentation rather than by the help of these branches of knowledge" (ibid.).

The degree to which workers could be molded to the needs of a rationalized work place was central. It was understood that some were more amenable to change, others less so. In light of these constraints, it may have been difficult to change people's behavior—to find just the right sorts of incentives and exercise adequate pressures to influence actions—but it was for this reason all the more worthwhile. Nonetheless, there were some forces that could not be overcome, no matter how hard one tried. Some workers were simply incorrigible. "As they say at [the German experimental farm] Pommritz, a lazy and bad worker would be of more use if he spent his time taking a walk in exchange for a wage, than working with the rest of the workers and just slowing them down" (Károly 1924, 16–17). Erdélyi also cited the significant barrier represented by the strength of tradition, especially found in those branches of the economy, such as agriculture, that have existed for the longest (1936, 113). "Enforcing the principle of rationality doesn't only encounter great technical and economic resistance; it encounters strong *emotional resistance* as well" (ibid.; italics in the original). This echoes the admonition from the German agrarian labor specialist Seedorf to sweep away all traditional modes of working to achieve a properly rationalized, well-structured firm (Reichenbach 1930, 232).

THE QUALITY OF MANAGEMENT

Since the mood of the workforce was crucial to improving productivity, managers were cautioned to take a careful and measured approach to supervision. Fischer cites Aereboe's maxim in this regard: "Nine-tenths of the results of farming depend on the selection of men and the way they are treated" (1936, 18). Vencel Nagy, the manager of the estates owned by the Order of Saint Benedict, made a similar point: "In the old days the value of land was measured by the number of workers' hands. Today a farm's value is raised by the quantity and quality of farm work. Whether full-time or occasional workers, it is extremely important that *our men be satisfied!*" (*Köztelek*, 27 Feb. 1944; italics in the original).

Károly defined the core management skills as understanding human character, a sense of justice, and self-restraint (1924, 77). Without these qualities, one simply could not expect to be able to improve productivity (ibid.). Sustaining the proper mood for work, and forestalling absolute penury, was absolutely central.

> The propertied class that works with agrarian workers, who lives among villagers knows best how different is the work of agriculturalists who are sound, healthy, able-bodied, satisfied, not bothered by family problems and worries from the struggles of men who have lost hope, are poorly fed, weakened, unsatisfied, and battling for survival. That's why the Hungarian propertied class cares deeply about the improvement of the living standards of the village people, with respect to body and spirit, morality and intellect, and not least material conditions. (*Köztelek*, 11 Feb. 1940)

To be effective as a manager, required, more than anything else, treating workers fairly (*igazságos bánásmód*), since fairness inspired trust (Károly 1924, 77). Drawing on mechanical metaphors, Kodar describes the qualities of workers and managers: "The worker is generally a very sensitive seismograph, especially in relation to injustices. He is able to judge whether his companion actually receives a well-earned honor or whether, through protection or without merit, he was preferred" (Kodar 1944, 24). A good manager was someone with a sixth sense for evaluating workers, someone who had "X-rays in [his] eyes" (ibid., 20).

An important technique to ensure fairness was to devise reasonable and fair wage schemes. Business economists were quick to point out that the perception of fairness and balance was crucial. Setting wages too low bred resentment among workers. Results would suffer, and trust would be lost (Reichenbach 1930, 190). Setting them too high raised suspicion that expectations of daily output would be increased. So a proper balance between effort and reward was not simply a calculation based on tangible results; it also ensured the proper emotional balance required to ensure that the outcome expected actually be realized. Of course, a just distribution of goods, as some authors phrased it, did not mean a fully *equal* distribution of farm income. Management's interpretation of fairness extended only so far as a small share of produce tied to results warranted. For example, Reichenbach conceded that the fairest system of payment would be to gauge wages to the amount a worker contributed to the overall earnings of the farm, but he thought such a system would be impossible to implement (1925, 66). Aereboe suggested that it might be possible on some manorial estates to establish a workers' cooperative and reward them with a commission from the firm's earnings. Reichenbach was less sanguine about the possibility of creating producers' cooperatives, citing the difficulties in calculations for judging income year to year and the problems that posed for employer/employee relations. His stronger concern, however, was the danger this would pose to exclusive authority of management to run the farm. "In this instance the workers surely would wish to meddle in management and the way income is calculated, but this reduces the employer's authority and discipline as a rule" (ibid., 67). His animus was no doubt increased by the Communists' promotion of cooperatives during the short-lived Soviet Republic in Hungary in 1919.

Managers were admonished to treat workers equally but also encouraged to come to know workers well enough to consider their personal traits (for example, see Fischer 1936, 18; Kölber 1932, 2). Working with a band of migrant workers from far away posed problems as their abilities and temperament were unknown to management. Therefore, it was better to send migrants into the fields to tend and harvest crops. Without additional knowledge management could not trust them with livestock or machines (Reichenbach 1930, 210). Familiarity with the labor force made it possible to weigh reward with punishment. "One can't treat workers in a cut and dried fashion. We only achieve true results if we treat them almost individually as far as possible. For one a good word, a reward is effective, for another strict orders and punishment work" (Reichenbach 1925, 17; see also Kálmán's comments about the necessity of knowing workers and their limits, *Köztelek*, 15 Aug. 1925). Familiarity did not mean surrendering one's position of authority, however. "The manager who speaks meekly and politely with his workers cannot maintain discipline. Workers take advantage of his weakness. On the other hand, in the presence of constant harsh and arrogant behavior they will easily lose their inclination to work, and indeed can become mean-spirited" (Reichenbach 1925, 17). The point was to be strict but fair (ibid.). In the final analysis, proper handling of workers made it possible for managers to "train workers to be willing to work" (Károly 1924, 76). As Reichenbach noted, reorganizing agricultural firms to improve productivity entailed increasing the numbers of office workers and supervisory personnel employed at the estate, as had been true in industry. He was confident, however, that the increase in cost would be covered by increasing output and reducing the number (and hence cost) of physical workers (Reichenbach 1925, 65). Reichenbach proceeded to argue that an increase in supervisory personnel was all to the good after so many qualified Hungarian estate managers lost their jobs in Slovakia and Transylvania and moved to Hungary when these regions were annexed to Czechoslovakia and Romania after World War I.

Finally, keeping workers satisfied also served to prevent the spread of radical ideas. With memories of the 1919 Soviet Republic in Hungary fresh on their minds, work scientists knew full well the dangers of extremism, to production as well as property. The close proximity of the Soviet Union, growing rapidly into a menacing presence, also played a role. "The true danger of Soviet propaganda does not reside in parades of drumbeats and red flags, or the open propagation of communism. Soviet propaganda is dangerous when they try to disturb state and social order *by taking advantage of unsatisfied and embittered elements who are cursed by the difficulties of economic life*" (*Köztelek*, 25 Jan. 1925; italics in the original). In support of his argument that agrarian wages had to be increased, Szeibert claimed that avoiding poverty should be a national priority, since squalor debauched workers and incited them to rebel (1939, 846). Steuer criticized his contemporaries for calling in police to restore order among discontented workers. Quoting a worker, he cited "the unfair wage, the empty stomach and the worker's child crying for bread" as the true agitators, arguing instead that the demand for decent wages was justified (1938, 602). This did not mean endorsing trade unions. Reichenbach was convinced that efforts to organize the rural work force, or worse, force them to move to the city for work, would drive them into the "arms of international socialism" (1930, 211; see also the editorial in *Köztelek* on "Alarming organizing," 25 Jan. 1925). As a solution to the emotional whirlwind of political extremism, Hatvani could offer the cold reason of science. "We mustn't pay attention solely to the economic advantages of some organization of work, but also to those factors that satisfy the highest social needs of the working masses, *because in all respects the satisfied worker is the most rational factor* in every aspect of economic and social life" (1935, 12).[14]

WAGES AND MONEY AS PRECISION INSTRUMENTS

The most powerful and effective instrument to alter working behaviors was a well-designed wage system. In most cases, this meant a monetary wage, but it could just as well mean benefits like keeping a cow or having a garden. Money was powerful, no doubt, but for that reason had to be handled judiciously. Fine distinctions between rates of output could be rewarded or punished with differential payment systems. If anything were to break the bonds of traditional production methods, national character, or inherent laziness, money was it. "It is a fact that an appropriate wage is the most suitable means—alongside a hygenic workplace and style of working—to increase the disposition to work" (Raith 1930, 256). Wages, in short, were an ideal disciplinary instrument.

The primary means by which workers were to be drawn into productive relationships was to have them take an interest in their work, make them appreciate their contribution. "We can only hope for effective work and work results if we make the worker interested [in his job] and connect his incentives to the firm's production" (Leopold 1934, 457). This was best done with wages. It is for this reason that work science paid extensive attention to deciphering various wage regimes and determining their value in agriculture. Inspiration came primarily from German agrarian specialists such as Pohl, Aereboe, Lüders, and Ries, though Taylor was ever present as a founding father (see, for example, Farkas 1941; Fischer 1936; Károly 1924; Kölber 1932; Reichenbach 1930). V. Fodor also made reference to American specialists in industrial management, mentioning Gantt, Halsey, and Emerson in his review (1925).

The foundational assumption behind the power of wages was that desire for material wealth was lodged in every man's soul. Hence promise of reward was highly effective. Károly writes of the "impulse for material gain" (1924, 79). Reichenbach tied material gain to aspirations for upward mobility: "*The material result of his work and the possibility of getting into higher social strata* can have influence over the worker's frame of mind and in this regard his productivity" (1925, 17; italics in the original). Quoting from the German journal *Landarbeit*, published by the Deutches Landes Gesellschaft, Reichenbach speaks of "material motives driving work" (ibid., 17–18), using the language of fundamental force (*indító ok*) he had deployed in discussing psychological factors in work. There was near unanimity among work scientists and managers that material inducements were the most powerful stimulus in their armamentarium. Poverty among agricultural workers only strengthened the dosage. "Only for a few people or in the rarest of characters is work pleasure, which we pursue for its own sake. Generally it is economic compulsion—

breadwinning in the true meaning of the term—and to a lesser degree the possibity to acquire some desirable goods in life" (Perneczky, *Köztelek*, 3 Jan. 1944). In his analysis of the Halsey and Bedaux systems, Keszler recognized their superiority to time wages, as they were premised on the ability of workers to expend the extra effort for higher reward. "Rather than prodding with words or worse—threatening to let someone go—this attempt means foregrounding the notion of agreement. This tool, however, only looks to be free from coercion in its external appearance; in reality it means hard constraint, which is hidden in the driving stimulus for survival and the strong urge to acquire a higher standard of living" (1941, 892). The absolute necessity of financial reward for prompting workers to produce was underscored by Fischer's conviction that the sign of a good manager was his ability to sustain workers' good humor without having recourse to additional pay (1936, 18).

Strangely, although the desire to improve one's lot was understood to be innate and natural, it was not present in all workers. Ignorant, humble workers consigned to their fate presented a problem, as they lacked any interest in a better life (Reichenbach 1930, 240). At a manorial estate in Eastern Prussia, Seedorf described a situation in which workers, neglecting new wage incentives, worked long enough to earn their customary day wage and then went home. Comparable sentiments could be found in Hungary, Reichenbach noted, especially among workers of Romanian (*oláh*) ethnicity. "Yet these days this kind of case is already rare. In most cases the wage of work completed has a stimulating effect. Farmers must train workers in work paid by performance" (ibid.).[15] It was not an easy lesson to master. One farm steward (*gazdatiszt*) shared his experiences in a *Köztelek* column about the difficulty of providing manorial workers new benefits, for example, new ways of ploughing private gardens, vaccination for livestock, and (most unusual) a nursery school. Under the title "Often little people must be helped against their will," he proudly declared that "all these innovations were greeted with loathing and opposition, but then there followed material betterment, joy, and satisfaction" (*Köztelek*, 3 Mar. 1940).

Making workers interested in their work (*érdekeltté tenni*) meant, first and foremost, to tie their pay to the results of their labor. Wages would increase, in proportion to their efforts. Of course, this was to be done while maintaining quality (Fischer 1936, 22). Since the majority of contracts in place were not structured around results, but defined as a temporal unit, new result-based wage systems had to be introduced. The first step in revising wages was to construct new contracts that stipulated the level of output at the outset. The second phase was to develop a series of rewards for output beyond that stipulated—bonuses, premiums, and so on.

The agrarian work science literature generally distinguished between two general wage forms. Though there were minor differences in nomenclature, the basic divide was between wage forms that were primarily temporal contracts, epitomized by manorial servants and day laborers, and wage contracts structured around results, intensity, and speed, which, in short, could be calibrated to task, such as job work or sharecropping.[16] The terms by which job work was set, and the means of judging output, could vary. "The basis of measuring the amount of work can be: (a) the amount of product manufactured, piece work (piece wage) or its weight and gauge, (b) the amount of material used, (c) the area worked, (d) the mass moved, (e) open space or hollow made" (Károly 1924, 78). The unit of measurement alone was insufficient, however, since quality was also a factor. Reichenbach refined this criterion by stipulating that one had to be able to redo the work until it was satisfactory (1930, 199). Examples he mentioned included shearing sheep, shucking corn, harvesting sugar beets, and digging ditches (ibid.). In his list of possible result-based tasks, Szeibert included activities like building maintenance (1939, 826). This would be the logic that pertained when result-based wages were designed for workers tending livestock (e.g., Kölber 1932, 52–56; Reichenbach 1930, 244–45). Unfortunately, much of the work done on farms was not easily chopped up into calibratable units. "Many kinds of work must be completed to produce the end result. The outcome is influenced by weather and physical conditions. In addition, the tasks necessary to reach an outcome in agricultural work often flow into each other and are regularly handled by different workers" (Szeibert 1939, 820–21).[17] One possible solution to this dilemma, Szeibert suggested, was to combine elements of both a time wage and result-based pay (*akkord*) (ibid., 821). Wages could also be moderated in light of the relative difficulty of the task, its pernicious health consequences, or simply how dirty one got doing it (Reichenbach 1930, 202). The move to result-based wages or sharecropping was considered inevitable, part of the shift to intensive cultivation. As such, payment by results was understood to be the most modern of all wage forms. "*The primary ambition of the employer farming rationally and intensively will be to increase the results of the most expensive labor power and human labor power to the highest degree.* Sharecropping or piece work remunerating results are the most suitable for this, while the least suitable is a time wage" (Schrikker 1942, 23; italics in the original). Using geographical shorthands for stages of modernization, Schrikker explained that the farther west one traveled, the more common it had become to replace time wages with result-based ones (ibid.).

Wage forms had advantages and disadvantages, as business economists and work scientists recognized. Increasing productivity by calibrating wages to effort was the most obvious advantage. Workers on a time-based regimen were nearly impossible to propel to higher productivity, as they had become indifferent to the presence of supervisory personnel (Károly 1924, 78–79). The tendency for result-based workers to speed up their pace was valuable, since it came in handy when workers were busiest (and scarcest) in the high seasons of the agricultural year (Reichenbach 1925, 69). Indeed, Rath argued that because agricultural products were perishable, rewarding prompt completion was more important than in other sectors, challenging those who believed that improvements in quantity and quality of production was only properly achievable in industry and office settings (*Köztelek*, 25 Feb. 1940). Moreover, working with result-based workers made it easier to figure out the labor needs of the farm reliably, since the effort expended could be accurately anticipated, unlike that of day laborers (Reichenbach 1925, 69–70). Of course, this calculation depended upon having a clear sense of workers' normal output (Reichenbach 1925, 71; see discussion of calculating normal performance earlier). Improving workers' output also meant cutting labor costs, as fewer workers were needed to complete tasks. Kálmán claimed to have achieved a reduction of 20–30 percent in manual labor needs when he experimented with Taylorism (*Köztelek*, 15 Aug. 1925). The disadvantages of some labor contracts also needed to be considered. Reichenbach ennumerated a number of problems in relation to sharecropping contracts: proper maintenance of the land, divergent yields, and quality of produce (1930, 207). The question of what is more profitable—contracting with specialized sharecroppers or growing certain crops in house—was an important consideration. To guide farmowners, Harmath published an extensive analysis of the range of profitability of sharecropping contracts (1937). His advice extended to farm owners who were not themselves trained experts, suggesting forms of sharecropping that did not require close, skilled supervision (ibid., 123–25). Even though time-based wages were considered the least productive, they also had their use. Day laborers could be employed only when needed, so they provided a helpful supplementary role in managing minor and unanticipated tasks at the estate. The value of manorial residents was their familiarity with the estate and local conditions (Világhy 1930, 16). Since manorial estate workers were primarily employed in tending to livestock, familiarity with the herd—cows, sheep, cattle, horses, pigs—could be as significant a factor as knowledge of local soil conditions or rain patterns. The value of migrant workers was their specialization in intensive crops (Schrikker 1942, 6). A disadvantage of migrant workers was that they were strangers to the farm, unfamiliar with local conditions, and unknown to management (ibid.). Some also claimed that migrant workers were morally questionable—conflating poverty with poor moral behavior—though Schrikker insisted that these claims were not generalizable (ibid.).

Managers explicitly targeted emotional states to encourage greater effort. In the context of wages and premiums, Reichenbach mentioned envy as a particularly good tool (1930, 241–42). Positive feelings of self-respect were also useful. Enhancing workers' independence from meddling managers strengthened their self-esteem and cultural level (ibid., 242).[18] These characteristics accrued to the advantage of the firm, since it was assumed that more culturally developed and skilled workers were more inclined to work harder. Indeed, Károly was of the opinion that "the freer and more independent the work and the more complicated the work, so the less possible it is to exercise influence on work by external pressure" (1924, 79). The best workers were put off by close and bothersome supervision (Reichenbach 1930, 266). The internalization of discipline—dependence on emotional states and psychological motives—was the hallmark of the good worker.

The hallmark of a good manager was his ability to learn how to supervise workers contracted to do result-based work. In short, supervisory staff were also called on to change their way of working with the introduction of new wage systems. Supervising result-based work demanded much greater attention and skill than time wages had in the past. Careful and conscientious managers were crucial. "[This is] one of the most important conditions of work supervision in the Taylor system. Without this the work results rapidly decline. When they notice the absence of inspection, the desire to reach great results involuntarily disappears in workers" (Fischer 1936, 21). The consequence of poor management habits could be easily calculated.

> The system where they shout at work from afar is wrong, because then they don't know who is being reprimanded and 50–60 workers will pay attention all at once, interrupting the work. If it only lasts two minutes, that comes to two hours of a worker's work and if it happens 10 times a day, then that means two day laborers. If we have something to say or show, we should do it up close, and if this doesn't work, then we should pull him aside, because a bad worker can detain the crew. (Kálmán, *Köztelek*, 15 Aug. 1925)

Reichenbach was convinced that by introducing payment by results, it would be possible to reduce the amount of time devoted to immediate supervision, and so reduce the frustration of sophisticated workers caused by meddlesome overseers. If the quality of the work could not be rectified later, however, then constant supervision was warranted. In those instances, the best strategy the manager could take on would be to devote his attention to organizational issues—effective division of labor, dealing with occasional problems, and supporting the workers. "He creates the impression among the workers that he is the manager, not the embodiment of uncertainty toward the farm workers" (Reichenbach 1930, 266). The German specialist Ries concurred with this judgement: "'The thorough scheduling of work and determining the most appropriate way of remuneration is more important than increased work supervision'" (quoted in ibid., 261).

An important innovation in agrarian wages was to provide bonuses to reward outstanding performance. Károly distinguished between two kinds of bonuses: the premium and the commission. "It is characteristic of rewards that they are negligible . . . in relation to the base wage. They are not based on measurement, but are approximate estimates, which are not actionable but carry the sense of a gift" (Károly 1924, 78).

In contrast, commissions or premiums were figured "on the basis of a rate determined according to the output reached with a number of interconnected jobs" (ibid.). While premiums were often paid in cash, they could also be distributed as payments in kind or as benefits. To cement the relationship between effort and reward, Reichenbach suggested that initially premiums should be paid daily, even if pay was usually distributed only once a week (1930, 242).

Premiums, indeed any result-based wage, had to be designed to reward individual work, not that of the group. Reichenbach considered this a "general rule of work science" (1930, 253). "The farm at Pommritz noticed with sugar beet cultivation that the amount of work completed increased to a great degree every time they worked alone and not in the crew. The work results of some piece work tended to be 30 percent more, and even a greater percentage more than piece rate work in crews or groups" (Reichenbach 1930, 252). Not all workers were capable of rising to the occasion of individual effort. Germans' experience suggested that women and youth should be assigned group work (ibid., 253). Reichenbach quoted Ries in this regard: "'if workers' character is less assertive . . . whether that be due to youth or low cultural level, then . . . we should assign them to a work crew, as opposed to working alone'"(ibid., 254). The only instance in which group work proved to be more productive than individualized contracts, according to Reichenbach, was when families were contracted as a group (ibid., 255).

How premiums were used varied. Discussing premiums introduced for ploughmen working with steam-powered ploughs, Kölber distinguished between premiums for a day's work, for a set amount of acreage, and for an entire season. In his opinion, the most useful of the three was the seasonal premium, which rewarded a progressively increasing premium for doing more than the stipulated acreage (Kölber 1932, 7). Workers feeding threshing machines were offered premiums for exceeding the amounts of grain to be threshed in a day (Reichenbach 1930, 243). Encouraging workers to complete their work earlier than was customary with day labor gave workers a shorter work day. Thus time won away from work became the reward (ibid.). The premiums Lüders used effectively overcame problems presented by the gendered division of labor. In one instance Lüders offered female day laborers an hourly premium if they would stay the full day at the estate, foregoing their usual habit of contracting for shorter days to give themselves time for housework. In another, he offered an additional hour's wages to women who would be willing to lead a team of oxen, usually a man's job (ibid.).[19] Fischer described an additional premium-task system (*prémium pensum*) in which the maximum amount of work was stipulated, an important feature to prevent workers from overtaxing draught animals in pursuit of a premium rewarding speed of completion (1936, 22). A variation on this theme for those working with mechanized equipment was to levy a maintenance fee (Kölber 1932, 8) or to reward workers who kept their machines working properly (Reichenbach 1930, 243). Lüders also mentioned providing manorial servants with an allowance for long service, as a means of keeping them from moving elsewhere (ibid.). Yet no matter how powerful incentives could be, they had their limits. V. Fodor had serious reservations about implementing Taylorism in agriculture. Not all workers could be prodded to work harder; the weak, less industrious, and malicious workers would not work harder, even if offered a premium for their efforts (V. Fodor 1925, 44).

MONEY VERSUS IN-KIND PAYMENTS

The dominant assumption among work scientists and business economists was that paying workers with money was a crucial long-term goal of rationalization. This was true not least because of their observations that more modernized countries had switched almost entirely to monetary wages (e.g., Reichenbach 1930, 190). But whether payments in kind should be eliminated in the short term was more controversial. The cost of replacing payments in kind with cash could be burdensome for manorial estates. Allowing manorial servants to keep a few animals, and work small plots of land, did not require any serious investment on the part of the estate. Indeed, the entire *kommenció* system required less working capital than a monetary wage system would have (Reichenbach 1930, 188). "When money is tight on the labor market one can see that sharecropping—and here particularly since 1930—gains ground to the detriment of cash piece work and day labor" (Leopold 1934, 475). Badics also approached the question in part by considering the general character of the agricultural sector and the broader economy.

When [farming] extensively under undeveloped economic conditions, where produce is cheap, money is relatively expensive, and workers are scarce. From the perspective of the farmer monetary wages constitute the greatest cost. Under developed economic conditions, where money is relatively cheap, the population dense, i.e., many workers are available, for whom the wages are relatively low, agricultural produce as a result is expensive—the same amount of monetary wages represents a much smaller cost. (1929, 21)

Of course, the value of money and the state of the economy changed over time, especially in Hungary early in the century. "Before [World War I] cash allowances were the larger portion of piece wages than today, since as the consequence of depreciation, money became worthless, which then was replaced—especially during communism [in 1919]—by increasing remuneration in kind" (Szalay 1931, 43).

A strong argument in favor of payments in kind was that receiving grains and other goods from the manorial estate ensured a basic standard of living (Badics 1929, 16–21; Kovács 1935, 214; Schrikker 1942, 21–22; Szeibert 1939). Szeibert noted that "it is our experience in general that transferring to a system of cash wages leads to the reduction in the workers' standard of living, quality and quantity of food" (1939, 820). The ability to keep animals was a particularly valuable element of an estate servant's benefit package. The one sure way that a manorial estate worker could accumulate a little capital was by selling animals at village markets. Just as important, perhaps, was being able to consume milk products. Not every estate permitted keeping cows, preferring to distribute milk to manorial families. Szeibert cautioned against this. "Allowing servants to keep cows is not only more advantageous for them, but also for estates, because farm servants who own cows tend to be calmer, more reliable, and more employable" (ibid., 823). Reichenbach voiced similar sentiments when he explained that while paying with money might be easier, it also introduced a level of freedom that was undesirable. "The servant paid in cash is not tied to the farm by any interest. He is quick to leave his place and so is not as reliable as the [yearly contracted] servant" (1930, 189). Protecting workers from the vagaries of markets—fluctuating prices, the costs of middlemen in commerce, and so on—was a central theme in discussions of payments in kind (e.g., Badics 1929; Reichenbach 1930, 189–90). But just as important to many observers was the need to protect workers from their own foibles. Workers had to learn how to deal with money. Leopold asserted that manorial servants simply didn't know how to deal intelligently with money (1934, 462).[20] Reichenbach claimed that unmarried workers were liable to spend their money on drink and unnecessary things (1930, 189). Marriage was understood to curb these appetites. Another reason offered for keeping in-kind payments was political. Blantz was convinced that the only way to prevent an organized wage movement was to pay workers in kind (1936, 102). Finally, the structure of the rural economy was also an argument against cash wages. In the absence of a well-developed consumer economy in rural communities Leopold asserted that "cash dries up faster and rots quicker than corn on the cob" (1934, 475).

Conclusion

The work of work science consumed hours and hours of time, though its tireless proponents considered it all well spent. Preliminary stages of ascertaining the value of labor in agriculture to scientific standards required agreement over what constituted a unit of activity, a standard workday, and a standard worker. With these in hand, they could proceed to figure out methods of commensuration to compile data that could be run through additional calculations to assess the costs of labor and the level of productivity of a farm. Only then would it be possible to begin to design wage systems that would reduce the cost of labor by improving productivity, if paired with proper attention by management to psychological dynamics. Ideally, the new systems would be subjected to field trials, demonstrating their efficacy conclusively. If good wage plans were successfully adopted, workers would eventually acquire an entirely new attitude toward work. They too would evaluate their tasks in terms of specific units of activity, define the workday in terms of a set number of hours, and learn how their employers wished them to behave. They would become modern wage workers.

Agrarian work scientists nourished a grand vision of scientific progress. A rigorous analytics of agricultural work and farm management would usher in a new era of empirically sound and conceptually sophisticated firm studies. Applying these insights to business strategies and government policy would improve the nation's fortunes, as it struggled to survive in intensely competitive international markets. Ultimately the transparency and precision of monetary wages would produce an efficient and docile labor force. Unfortunately, work scientists' trust that money would solve problems of productivity and discipline was illusory. In the period between 1918 and 1946, government-issued money was anything but reliable and at times entirely useless, repeatedly shaking people's faith in cash and coin. It is to this complex history I now turn.

4

The Problem with Money

In the century following the abolition of serfdom in 1848, money was many things in Hungary: powerful, fertile, unreliable, fickle, destructive, and dangerous. Money was the source of many problems and, in some cases, their solution. For agrarian business economists, money had the power to transform the economy and rejuvenate work. Others acknowledged the power of money but were reluctant to see its destructive force decimate the Hungarian nation. In short, the cultural sentiment of money in Hungary was a mixture of fascination and ambivalence. Complicating this already complex picture were the recurring difficulties the nation faced in maintaining a stable currency. Money was construed as a morally contradictory force, capable of eroding social bonds even as it invigorated the nation's economy. It is the purpose of this chapter to discuss a range of attitudes and variety of episodes surrounding money to illustrate this strange history. Doing so will also shed light on the imperative to calculate labor value in abstract units. In an agrarian economy dominated by payment in kind and a lingering suspicion of the national currency, figuring the value of labor in terms of fixed monetary wages was ill advised.

What is Money For?

In Simmel's famous formulation, money is itself a means, the true cipher of the modern economy. To confuse it with its physical quality, or even with the valuable metals it represents (gold, silver), is to misunderstand its primary function. Marx described the misplaced concreteness money represented as its fetishization, that is, seeing it as an object when in fact money was first and foremost a social relation. Both of these views inform the analysis of money to follow, although the manner in which I will deploy these ideas is meant to historicize the processes whereby money becomes a means, the ways that money constitutes social relations over time, in varying conditions and contexts. While Marx and Simmel developed their insights into the strange phenomenon of money in the process of analyzing the changing social relations characterizing modern society, many of the specific features they describe in the late nineteenth century had yet to accurately portray how money lived in Central Europe—even in their own territory—much less elsewhere around the globe. As the history of modern monetary instruments tells us, creating and eventually stabilizing national money was a long and arduous process, which proceeded at a different pace in various regions dependent on far more than simply issuing a national currency. The reliability of exchange using monetary instruments could wax and wane, as the fortunes of the national (or local government) shifted over time. National money was completely unreliable during inflationary spirals, economic crises, and war. Therefore, institutionalizing a single and exclusive national money confronted obstacles far more substantial than the political and economic logistics of founding national banks and convincing citizens of its value.

In Hungary, large segments of society had moral trepidations about the use of money well into the twentieth century, though the reasons for this varied. In the decades following the abolition of serfdom in 1848, money was often thought to cut the bonds tying servants to masters, eroding habits of deference and authority necessary for civilized society. Aristocrats were well known to disdain handling money.[1] Their disdain extended to any association with business as well, a phenomenon complicated by anti-Semitism and Catholic sanctions against usury. This history is a familiar one, played out in other regions of Central and Eastern Europe. Among the ranks of agrarian socialists gaining ground in the 1890s, money epitomized the ruthless forces of dehumanizing exploitation imposed by capitalist relations of production. Yet voices advocating alternative conceptions also were raised. Money could be productive and useful. Money could entice workers to a job that they would otherwise shun. Creating markets for Hungarian products would strengthen the economy and improve the fortunes of farmers and workers. So, too, money was key to the success of modernization promoted by agrarian business economists and work scientists, since its power could only be exerted if it were the dominant means of compensation.

Hungary had its share of wars, revolutions, and economic crises. In the course of one generation, Hungarians experienced a war, the dissolution of an empire, the birth of a nation state, a brief socialist revolution, several years of inflation, a decade of economic depression, another war, and the greatest inflationary spiral in world history.² In every instance, the ability to use money—as cash, as credit, and as capital—was compromised; time and again the ostensibly natural fungibility of money was frustrated. And when money could not move around easily, the ability to extract value—as profit or as state revenue—was seriously hampered. New currencies had to be created, more printed, monies recalibrated, and finally, as a last resort, abandoned, for example, when taxes were collected in agricultural goods rather than in cash. In short, the possibility of measuring value was constantly frustrated because the instrument was faulty. In the absence of a reliable currency, it was hard to determine the monetary value of objects and people on the open market. So despite the extensive effort devoted to figure out how to measure value—the value of labor in this case—the means to stabilize value, to "domesticate" it, to harness it scientifically were inadequate to the task. The destructive and erosive quality of money was an ever-present issue, as was the attempt to assuage these concerns in the interest of making money a simple tool, a means toward another end. It is a testament to the conceptual power of money that the recurring problems Hungarians faced with fiscal instruments did not dissuade work scientists and agrarian engineers from believing that the best policy for modernizing labor organization and increasing labor productivity would be found in transforming abstract units of value determined by science into monetary wages. This confirms Simmel's and Marx's ideas about means and about fetishes, since even in the absence of reliable money, the notion that money would be the savior of capitalist productivity still held sway.

In the first half of this chapter, I will rehearse several debates over the use of money voiced in the latter half of the nineteenth century. At the outset, I would like to emphasize that I will not be offering a simple unilinear trajectory from discomfort with money to its complete acceptance. This is not a story of unidirectional movement, but a story about the conflicting attitudes and contradictory consequences of money. In the second half of the chapter, I will describe the recurring crises around money Hungarians faced between 1918 and 1946.

The Inherent Value of Things³

In the latter half of the nineteenth century, value was broadly believed to reside in things, qualities that were unchanging. This could be expressed in relation to other goods or actions, but only insofar as it assisted the conceptualization of value as fixed in material properties. Accordingly, valuing a farm's assets entailed estimations based on the average output of land and the quality of produce. Pastures were evaluated in terms of the quantity and quality of feed, whereas arable lands were defined in terms of wheat cultivation. All crops were assigned a numerical value in relation to one hundred pounds of wheat, for example, rye was valued at 120 and oats at 150. The numerical indicators permitted landowners to determine the costs of buildings, tools, and land in more reliable indices than the prevailing price.

> A suitable ... estimate is arranged such that the current and lasting value of the land figures prominently, and the estimate should be as valid centuries from now as the time when it was done. This, however, is not feasible if the value is ascertained in money, since the value of money being changing, estimates made in monetary value are only valid for the present, yet in the future they do not offer certainty. The proportion that stands between the values of plants usually produced for human and animal food and to the value of land is more lasting. Consequently, money should not be taken as the symbol of value; instead it should be the primary crop for bread of the respective country or region, hence in our country wheat or rye. (*Gazdásagi Lapok*, 29 Jan. 1849)

Natural products embodied value in ways money could never do. Indeed, elevating money as the medium for setting value above the constancy of natural products was seen to be abnormal, hence demoralizing and dangerous.

With the abolition of serfdom, aristocrats could no longer demand labor service to work their estates.[4] The possibility of having to consider paying for labor was a difficult pill to swallow.

> The [feudal] estate, which brought income to the landlord without any work or industry, has shrivelled, and the source promised in its place has not yet presented itself. Since seignoral property had been cultivated primarily with serf labor, it has now been deprived of the terms of its profitability. [Seignoral property] has earned less than nothing in many places, because one must pay the worker so much that the cost of labor already consumes the potential benefit of the estate lands, [leaving them] in miserable conditions. (*Gazdasági Lapok*, 1 Jan. 1849)

It was becoming clear that the freeing of the serfs meant the repudiation of the peasantry's dependence on the lord for sustenance. In fact, the tables were turned, forcing lords to become increasingly dependent upon the peasantry, who could then demand a wage more to their liking. "Where the worker has an increased consciousness of his indispensability—be that a worker from any part of the world—he will increase his demands and claims, namely day labor will be expensive" (Kenessey 1868, 13). Indeed, many peasants refused in the immediate postfeudal years to work for money at all. For many estates, little or no cash reserves limited their ability to shift to a nonfeudal economy. But even for those estates in which money was available, the difficulty of recruiting labor was still quite great. László Korizmics made the following comment in the late 1860s:

> As to ... conditions ... on the Great Plain, in our opinion, it would surely be good for larger estates, even if they have the financial means to buy equipment, to think twice before setting up their farms to be managed by them alone. The major reason for this is that although it is difficult, one can still get labor power for land; on the other hand, for money one cannot get labor power in many cases for any price whatsoever, especially in the amount that immense estates would require. (Vörös 1976, 68; quoted in original text)

All sorts of schemes to institute sharecropping arrangements were contemplated. One author advised landowners who did not want or could not afford to pay money to lease lands for sharecropping row crops or feed under the stipulation that the sharecroppers provide a team or two workers for each cadastral acre of land. "In this way, without ploughing or sowing, the landowner gets pure profit, and sometimes perhaps more than if he had cultivated the land with his own labor. But he also obtains readily available draft power and manual labor to cultivate his other lands, and at a time when otherwise one could not even obtain it for a high price" (*Magyar Gazda*, 12 Jul. 1859). Lords obviously resented having to pay any wage, much less a decent one. As a consequence, peasants' refusal of an insufficient wage was taken as an indication of sloth. "[T]he curse of the life of the common people is this: *our people's inclination that if they have something to eat, they would rather laze around home than*—in their estimation—*go to work for measly wages*" (Kenessey 1868, 47; italics in original).

Money was believed to corrode relations between masters and servants, farmowners and laborers. Noble sentiments were being eliminated by greed and the pursuit of monetary reward. Money had demoralizing qualities; being enslaved to money was a frightening prospect.

> We liberated our serfs, but we established a yoke far heavier, because today the great mass of the commercial class does not work for itself; it works for the master of money. The modern economy proper is not composed of true values, nor owning precious metals, but from laying claim to the proceeds of others' work. The state, town, official, landowner, manufacturer are all indebted and they are forced to relinquish the fruit of their labor to financial capital. (*Magyar Gazdák Szemléje*, Jan. 1896, 7–8)

Money had a pernicious influence on a master's influence over the help, as it rendered servants "familyless and homeless." [5] It was contaminated, associated with less cultivated or upright people, and wont to bring out the worst qualities in individuals, associated with poorer work habits and a less moral community. In the classic sense, it was polluting to the social fabric.

> Above all I recommend that next year we have the harvest done for a fair portion, not for money or by the job, as has become common in recent years. Payment in kind commits the harvester much more than money does: first because he comes out much better when there is a fair division, and secondly, because harvesters have a penchant for shares, a custom inherited from their fathers.... The share-harvester comes usually from the more respectable strata of the people as a rule, while the man who harvests for money, except for the harvest bosses, comes almost entirely from the lowest strata, the rabble, which hardly has any notion of law and obligations, and is more inclined in every way than the previous group to violate discipline.... Most landowners, prompted by growing state and individual liabilities, are forced to ... hire harvesters for cash; ... this innovation has cast aside the old, patriarchal character of harvesting work, has severed the cordial tie that existed in the past between the landowner—usually the peasant—and the harvester. (Kenessey 1868, 43–44)

The cordial relationship between master and harvester guaranteed the proper fulfillment of duties, while the calculation of cost and profit hastened the demise of propriety and respect which ensured a decent harvest for the lord.

On the other hand, there were those who unabashedly called for higher wages as a solution to the problem of intransigent laborers.

> There is still a great obstacle, which with the establishment of our provisional system must be overcome like a man, be it in the interest of regular pasturing or stabling: the servants and herdsmen's lack of willpower, thieving by outlaws and fencers wide-spread over the broad region of the Great Plain, also the crime of torturing animals, which in specific regions of our homeland has become frequent due to the moral depravity of our people; ... one can only help this deep-rooted trouble with *money*, according to the Hungarian saying, *one can even buy the devil with money.* Namely, one could exert an influence by offering prizes and awards, a percentage after the fertility of our animals, etc. But if we cannot help things this way either, then let us leave the old herdsmen with their fixed ideas; let us direct our attention to the young generation, our first task being to ensure their financial circumstances. (*Falusi Gazda*, 13 Sept. 1865; italics in the original)

The author shares with his fellow critics all the prejudices against the intransigent worker, painting an image of filth, impurity, immorality. And it is precisely for this reason that he believes money—with all its unfortunate associations—would be an appropriate tool to sway their hearts, "for even the devil can be bought." Hence the author's willingness to introduce wages is a concession to the immorality of the worker, not the expression of new attitudes toward money as a morally neutral instrument of economic growth.

If wealthy landowners were ambivalent about the role of money in agricultural affairs, they were completely clear about the inability of the lower classes to use money wisely. Describing practices employed at the manors of Archduke József, the author states unequivocally:

> Manorial workers' pay is based mostly on remuneration in produce, which according to the experience of the estates has proved to be the most correct principle, because they cannot estimate the value of money as much to manage to farm with it as they do with payment in kind. Where manorial workers have begun to turn away from the old accepted custom—that is, they do not have their grain ground, but sell it and buy flour—there they are not as frugal as they were before. (*Magyar Gazdák Szemléje*, May 1896, 157)

The author identifies the destructive role of money by showing how it undermines habits of frugality: rather than processing wheat locally, manorial workers sell it and buy flour. A truly frugal economy, in this perspective, is one that avoids money entirely and builds on the subsistence economy of natural products and local relationships.

Left-leaning spokesmen also worried about the corrosive effects of money, but for different reasons. "The last spark of human feeling has died in wealthy people. Only their heartlessness is greater than their desire for profit. They consider the worker to be a tool, which provides a means for their carefree and lazy life. . . . They treat the worker like a draft animal. He is required to make do and not have any expectations" (Csizmadia 1896, 25, 28). Workers experienced capitalism as dehumanizing. They were being transformed into another factor in capital investment, another component of the economic unit.

The sullied reputation of commerce accounted for much of the discomfort voiced in the latter half of the nineteenth century toward the increasingly crucial role of money in the economy. Much of the discomfort surrounding commerce was the perception that it interfered with the natural expression of value. By their manipulation of markets and of money, mill owners and other commercial interests could inflate prices, depriving peasants of a decent price for their wares, contributing to their wretched poverty. "Money has stepped in in place of contributions in kind, opening the gate to the ravaging effect of middleman's business" (*Magyar Gazdák Szemléje*, Jan. 1896, 9). The attitude that market actions were essentially artificial is well illustrated in the following quote, in which commercial interests literally "make prices": "Free competition makes way for the domination by the strongest. The host of price makers—resourceful, agile, and constantly getting stronger—intrudes between the sweating farmer and the modest consumer, getting the better of the producers and holding the consumers in their hands" (*Magyar Gazdák Szemléje*, Aug. 1896, 83). Agrarian spokespersons resented the ascendancy of shopkeepers, industrialists, and merchants over the worthy peasant. The values of national identity and civilization were undermined in this new regime of money and power.

> No one can consider them proper proportions if the younger salesman of some merchant enjoys a salary of 800–1,000 forints, while a teacher is forced to consider a minimum stipulation of 400–500 forints as an achievement. The stingy settling of a worker's wage stands in stark contrast to this unjustified munificence. The wife and children of a worker are also forced to work, their housing is not fit for human inhabitation, their diet is wretched and insufficient, teaching and nurturing are suspended. The worker does not even know his boss. This is the situation of the worker today. (*Magyar Gazdák Szemléje*, Jul. 1896, 9)

The notion that forces intervene between producers and the market is especially well articulated in discussions about wheat futures and the stock exchange. Many clearly saw these institutions as inappropriate, finding the sale of futures totally exploitative. The danger of abuse was certainly there, prompting the minister of agriculture to issue a decree to all agrarian associations to be on the lookout for merchants conning peasants into selling their wheat early, usually at a loss (*Földmívelési Ertesitő*, May 1898, 413). Futures were often discussed as paper wheat, an image that jars the very tangible sensibilities of the Hungarian agriculturalist.[6]

> For [speculators] wheat, the dear fruit of the farmer's toilsome work, is only a pretext, a reason for the game. While they live from the *pretext*, from the thousands and thousands of quintals of paper-wheat bought and sold, in the meantime the plough handle falls out of the hands of the cultivator of the soil, and land ceases to provide an income. . . . The firm bases of our economic life have been demolished. . . . Professional *gamblers* have filled the rooms of the exchange. (*Magyar Gazdák Szemléje*, Aug. 1896, 85–86; italics in the original).

The whole notion of paper wheat was clearly anathema and bred disdain. Agrarian writers derided the notion that the way that to make money was for commercial interests to play the market. Middlemen and stock brokers, merchants and futures buyers were all classed as usurious hangers-on, living off the humble populace in villages and towns. Speculation, futures trading, the stock market all came in for criticism, representing to the authors underhanded and inappropriate means of acquiring wealth. These shady dealings were opposed to the integrity of agricultural work and land ownership. István Bernát made this position very clear in his call for outlawing the futures business.

> Once again we must advance ourselves with productive labor and the tools of national prosperity.... Reliable commerce has as little *interest* in feeding and maintaining parasites as producers, and what use are low prices to consumers if they destroy producers, who are the consumers of their products? For a certain time it is possible to eclipse [overshadow] the laws of economic life, as with moral order, but not for long *with impunity*. (*Magyar Gazdák Szemléje*, Aug. 1896, 93; italics in the original)

This one-sided view did not go unchallenged. Pap pointed this out delicately in his critique of agrarian economic reasoning:

> It's true, the state did not eliminate the stock market or the futures business. And so—say the agrarianists—this made it possible for the stock market to pilfer the price of wheat. Well, I do not want in any way to write here a detailed defense of futures business.... I just want to ask one thing. Who believes that a merchant could be found who would buy hundreds of thousands of tons of wheat after the harvest, if the possibility were not there for the merchant who was actually buying [wheat] to insure himself by futures sales against the fall in prices that could occur between the actual purchase and actual sale? (*Magyar Gazdák Szemléje*, Dec. 1896, 439)

No matter how sensible the author's reasoning, it did not alter the view that paper wheat could never substitute for "real wheat."

Throughout these acrimonious debates we find hidden, or only thinly veiled, allusions to outsiders—foreigners and Jews—as the culprits in economic disarray. The prejudice amongst the aristocracy against engaging in monetary transactions and financial dealings was fed by their disdain for lower classes (merchants) and their fear and hatred of foreigners, especially Jews. The association of Jews with money and even speculation—the unbridled and irresponsible use and abuse of money—was widespread. To distinguish themselves from merchants, proud proclamations like "Hungarians are not a commercial people" justified a dismissal of monetary concerns (*Magyar Gazdák Szemléje*, Aug. 1896, 82; see chapter 2 for a discussion of prejudices against business). But circumstances had changed by the end of the nineteenth century, when commercial interests threatened the hierarchy of wealth grounded in aristocratic privilege. "As the masters of factories exploit consumers with the help of *cartels* not the smallest thing occurs to inhibit these machinations. They imprison small thieves, and yet those who pocket millions year after year because they succeeded in undermining the healthy foundations of economic development by eliminating free competition are raised to noble rank" (*Magyar Gazdák Szemléje*, Jun. 1896, 223; emphasis in the original).[7] This merely exacerbated the ambivalence aristocrats felt toward the power of money, especially as it had become a direct cause of the erosion of their privileges. Wealthy landowners attributed social unrest to the policies of mercantilists, their catch-all term for commercial interests. "The proponents . . . of the mercantile school . . . believe that with money, legal articles, gendarmes, and prisons one can easily replace that cementing and conciliating force that is found in moral sentiments and in the loyal and selfless fulfillment of duty" (*Magyar Gazdák Szemléje*, Jun. 1896, 222–23). In the new world of money and capital, natural relationships were doomed, be they in materials like gold or wheat or relations of deference, like the patriarchal view of master-servant obligations. Others had a different conception of value and saw money as a positive influence in economic development.

Redefining Value

The publication in 1830 of Széchenyi's treatise *Credit* (*Hitel*) constituted a founding moment of the progressive reform period in Hungary. Széchenyi's manifesto spoke forcefully on behalf of a money economy. Széchenyi argued at length that "the hand compelled to work grows only colorless and bad smelling flowers" (Orosz 1962, 63). Széchenyi's thesis was simple: Hungarian economic development required that relations of property and production be radically restructured, a process that could be initiated by the infusion of foreign capital. Central to the restructuring of agriculture would be the elimination of feudal service contracts (*robot*). Many of his ideas about the elimination of *robot* were not new and had been expressed among the flurry of voices of the Josephine period of enlightened criticism in the eighteenth century. His was a systemic critique of feudalism. "The deficiencies of the 'systhema' must be eliminated, and Hungary 'must extricate itself from the hideous confusion of degrading feudalism'" (ibid., 59). Széchenyi's agrarian program contained radical elements. In addition to promoting elimination of feudal labor service and the tithe, he also called for the separation and privatization of all properties (fields and pastures) and the introduction of relative prices to replace price fixing of certain products by guilds (Széchenyi 1830, 92–107). The danger of useless action, of undirected purpose, underlies Széchenyi's criticism of feudal labor service. "The reason the homeland is covered by swamp and wasteland that forms an eyesore is not because there are no people and no hands, but because so much work is lost without result" (ibid., 101). He goes on to explain:

> "Work, work is the foundation of national economy!" Thus do so many lead off, relying with commendable zeal on the names of Say, Ricardo, Malthus, and Sismondi. Yet it is not work, but well-ordered work, in short, reason, that is the foundation of the national economy. If in 1830 I devote all my energy with the greatest strenuousness and tenacity to digging a great hole that in 1831 I fill up again, and so on, then my work has been entirely lost; if on the other hand I were able to command my men not to do anything without benefit, then all of my work would be profit. (ibid., 102)

Széchenyi's goal was to invest work with purpose. These considerations were entirely absent in the feudal economy of service and fees. The tithe was just as poorly conceived as *robot*, because it undermined the relationship between effort and reward. "[The English publicist Arthur] Young . . . has said that a 'more successful means for the demoralization of agriculture than the tithe cannot be found.' The more industrious a serf is, the better they punish him, because the more he grows, the more he has to submit . . ." (ibid., 104). So, too, he argued against the ability of guilds to fix the price of meat. "Good meat is not more expensive than bad and so why should the farmer attempt to produce better meat?" (ibid.).

The value of money, its moral force, must be recognized as the propellant to economic growth, national prosperity, and individual freedom. Speaking to an aristocratic audience, he was at pains to dispel attitudes that considered credit, and any financial dealings generally, as sullied, dirty, and disreputable. It was precisely the absence of money that crippled society, morally as well as economically. Contrary to popular opinion, money did not pose danger to the social fabric, but it was rather the inability to acquire it, to have a free flowing, open market in money that was pernicious, indeed, the work of the devil. Wealth was not a fixed property nor did it inhere in certain objects. In the chapter entitled "The Hungarian Property Owner is Poorer than He Should be in Relation to His Property" (*A' magyar birtokos szegényebb, mint birtokához képest lennie kellene*), Széchenyi began by discussing different meanings of property or wealth.

> The two sects most numerous in their agreement are those, some of whom consider gold and money to be property and wealth, the others who believe it to be land and livestock. In my opinion it is neither treasure nor meadow; rather it is the use obtained by such things. What good would gold have done Robinson on his island, and what are the vast fields, forests, and wilderness of some Hungarians worth to them? (ibid., 41)

The concept of benefit, of use, of profit was central to Széchenyi's understanding of a modern economy.

Széchenyi explained that to pursue profit—always keeping one's eye on distant goals—could only be achieved by tooling one's actions to the immediacy of the moment, that is, by recognizing value as fleeting and mobile.

> One only has the greatest possible benefit of money, land, or anything else if at every moment one can apply one or the other as one wishes.... *Value is closely tied to the moment.* For the thirsty man a fresh drink, the tired a bed, the freezing man a fire, a coast for he who struggles at sea, the timid a fort or the side of a hill is worth more at the right moment than every other treasure in the world. And that is why we often see a *wealthy* man with little property, and so too a truly *poor* man with tremendous wealth, because negligible wealth can normally produce the very best of life, and Croesus's treasure is insufficient, even to satisfy basic needs, if we cannot avail ourselves boldly and according to our pleasures. (ibid., 42; italics in the original)

Economic development lies, finally, in drive for gain. "Forward then with cold-blooded and clear-headed *counting* (*számlálás*), because in farming and commerce only *the hope of gain or profit* excites" (ibid., 77; italics in the original).

Széchenyi's was not a lone voice in the woods. Other authors proposed the idea, apparently inconceivable to most wealthy landowners, that the judicious use of money may have positive consequences for agricultural production and worker morale. Writing in response to Gaál's plea for renewed patriarchialism in social life, one author described practices implemented at manors owned by the Roman Catholic hierarchy, in which dividends were distributed among manorial servants.

> The worker dividend system, the so-called *perczentuatió*, has only been introduced in a few places and in those cases mostly stewards come in for a share. It has also been extended to shepherds at various sites in the Jaszó estate. For years now it has been in force at the bishop's estate of Pécs that manorial workers receive ten percent of the officer's share; not only the overseers (bosses) receive it, but the more diligent manorial servants do too who have been in service for a longer time and have larger families, in seemly proportion. Its moral and financial influence has been excellent. The manorial work force is much more diligent and permanent. (*Magyar Gazdák Szemléje*, Sept. 1896, 210)

Schemes for dividing up profits among the work force enhanced productivity and bred loyalty. In the author's mind, no questions remained; money could be used to encourage better work habits on the part of manorial workers.

Gusztáv Leopold also argued that higher productivity was tied to worker satisfaction. "*It is not possible to farm intensively with serfs. For that one needs worker elements who are more intelligent and who are more valued—in pay and in the way they are treated—than for extensive farming*" (Leopold 1911, 540; italics in the original). But for most, humane treatment was not a goal, except insofar as it was realized through the paternalist relationships of master and servants, in which workers were grateful for whatever wages they received. Even so important a topic as the health of the nation's workers could become a concern primarily as it assured greater economic prosperity. "Improving health means more workers' hands and more consumers, of which Hungarian agriculture absolutely has need to advance" (*Magyar Gazdák Szemléje*, Dec. 1896, 460).

Csizmadia's solution to the dehumanizing consequences of capitalist relations differed from his wealthy compatriots. He did not draw the conclusion that workers should be excluded from a monetary economy. On the contrary, he argued specifically that workers should no longer be paid in kind, but in money (Csizmadia 1896, 40). In fact, his view was not that workers had a poor sense of money use but that they did not have high enough expectations for their own lives.

> They consider the reason for the problems of the agricultural work force to be that they live and dress in a gentlemanly fashion. But this is not true! Rather, the cause for their problems and misery is that their demands are not great. If the needs of agricultural workers had been more extensive, then they wouldn't have reached this point. They should improve their demands so that they call for what is necessary for one man to live. (ibid.)

Csizmadia was calling for Fordism, a bit ahead of his time.

The market was seen by some as a valuable tool. Echoing Széchenyi's argument about value in use, Peterdy asked, "What is the greatest treasure worth, if it cannot be sold? What would gold and silver be worth if they couldn't be used for our basic needs?" (*Magyar Gazda*, 6 Dec. 1859). He proceeded to admonish the farmer who was willing to go to any lengths to improve cultivation but who would not take his goods to market.

> Since when is it a shameful or insensitive thing to sell the produce we have amassed honorably? Are we swimming in prosperity and drowning like flies? No, sirs. As long as we are not able to market our produce, we are not a resourceful nation.... To market!... To cultivate!... [is our] civil duty!... Domestic industry and export! The basic source of the well being of the Hungarian state! (ibid., 518).

If farmers were being exploited by middlemen whose only interest was to push prices down, then they needed to take things into their own hands. They should establish their own association, to market their crops abroad and raise the profile of Hungarian products overall, by attending to the quality of their goods (*Falusi Gazda*, 11 Oct. 1865).

Agriculture would flourish when estate owners and peasants would learn to make their farming more businesslike. This entailed not only better accounting procedures but also investing more directly in the processing of raw goods and materials, to sell directly off the farm. "The farmer [*gazda*] must aspire to achieving two primary traits: he must be more of a manufacturer and more businesslike than he has been until now...." (Gaál 1885, 86). One may wish to improve one's farm, but as one observer opined, "Hungarian farms generally reward drudgery poorly" (*Gazdasági Lapok*, 8 Jun. 1854). Becoming more businesslike involved, then, a rethinking of the use of one's produce—processing it for greater profit—and the use of other components of production, capital, money, and most importantly, one's own actions, for example, time. "Work is the sort of capital that one should not let lie about without investing, neither society should nor the individual" (Kenessey 1868, 47). "The *English* proverb is properly applicable as *time is money, conversely money is time*" (*Falusi Gazda*, 19 Jul. 1865; italics in the original). Phrased more colorfully was the comment that "the lack of markets is a tormenting worm that gnaws at the Hungarian producer's heart" (*Magyar Gazda*, 6 Dec. 1859).

Investment was key. Capital in all its forms had to be protected and increased. Feeding cattle and keeping them well tended was absolutely necessary, "since the petty livestock we cling to is fake capital . . . our cattle's numbers form a living savings bank, on which we may draw at all times in the days of need and tribulation" (Kenessey 1858, 20, 11). Using manure constituted a crucial investment, as "*manure is money*, because without manure even the richest land becomes depleted" (*Gazdasági Lapok*, 1 Jun. 1854; italics in the original). To facilitate access to capital, proposals to establish new credit banks grew more numerous. The absence of money in the economy was felt strongly in particular by aristocratic landowners, who as of 1848 were paying taxes for the first time. No less urgent was the need for monies devoted to recompensating the nobility for the loss of their serfs. "Naturally lack of money brought in its train the stagnation of the economy, which furthered the lack of money..." (*Gazdasági Lapok*, 6 Jul. 1854). Wealthy landowners looked for the appropriate body to initiate these efforts, such as the Hungarian Agricultural Association (Magyar Gazdasági Egyesület) (ibid., 336); others spoke in favor of a national bank, managed by the state. With time, attention was paid to the need of smaller farms, which were to be assisted with credit from "folk banks" (*népbank*). One author even argued that by establishing savings banks and pensions, manorial servants could be drawn into better work habits and a greater sense of self-reliance (*Falusi Gazda*, 13 Sept. 1865). Perhaps the most encompassing effort to improve the commercial interests of farmers was the establishment of the national cooperative network, Hangya (Consumer and Marketing Cooperative of the Hungarian Landowners' Association; Magyar Gazdaszövetség Fogyasztási Értékesítési Szövetkezete) in 1898. A part of the international movement to establish retail cooperatives, Hangya aided producers in marketing their goods.

Ambivalence surrounding the use of money continued into the twentieth century. Even though the use of money was common in business transactions in city and country after the First World War, a significant proportion of the agrarian workforce was still paid in kind, a barrier to productivity work scientists hoped to eliminate. Greater barriers were constituted, however, by the precarious nature of Hungarian currency in the years between 1918 and 1946. It is to these difficulties I now turn.

Emergency Currencies, Temporary Cash, and Alternative Monies in the Twentieth Century

Widely varying opinions on the value of money and the probity of monetary transactions has complicated the use of money in Hungary. Historical experiences of inflation after the First and Second World Wars worsened people's trepidations and eroded confidence in the National Bank and the money it issued as a tool of everyday interaction. When cash evaporated, and alternative currencies failed as substitutes, Hungarians retreated into bartering in goods and services as the only sure means of getting by.

AN EMPIRE AND ITS CURRENCY COLLAPSE

The fiscal landscape created by the dissolution of the Austro-Hungarian Empire was slow in taking form. During the war, the government had issued unsecured banknotes to finance the increasingly costly disaster, increasing the number of banknotes by 334 times the amount in 1914 (Ambrus 1979, 15). Facing insufficient capacity to print the so-called "blue" money used in peacetime, in October of 1918 the Austro-Hungarian Bank released what came to be known as "white money," which had been printed quickly on offset machines (Leányfalusi and Nagy 1997, 13). Though meant to be a short term strategy (ibid., 16), white money was used in Hungary until November of 1920. It never enjoyed the full confidence of the citizenry, despite repeated proclamations by the government that it was fully legal tender (Ambrus 1979, 34).

As a condition of the peace treaty at St. Germaine signed by the Austrians, all successor states were required to establish their own currencies within a year. The first order of business was to create hybrid forms of currency, that is, the new states had to stamp bank notes previously issued by the Austro-Hungarian Bank, thereby distinguishing them from each other (Rádóczy 1984, 20). This task was to be accomplished within two months. While the procedures for establishing new currencies were straightforward, the actual process of transforming old banknotes into new currencies was not. Ink stamps devised to "nationalize" banknotes were easily copied. The Kingdom of Serbs, Croats, and Slovenes initiated their overstamping in January 1919, but "[t]he forgeries of [the ink] stamps became so numerous that the officials themselves could no longer detect whether the stamps were genuine or forged, so that the Government was compelled to accept large quantities of these falsified notes which were presented for the second stamping" (Garber and Spencer 1994, 9). On the second round, stamps bearing the new national emblem were affixed to the notes. In the meantime, unadulterated banknotes—white money—remained legal tender in Hungary, at which point the country became a dumping ground for any banknotes not yet stamped by the new governments in the making, adding to the already worsening postwar inflationary conditions (Leányfalusi and Nagy 1997, 13). Policing the flow of cash in the contentious atmosphere of postwar territorial squabbles was a near-impossible task.

Relations between the Hungarian and Austrian officials in the Austro-Hungarian Bank were strained in the postwar climate. Figuring out the relative burdens of the war debt and the credit/debt balance between Austria, Hungary, and the successor states took a while; the Austro-Hungarian Bank was fully dissolved in 1922. In the absence of its own national bank, fledgling Hungarian governments fought with Vienna for months over printing money. In the months following the war's end, the Hungarian branch of the Austro-Hungarian Bank had its hands tied by Vienna, which still held the patent for issuing banknotes. Hungarians were cautioned not to print money, while at the same time Austrians did not provide Hungarians with the banknotes they needed. Over time, the Austro-Hungarian Bank's control over printing of money could no longer be tolerated. When the Hungarian Soviet Republic was declared on March 21, 1919, the new government took charge, directing the Postal Bank to print several different denominations of white money. Plans to mint coins never came to fruition, as the republic only lasted 133 days (Leányfalusi and Nagy 1997, 13). The availability of sufficient banknotes and coins was a problem in the capital city, but much worse in the countryside. As a result, a number of towns, cities, and even factories printed their own emergency currencies (necessary money, *szükségpénz*) in the fall of 1918 and the first six months of 1919 (ibid., 112). Businesses, factories, publishing companies, banks, and even religious dominations issued emergency monies, but so too did the authorities—city and town councils, including the City Council of Budapest.

A major problem with white money was its limited denominations. White money was difficult to cash, so completing transactions in the marketplace or paying wages became a problem. Workers in Nagykanizsa threatened to close down factories if they weren't paid at least in part with change (*aprópénz*) (ibid., 107). Everyday transactions in stores and shops were stymied, as a letter sent in May 1919 to the editorial board of the *Red Newspaper* (*Vörös Újság*) describes:

> Whether in shops, coffee-houses, restaurants, laundries, or even amusement halls, that is, wherever one shops or pays [cash], we receive a small piece of white paper, which marks in pencil the amount we would have received in change. This is the newest money. Shopkeepers have prevented the elimination of small change before the government has. Today they issue money. Shopkeepers exercise the franchise of the defunct Austro-Hungarian Bank. (ibid.)

To deal with these impediments, the Revolutionary Governing Council encouraged the use of checks in place of cash. The innovation required the government to mount an informational campaign: to explain the differences between banknotes and checks and to promote their use (ibid., 108).

Suspicions about the legitimacy of white money eroded confidence. White money had a hard time competing with blue money. It was shoddy and often lacked printing on one side of the bill. It was easier to counterfeit than the copperplate-printed blue money (Ambrus 1979, 93). Since the white money that had been printed in Vienna carried an expiration date—it was to have been replaced with new banknotes by the end of July 1919—it was more and more difficult to convince people that white and blue money were of equal value (ibid., 34). Eventually, in July 1919 the Soviet Republic declared blue money illegal, so that the currency being printed by the Postal Bank would take hold. Hungary stumbled on for several more years without a stable post-empire currency, beset by unprecedented levels of inflation. A national bank was finally established in 1924, fully equipped with patents to issue money. Nonetheless, it took three more years for the country to switch to a new currency, the pengő, to replace the crown.

INFLATION FOLLOWING WORLD WAR II: DÉJÀ VU ALL OVER AGAIN

In the year following the end of the fighting in May 1945, the country experienced a period of rapid inflation. This is a common occurrence in postwar economies. What sets this case apart is that no other country has exceeded the rate of inflation Hungarians witnessed between June 1945 and August 1946.[8] While the explosion of the post–World War I German inflationary spiral is well known, it cannot match the rocketing prices in Hungary after World War II. Nötel compares the two inflationary episodes in terms of the exchange rate of the dollar: 4.2 billion (10^{12}) marks in Germany post–World War I versus 5 quintillion (10^{30}) paper pengős in post–World War II Hungary (Nötel 1986, 538). This tragic turn of events was true despite measures taken after World War I to prevent its reoccurrence.

In 1924 when the National Bank was founded, the government pledged not to finance its expenditures by issuing unsecured banknotes. The law forbade the new institution from bankrolling the government. Taxes were to be levied to cover the state's expenditures. This principle held until the pressures of military rearmament in the late 1930s increased. The independence of the national bank was abrogated with the passage of the civil defense law in 1938, opening the bank's coffers once more to the hands of government officials (Siklos 1991, 60–61). Pressures to monetize the national debt were exacerbated by Germany's indebtedness to Hungary throughout the war, a result of its policy of maintaining a balance of trade (an exchange clearing account system) rather than paying for goods outright; the debt grew from 326 million pengő in 1941 to 2,918 million P in 1944 (ibid., 51).[9]

Irresponsible officials in the last days of the Nazi-allied regime issued nine billion P notes and shipped them westward as they fled the invading Soviet troops, having destroyed the printing presses before their departure and taken the banknote plates with them (Siklos 1991, 4; Rádóczy 1984, 84–85). To add insult to injury, government officials also absconded with the nation's gold reserves, which they then surrendered to the US military in Germany (Siklos 1991, 250). With insufficient foreign currency left in the bank's coffers, the Hungarian state had to borrow money in order to pay for the ink needed to print currency (Bomberger and Makinen 1983, 807). To muddy the fiscal waters even further, the Soviet army also issued banknotes—to the tune of 4,800,000 P—between the end of 1944 and early 1946 (Botos 2006, 170). Clearly marked as issued by the Soviet Army Command, these notes were fully legal tender. That did not allay concerns about their legitimacy;

TABLE 4.1 AVERAGE DAILY PRICE INCREASE (%)

		On the last day of the month	
Date	Percentage increase (%)	Price index August 26, 1939 = 1	Dollar rate of exchange on the black market (pengő)
July 1945	1	105	1,320
August	2	171	1,510
September	4	379	5,400
October	18	2,431	23,500
November	15	12,979	108,000
December	6[1]	41,478	290,000
January 1946	2	72,330	795,000
February	18	435,887	2,850,000
March	11	1,872,913	17,750,000
April	60	35,790,361	232,000,000
May	1,012	11,267 million	59,000 million
First half of June	504	862,317 million	7,600,000 million
Second half of June	8,504	945 billion[2]	42,000 billion
First week of July	45,904	3,066,254 billion	2,000,000 billion
Second week of July	53,214	11,426 trillion	481,500 trillion
Third week of July	45,014	36,018,059 trillion	5,800,000,000 trillion
Fourth week of July	158,486	399,623 quadrillion	4,600,000 quadrillion

Notes: Daily price increases during the inflationary period from July 1945 to mid-July 1946.
[1] The slowdown was caused by the currency stamping on the 19th of December, 1945
[2] There are twelve zeros in a billion, eighteen in a trillion, and twenty-four in a quadrillion (Pető and Szakács 1985, 61)

"farmers, in particular, [were reluctant] to accept such notes in payment for agricultural produce" (Siklos 1991, 252n2).

The jump in the rate of inflation was not immediate; in the latter months of 1945, prices showed an average daily increase of 2 percent, 4 percent, 18 percent, 15 percent, and 6 percent. Acceleration truly gathered steam in February 1946. At this point, peasants refused to take pengő in payment for their goods (Siklos 1991, 5). In the city, people left work as soon as they were paid to shop, fearing any delay would eat into the value of their wages (Nogaro 1948, 529). Initially, people were paid twice a week, but as the inflation worsened, they were paid every day to accommodate the rapidly changing value of cash (Eckstein 1952, 237). Some commodities were still considered valuable and could be used to acquire other goods: tobacco, cigarettes, rice, scarce items of clothing and industrial goods, raw materials, and gold (Botos 2006, 178).[10]

CALORIE MONEY

In the early months of the inflation, there was no scarcity of ideas about how to stave off the worsening fiscal situation. In September 1945, the Hungarian National Bank made the simple suggestion that the government make a case for postponing the costs of reparations imposed by the international community (Botos 2006, 186). Nothing came of this.[11] At the end of October, the chief counsellor of the bank argued in favor of strong government intervention—stabilization of wages and prices, state-directed production, and redistribution of goods—as it was now possible to foresee the complete collapse of the pengő (ibid.). He was not loath to advocate borrowing money from abroad, as did others in the months to come (ibid., 187).[12]

By February, difficulties with securing and transporting sufficient foodstuffs to town forced many to part with valuable heirlooms or other disposable household goods, simply to get enough to eat. In December of 1945 the daily bread ration had already been cut back: from forty-five to twenty-five decagrams for manual laborers with the hardest jobs, thirty to thirty-four to fifteen for other manual workers, while office workers and children were assigned an extra ration of five decagrams (Pető and Szakács 1985, 45). Those without these resources were in serious trouble.

> At the beginning of 1946, the average food ration of the around 800,000 adults living in Budapest did not reach 480 calories a day. In the first three months the public supply apparatus could only ensure on an average 1,000–1,200 calories for the 135,000 heavy manual laborers of Budapest, 820–980 calories for the 160,000 other physical laborers, 600–650 calories for the 115,000 office workers, and 670–830 calories for the roughly 220,000 children under the age of 12. (ibid., 46)

Recognizing the dire straits families faced, city officials in Budapest set up people's kitchens on the Pest side of the Danube as soon as the fighting between the retreating German forces and the advancing Soviet army was over (end of January 1945). By the summer there were 124 official soup kitchens and 278 workplace canteens in Budapest, feeding 137,000 adults and 7,800 children (ibid., 45–46).

In response to the worsening situation, the government and the Economic Supreme Council (Gazdasági Főtanács) crafted a new policy in February of 1946, the so-called "calorie pengő" (*kalóriaPengő*). The idea was to have factories provide workers with food, simply replacing a monetary wage with goods in kind determined by their caloric value. The daily caloric needs of workers and civil servants were set at 2,840 calories for (heavy) manual workers and 2,470 for all other categories of workers. The size of the worker's family was also to be counted in the wage; family members were to be allocated 1,200 calories (ibid.).[13] Each week, the Economic Council priced a food basket with different contents, figuring the prices on Monday and issuing them officially on Thursday in order to ensure the information be available by payday on the weekend.

From the outset, the purpose of calorie money—provisioning the labor force to ensure that reparations and reconstruction would continue apace—was constantly frustrated. Factories had just as much trouble acquiring the foodstuffs necessary to supply their workers as did the workers themselves. In some cases, the enterprises producing the sought-after goods—following a practice already in place during the Second World War—organized their own purchasing groups, which travelled to villages and rural towns to buy up food in exchange for manufactured product (Botos 2006, 178). These problems were far greater for small industry as they lacked any comparable infrastructure.[14] Recognizing the difficulties employers faced in finding enough food to pay their workers, the government relented and made it possible to pay the pengő equivalent of the "caloric wage" (Pető and Szakács 1985, 46). To assist this transaction, the government devised a calorie price index to ensure that workers' wages would keep up with inflation. In some instances, workers simply refused to be paid in anything else but in kind, as was the case with miners in Tatabánya, but foodstuffs were more available in town, making it possible for the factories to comply most of the time (Pittaway 2012, 60–61).[15]

At the same time that the calorie pengő policy was adopted, the government also standardized wages within trades with a change in industrial workers' contracts. Since only a small percentage of workers' pay would now be represented in the base wage, expectations were that differences in wages between categories of workers (skilled, unskilled, etc.) would disappear.[16] These hopes were not fulfilled. As the report prepared by the Hungarian Economics Research Institute issued later in the year stated, when workers were paid in food, wages were fairly stable over time.[17] But as a result of inflation, the pengő equivalent of the caloric wage constantly eroded.

> During the four weeks from April 20th to May 20th prices showed an 80-fold increase, while in contrast the increase grew 3,350 times from May 20th to June 20th. The official index of the consumer prices of food was 3.1 million, as opposed to the 6 billion free market food index. On the other hand, during the four weeks at issue wages only increased 1,250 times, inclusive of calorie conversion. This demonstrates that the increase in wages only offset the increase in prices by a third, while in the previous four weeks wage rises were more than 60 percent (Büky 1946, 6).

Regulations were broken in the interests of keeping workers going. "It is inconceivable that someone would be capable of working, much less stay alive, in the midst of this degree of decrease in real wages in June 1945."[18] In light of the radical devaluation of wages, workers were regularly paid far more than officially sanctioned. "In practice the cash wages set in the collective contracts were generally overpaid by many times, and the drop in the standard of living relative to June of 1945 was in reality no more than 40–60%."[19] The authors of the report didn't even try to calculate what percentage of wages were illegal, mentioning the wide differences between factories and branches of industry, as well as over time.[20] In small industry, where cash wages were common, illegal wages were assumed to be very high.

Within two weeks of the new policy, workers were already voicing their dissatisfaction. Workers from Makó marshaled their own evidence and submitted their results to officials in the Social Democratic Party.

> Allocations in kind as a consequence of the possibility of converting calories only exists on paper; it has completely failed. It is bankrupt because the conversion price established by the Economic Supreme Council stays far lower than market prices. Let the prices below serve as verification:
>
> the weekly allotment for employees:
> two eggs
>   conversion price (47 calories ×18 P.)   P. 2,646
>   market price   P. 12,000
> 20 decagrams of cheese curd
>   conversion price (220 cal. × 18 P.)   3,960
>   market price   12,000
> 1.40 kg of cabbage
>   conversion price (420 cal. × 18 P.)   7,560
>   market price   84,000.[21]

In the face of these difficulties, the workers in Makó called on the government to regulate the market and to impose more stringent oversight of commerce, the post office, and the railroad. Banks were to be closely monitored, at least until they were nationalized. (Unsurprisingly, banks and black marketeers were equated throughout the document.) The country needed a police force dedicated to overseeing economic transactions (*gazdasági rendőrség*). And last but not least, the workers insisted that they be paid in kind. If not, then the prevailing prices at the local market had to be the basis for wages calculated in calorie money. Otherwise, "if a noticeable improvement does not appear in a short time, then the executive committee cannot be held responsible for the consequences."[22]

In April, people had given up quoting prices in numbers, preferring to refer to the color of banknotes (Siklos 1991, 6). "The calculating abilities of people could no longer take the astronomical rates. Confusion arose in official mathematical circles whether a period should be inserted between million and billion or billion and trillion" (Büky 1946, 5).[23] Needless to say, the crisis in the national currency strained more than people's skills in multiplication. Black marketeering and price gouging—familiar postwar hobbies—only added fuel to the fire. The rapid deterioration of the value of banknotes created incentives—for those still with money in hand—to spend recklessly and invest haphazardly (Nötel 1986, 541).

As the inflation worsened, word spread that a stabilization program was in the works, rumors substantiated by wooden sloganeering in party organs. In May, the catchy phrase "All power to creating good money" graced the pages of *Free Land* (*Szabad Föld*), the Communist Party weekly newspaper for cooperative farm members.[24] Peddling hope was a lost cause; memories of the inflationary catastrophe following World War I were too fresh and skepticism of government too high. "By June 1946, it was virtually impossible to find regular Pengő currency in circulation in Budapest and other cities, especially after 2 p.m., the hour the banks closed. Businesses and individuals would deposit practically all their currency in banks and withdraw a scaled-up sum the following morning with which to conduct business" (Bomberger and Makinen 1983, 808). For most, however, recourse to barter was the only means of getting by.

Dire straits led some not sitting in banks or ministerial chairs to propose alternative monetary metrics. In an attempt to stabilize the value of goods and labor, several proposals surfaced designing alternative systems of metrication, usually taking labor as the basic unit of value. While primarily designed to fix the relationship between labor and goods—to make it possible for the working person to acquire basic necessities—these plans also had strong sanctions built in against price gouging and black marketeering. The dignity of labor had to be preserved; the abomination of becoming rich at the expense of others had to be outlawed. It should not go unnoticed that these plans shared crucial features of the metrication projects agrarian work scientists developed seeking to standardize the value of labor. Unfortunately, the wealth of ideas on the table did little to stem the tide. People were forced to make do, in the process bequeathing us colorful stories that mask painful deprivation.

The Path to Stabilization

The forint was introduced on the first of August, timed to coincide with the wheat harvest, an occasion traditionally associated with bounty and celebration.[25] The nation's fiscal savior had been born. The rapacious pengő—powerfully depicted in a poster as a child eating away at its mother's breast—would be banished to the dustbin (see figure 4.1). The Communist Party did its best to take credit for crafting the new money, though its claims were not unchallenged. A satirical cartoon that appeared in the joke magazine published by the conservative Smallholders' Party shows the Communist leader Rákosi parading around the newborn coin, arm in arm with his wife, a happy peasant woman named Democracy. On the sideline, figures representing the other major political parties remark on the charade (see figure 4.2). One of them cites a Hungarian proverb: "You know. When a child is born, you can only be sure who the mother is."

Anticipating difficulties ahead, government officials were keenly aware of the need to cushion the forint's arrival. To do so, they stockpiled goods in the months leading up to the stabilization in order to flood stores with long sought-after goods. And the manner of acquiring these goods showed the officials to be very clever in manipulating markets.

> To obtain some of these goods, Hungary concluded a series of barter treaties during early 1946 . . . transactions which involved a clever arbitrage of markets. Large quantities of Hungarian tobacco products, e.g., were sold in Vienna, where their price was high in dollars. The dollars were used to purchase cheap Polish sugar. The sugar was sold in Bucharest where its price was high in broken gold. The gold was taken to Budapest where it was struck by the Hungarian mint into gold Napoleons. These gold coins were then used to buy imported goods from Switzerland. (Bomberger and Makinen 1983, 818n28)

Word of bounty spread quickly. Farmers flocked to the city, exchanging their foodstuffs for forints, which would then be spent in the store down the street offering industrial goods (Eckstein 1952, 239). Yet the process of shifting to a new currency took time. As late as December, agricultural workers were being paid in kind, as forints were hard to find in the countryside. Receiving their wage in kind was also preferable, since in the absence of a surplus, it was not possible to buy grains for money. Efforts on the part of the National Board to Set Agricultural Wages (Országos Mezőgazdasági Munkabérmegállapító Bizottság) to figure out the forint value of agricultural wages in kind were slow.[26] State budget deficits continued for several months, as had been anticipated in

FIGURE 4.1. Poster entitled "Money" depicting a rapacious child (pengő) devouring its mother (1945).

FIGURE 4.2. Cartoon from the the satirical newspaper published in early August 1946 by the conservative peasant party. Mocking the headline from the Communist newspaper declaring Rákosi the father of the forint, the picture shows three prominent politicians—Nagy from the Smallholders' Party, Veres representing the National Peasant Party, and Szakasits from the Social Democrats—watching as Rákosi escorts Democracy with their child in a pram. Citing a common proverb about never being sure of paternity claims, the caption reads: "As we know, we can only be certain of the mother."

the plans to restructure the taxation system designed in the spring, but with the nationalization of industry that soon followed collecting taxes no longer posed problems (Siklos 1989, 137).

Conclusion

I began the chapter by noting the discrepancy between Marx's and Simmel's theorizing about money as a means and a fetish on the one hand and the recurring barriers to its use (as banknotes or cash) at the time in Central Europe on the other. Yet the core of Marx's and Simmel's insights into the magical properties money exercised on the imagination held true. Agrarian work scientists and business economists aimed their sights on monetary wage schemes as the key to increasing productivity and profitability. In light of these difficulties, all the work that economists and work scientists devoted to crafting an abstract "standard daily labor unit" in the 1920s and 1930s can be seen in a new light. Specialists were not simply aiming for a rigorously defined metric of labor value to deploy in studies of productivity or commensuration exercises to rationalize agricultural production scientifically but in fact were pushed to wax abstractly precisely because of the absence of monetary prices for labor. The new work units experts designed bore all the features of a commodified wage: discrete units of activity of a specific duration performed by specific categories of social actors with certain skills or physical attributes. Happily, these abstract metrics served just as effectively in commodifying labor even though what we assume to be the usual motors driving commensuration and comparison—labor markets—were absent. The abstract formulation of labor value, a scientific goal on its own terms, would serve the practical needs of business owners and accountants until a fully developed market infrastructure was in place. Those days would be long in coming.

PART TWO

5

State Matters

On Sunday the 25th of July, 1948, the following headline greeted readers of *Free People* (*Szabad Nép*), the Communist Party newspaper: "Embezzling, Sabotage, Spying: The Non-Party Block in the Ministry of Agriculture." This shrill charge was the first indication of the Communist Party's public assault on the Ministry of Agriculture, which became known in the course of the trial mounted in September as the "citadel of feudalism" (*a feudalizmus fellegvára*). Eighty-four people were charged in the trial, approximately one-third of whom were high-level bureaucrats. The rest of those targeted in the trial were personnel working at state-owned farms, charged with corruption for shady business dealings or incompetence. The upper strata of defendants, the bureaucratic elite, were charged on three counts of organizing to overthrow the state, one count of obstructing the land reform, and two counts of sabotage.[1]

The frontal attack on the Ministry of Agriculture was the first major attack on a government ministry by the Communist Party, but it was only one of several crucial trials mounted by the Communist Party as it strengthened its grip on power in 1947–48. These targeted significant sectors of Hungarian society and politics: the 1947 trial of the leaders of the Smallholders' Party, an erstwhile coalition party in the government; attacks on foreign business interests, such as MAORT (Standard Oil); discovery and expulsion of foreign agents, most of whom were British or US citizens; and the famous trial mounted against the Archbishop of the Hungarian Catholic Church, Mindszenty, in 1948–49. Citizens learned of these transgressions in the colorful prose and hyperbole of the *Free People*. Commentary on the Ministry of Agriculture trial, like the others, was marked by tirades against fascist reactionaries and shoulder-slapping accolades for the vigilant secret police who uncovered the nest of socialism's enemies. Purging political rivals and disloyal bureaucrats from the government and ranks of the Communist Party was a crucial move in securing control over the entire governing apparatus.

Portraying the Ministry of Agriculture as an obsolete feudal bastion of aristocratic privilege was a convenient strategy deployed by the Communist Party to justify taking over the cumbersome institution standing in the way of socialist modernization. But the Ministry of Agriculture was far from being a feudal remnant; it had been frequently reorganized in the four decades of the twentieth century in response to shifting administrative priorities. Between 1890 and 1943, the numbers of personnel in the ministry increased from a modest 148 to 1,176, expanded primarily by increasing the staff of professional experts and accountants (Estók et al. 2003, 179); the number of research institutes and experimental farms the ministry sponsored also increased. Manorial production was also depicted as being mired in an antiquated past, an impediment to fully modernized agriculture. Work scientists and agricultural economists may have agreed with the Communist Party's assessment, but it nonetheless misrepresents the capitalist imperatives wealthy agrarian elites grappled with during the interwar years.

The depiction of radical change brought about by the Hungarian Communist Party has been a given in studies of the region. In the following account, I will offer evidence to challenge this view, showing far greater continuity in administrative procedures and regulations than has been acknowledged. Denying continuities in institutions and practices is a strategy Gabrielle Hecht has described as "technopolitial rupture-talk," or "the rhetorical invocation of technological inventions to declare the arrival of a new era or a new division in the world" (2003, 2). Hecht describes the French nuclear program as an originary technology, that is, developing nuclear technology is portrayed as a feature of France's advanced technological prowess in contrast to the underdeveloped character of its colonial possessions, even though the nuclear program relied on important contributions from African colleagues. Denying Africans' participation reinstantiated the boundary between developed and underdeveloped nations so crucial to sustaining European hegemony over the global south. In this instance, the "originary" technology is the socialist state, as evidenced in numerous studies of the transition to socialism—whether penned in the East or West—that depicted the rise of the Communists to power as a radical departure in political institutions and economic policies (see the introduction). Unfortunately, this approach ignores the history of planning and state intervention in economic affairs that characterized the Hungarian nation-state all throughout the 1930s and 1940s, a history I have described elsewhere as the "technopolitical lineage of state planning" (Lampland 2011, 156). The socialist party/state was far from being an originary technology, since Communist Party institutions were built on existing structures of a planned economy. Even if this had not been the case, it would have been foolish for the Communist Party to demolish the existing administrative capacities of the state precisely at a time when the role of the state in managing the entire economy was about to grow exponentially. Fiscal authority, administrative regulation of commercial activities, professional certification, public health functions, and so forth could not be restructured overnight, and in fact there was little reason to do so. Knowing that the new Communist regime inherited a substantial administrative infrastructure from previous governments also means that the limitations of political institutions would also continue to hold—the oft bemoaned inadequacy of government agencies in fostering modern, rationalized business practices so familiar from the pages of *Köztelek* (see chapters 2 and 3). How would the new regime go about expanding the capacities of the party/state to address these problems in ways that served their new political and economic vision? Acknowledging the continuity of many political and economic institutions into the early 1950s allows us much greater analytic purchase on what in fact the Communists were forced to confront as they pursued an alternative political agenda. What would they see fit to change in economic, political, and academic institutions, how would their efforts fare, and why?

In the following chapter, I will focus on the crucial role of expertise in the first years of the transition as a means of tracking how political, academic, and economic institutions were being redesigned to alter state capacities and strengthen Communist Party control. Initially, the Communist Party would face limited options in altering the political and economic landscape. Realizing the goals Communists set for themselves was consistently frustrated by the inadequacy of government bodies delegated to the tasks and the difficulty of resolving the question of professional expertise and political loyalty.[2] Moreover, the question of expertise in the mid-1940s was politicized in new ways by an expanded public arena. When agreement could be reached, or at least a compromise struck, about the qualifications of new personnel, finding people who satisfied the criteria was difficult. All of this was complicated by ugly political squabbles and vociferous public attacks. Reforming the state did not entail simply discarding the dessicated prewar system of prestige and power; it also required staffing government with qualified experts able to craft regulations and implement laws. The thorny problem of what constituted expertise was left unspoken, even though there were wide-ranging attitudes about the relative importance of book knowledge, expertise, and practical experience. Not surprisingly, debates over expertise were wound tightly around problems of interest representation. The broadest outline of these divides is represented by the renowned anti-intellectualism of working-class cadres in the Communist Party, the usual image of which pits the solidarity of class ignorance against the sophistication of an educated elite.[3] This is an oversimplification. How and why individuals ended up assigned to posts throughout government was a complex process, particularly in the years between 1945 and 1948. We know the shorthand of this story in terms of the jockeying of political parties for power. Yet the project of restructuring the state in the mid-1940s was as much a public debate over whose knowledge counted as it was over the structure and distribution of offices within the government.

These problems were not restricted to government offices. Higher education was also subject to Communist Party scrutiny, becoming an additional dimension of the reforms initiated in 1945 to restructure technical and university-level training. As the Communist Party grew in influence in 1947 and 1948, the focus shifted away from the administrative structure to the character of the faculty and student body. Faculty were vetted in terms of their commitments to "bourgeois science"; students were admitted on the basis of their class identity as much or more so than their intellectual aptitude. The curriculum was rearranged, adding new courses on Marxist-Leninist economics, pride of place being devoted to the invaluable example of Soviet science and political history. These changes were instituted despite a significant absence of qualified faculty to teach these courses, a problem that also affected staffing and productivity at crucial research institutes.[4] As a result, substantial changes in pedagogy and research were hampered by the scarcity of qualified experts, leading to the continued employment of "bourgeois" faculty.

Immense efforts were devoted to altering the political, academic, and economic landscape. Substantial changes in legal structures were instituted, economic regulations passed, and new ministries and research institutes founded. All of them foundered on the inability of the new institutions to fulfill their tasks for lack of qualified personnel, or at least the right class-identified personnel. This proved to be a serious impediment to the new government, since changing attitudes was far more difficult than writing new laws or establishing new bureaucratic entites.

Reconfiguring Government

The mood of the country after the cessation of fighting in the spring of 1945 was a heady mix of exhaustion, elation, trepidation, optimism, fear, and hope. The limits of the possible had shifted. Long overdue social reforms could proceed, while visions of economic renewal were held hostage to the exigencies of war reparations. The limits of the possible in politics were far less clear. Debates over the future of Hungarian politics flourished: What should democracy look like, and how could it be achieved? The provisional government cobbled together in late 1944 stumbled along, helpless in the face of land grabs preempting land reform legislation and powerless over Soviet troops stationed across the country. Elections in 1945 set the stage for the new coalition government, introducing a modicum of political stability, barely tolerated by the Communists, who were disappointed with the results. The ensuing two years of haggling over different parties' control of government bodies kept several political parties in the public eye, but the Communist Party's sustained efforts to take over crucial functions of the government were unrelenting. Paired with a merciless tirade against reactionary elements, leading a number of important politicians to leave the country for the West, the Communist Party was able to sweep several parties off the books, culminating in a largely sham election in 1947 consolidating Communist Party political dominance.

A provisional parliament and executive body was established in December of 1944 in Debrecen, a city near the eastern border.[5] Four political parties were represented: the Hungarian Communist Party (MKP), the Social Democratic Party (SZDP), the National Peasant Party (NPP), and the Smallholders (FKgP). Elections were held immediately. To cobble together a group of representatives for the assembly, the committees charged with the preparations borrowed trucks in early December from the Red Army to reach communities behind the front (Palasik 2000, 30). "In retrospect, it is remarkable that in a little more than a week more than a million and a half people participated in the elections at a time when hardly more than half the country was freed and there existed no system of communication and transportation" (Kenez 2006, 30–31). The Soviets did more than lend trucks. They were directly involved in selecting specific individuals for positions within the governing executive. Yet as Kenez points out, the composition of the initial government was very conservative, evidence, he surmised, that the Soviets did not expect this group of leaders to be around for very long (ibid., 26).[6] In fact, a full-scale national election was not held until November of 1945, setting the stage for wrangling over specific ministerial portfolios among the four parties elected.

In the meantime, the administrative bodies taking care of everyday tasks at the local level—policing, public works (which essentially meant fulfilling the Red Army's demands for food and labor), scheduling market day, etc.— had to resume their work as soon as possible. Officials strongly supportive of the prewar conservative government had fled, leaving citizens to fend on their own. Soviet military authorities were quick to urge communities to take charge, posting information and lists of tasks required to protect the interests of the civilian population and the Red Army, often within two to three days of the fighting moving further westward (Somlyai 1985, 65). As in the case of the national assembly, the Soviets did not insist on a radical restructuring of local administrative units but did participate in encouraging "democratically thinking" elements to step into the shoes of the now absent bureaucrats (ibid., 66). In the early period reinstating bureaucratic capacities, political parties per se were not significant actors. In some communities, in particular those with a leftist political history as in Békés County, local communists attempted to radicalize municipal offices by creating directorates, or people's councils, to replace municipal associations, but both the Soviet military authorities and Muscovite Communist leaders quashed these efforts (ibid.).

As the fighting ceased, and a postwar calm spread, administrative reform was now on the table. Debates over how the state should be administered were an interminable feature of public life in Hungary, each decade from the *Ausgleich* until the end of World War II a witness to new reform packages and tinkerings with the law (Csizmadia 1976; Gyarmati 1989, 4). Whether in 1876 or 1946, the status of the county occupied center stage, respected in some quarters as a venerable institution with a proud history and in others vilified as an impediment to a smoothly functioning apparatus (Csizmadia 1979, 564). In 1943, Imre Némethy, administrative judge and editor of the journal *Hungarian Administration* (*Magyar Közigazgatás*), spoke in glowing terms about "our ancient system of regional self-governing bodies" (ibid., 525). "I would consider it an ideal outcome if the members of the municipal authority and the bodies representing towns would personally take part in the proper work of the self-governing bodies of their county or town (city) in the traditional spirit of *nobile officium*" (ibid., 526). This benevolent view of county offices was not shared by those on the left. In a 1948 essay, Beér, who was a central player in writing the socialist constitution, described the old county system in harsher terms.

> The county is a body of privileged aristocrats and the embodiment of noble perogative. It is for this reason that after the liberation of serfs [in 1848] the county could never become an instrument of embourgeoisement for the masses. Besides the social base, material conditions also prevented [embourgeoisement], since the territory developed on the basis of landlords' property and the organization of military units. The economically and regionally perverse borders of these areas ignored economic connections and the conditions of settlements and that is why those who settled [on the landlords' property] could not create the economic solidarity indispensable to a true self-governing body. (Beér, quoted in Csizmadia 1979, 566)

Béla Grünwald's characterization from 1876 was more succinct: "'... bad administration and the county are as closely connected as cause and effect'" (ibid., 564).

Prominent political and intellectual figures across the political spectrum put their energies into reforming the institutional structure of new governmental offices.[7] Plans emerged from several parties—first the National Peasant Party, then the Smallholders' Party, and later the Social Democrats—outlining new administrative structures, streamlining state practices, and introducing greater decision-making power and independence to local community governance. The Communist Party waited patiently in the wings, appointing various exploratory committees to consider all the issues, a tactic intended to delay having to publish its vision for reforming the state. As the reforms were debated, the Communist Party focused its efforts in party deliberations over the most effective means of seizing control of crucial functions of the government.

Once fully ensconced in power, the Communist Party set out the principles of the new council system in 1950. Strongly modeled on Soviet practice, the broad features are well known: democratic centralism, complete control by the Communist Party—as representatives of the new ruling class of workers—and the complex interwoven character of government and party offices described in the common shorthand "party/state." The opening paragraph of the law describes the new state in glowing terms: "Progressing on the road to socialism, the Hungarian People's Republic constructs the kind of state structure that ensures the worker's active and constant participation in exercising state power and in the work of state administration. [The state] brings its concerns closer to the working masses, and consistently realizes the principle of socialist legality" (1950.évi I. törvény).[8] Every administrative unit—counties (*megye*), districts (*járás*), cities (*város*), communities (*község*), and the boroughs of larger cities—would be represented by a council elected by local residents. All councils were subordinated finally to the parliament and the Presidential Council of the People's Republic (Népköztársaság Elnöki Tanácsa). "The work of the councils concentrated narrowly on seeing to economic tasks, essentially focusing on executing forced deliveries, tax collection and parcellization, and lessening difficulties with public supply. Higher administrative agencies flooded local council authorities with a deluge of regulations and orders and effectively crippled their work" (Balogh et. al. 1978, 166). Needless to say, councils at every level failed miserably at engaging the people in decision making. They were successful, however, in bringing bureaucratic practices "closer to the working masses." "Since the councils dealt solely with the inhabitants' obligations, they became the sole vehicle of unpopular orders from central authorities, so that inhabitants considered them primarily as despotic, [bent on] punishing and harrassing them" (ibid., 167). Of course, these agencies were subordinated to the party, whose administrative structure completely paralleled the full hierarchy of the government, from top to bottom.

> In regard to the workings of the power structure, the organizations and leaders of the party apparatus stood higher in the hierarchy than comparable level leaders and administrative organizations. So the department head of the county party committee was a person with greater power than the head of the agricultural department of the county council. This meant that in everyday practice the actual decisions were made in the party committee and its appropriate departments, and the comparable level of the administration primarily played the role of the executive branch. (Valuch 1996, 14)

This standard view—that the Communist Party ruling elite dominated both branches of the ostensibly parallel offices of party and state—needs to be qualified. While it is true that even the most mundane decisions were discussed by the Central Committee, many of the day-to-day tasks of actually drafting policy and administering state affairs were left to lower-level government bureaucrats. The degree to which the Communist Party would succeed in maintaining control over the state apparatus (much less its own affairs) was a questionable proposition from the start, its folly amply demonstrated by the empirical record. It is a testament to the the lingering power of Marxist-Leninist state ideology that the view of Communist Party dominance still holds sway.

THE POLITICS OF STAFFING

Months before the Germans' defeat, the provisional government initiated denazification proceedings against members of the Hungarian government. As a German ally during the war, the Hungarian government bore responsibility for implementing policies to support the Nazis' military objectives and for actively engaging in war crimes. A screening process, referred to as the B listing, was initiated on the 4th of January 1945 to review the actions taken by government officials during the war (Gyarmati 1989, 6). This was augmented by a second regulation issued in early May. "The first decree treated a public employee as a person employed in the administrative, jurisdictional, education or economic organization of a municipal authority, city, or village. The second expanded the screening process also to officials of 'industrial and commercial enterprises enjoying state support, as well as social organizations and public welfare cooperatives'" (Palasik 2000, 71). The politics of staffing new government agencies was held hostage to the proceedings, which were anything but transparent.

In the emotionally charged postwar atmosphere, it was extremely difficult to establish the veracity of one's loyalty to the past regime. Suspicions were piqued. So who actually ended up on B lists and why was a complicated issue. The process required officials to submit a declaration, after which their names would be made public, in order to solicit commentaries on their past behaviors.[9] Provisions were made to protect from unfair treatment those who had been compelled to take an oath to the previous government in order to keep their jobs. Consideration was also given to those who fled the Red Army out of fear rather than specific political motivations, such as choosing to move west with the retreating German army (ibid., 71–72). Kenez cites sources claiming that between 60,000 and 100,000 people lost their jobs for ostensibly political reasons (Kenez 2006, 131). Palasik cites data specifically on the number of screenings conducted on public officials, limited to Budapest in the absence of data from elsewhere in the country.

> According to a summary for Budapest, between March 26 and October 26 41,602 legally binding decisions were delivered, of which 34,019 were "exonerated." The summary from January 10th of 1946 substantiates this trend: among the 42,136 employees and officials undergoing the screening process, 37,351 (88 percent) were exonerated. Among the 4,785, however, who were not cleared only 825 persons were let go and only 264 sent into retirement. (Palasik 2000, 73)

Extrapolating from Palasik's evidence for the entire country—remembering, of course, that Budapest was the capital city where the overwhelming majority of government offices were found, and so would have constituted the primary site of the vetting process—far fewer people may have been let go or sent to trial than Kenez suggests. That does not lessen the fear and anxiety many felt, knowing full well that charges could be fabricated to settle personal scores, as had been true during the war.

Not surprisingly, screening of public officials became embroiled in turf wars between the parties over government positions.

> In addition to the desire to ensure "democratization" by screening administrative personnel, the parties also had to present their own staff of competent cadre. All the while they had to prepare for national elections that would eliminate the provisionality of the legislature and government. The principled objective of democratic transformation and the power struggle between the coalition parties charged with achieving this created a peculiar catch-22....

> In a screening process that would decide the livehood of tens of thousands, there were committtees composed of the delegates of the affected coalition parties on the one hand, and on the other, these self-same parties were taking steps to increase their membership. This was done partly in the interest of a greater political presence in the administration and partly with a mind to success in the elections. (Gyarmati 1989, 6)

Moreover, as Gyarmati points out, not every party had enough qualified members to staff the administration to the degree they anticipated. In other words, procedures designed to rid the bureaucracy of the old guard and streamline the bureaucracy were hostage to the ambitions of the parties to expand their base within the government and forced parties to intervene in the screening process on behalf of their members no matter how qualified (ibid., 6). Even the leaders within a party could find themselves working at cross purposes in the reform. Setting the stage for a planned economy led the Communist Party to advocate reforms that would come to haunt them once they took power. "During the coalition period, the Communist Party supported the expansion of various levels of professional bodies in certain ministries even though . . . [the Communist Party recognized] this kind of overorganization . . . produced haphazardness, supervisory anarchy, and prolific bureaucratism" (Gyarmati 1981, 183).[10] The countervailing tendencies of reducing the size of the bureaucracy while securing as many jobs for party members as possible (and purging the bureaucracy of one's political opponents) elicited contradictory responses: "they accused them of leniency and inhuman severity at the same time" (Palasik 2000, 73). Bibó described it in the following terms: "'we wait in vain for a process that is built on denunciations, accusations, witnesses, and pleas to purge the civil service of hidebound, punctilious, incompetent, and caste-minded officials. If this is what we expected, then indeed it was too lenient, because by and large they didn't consider the officials' personality and professional value, but looked at the information produced against them. . . .'" (ibid.). He continued: "'we cannot expect effective work from a committee or other body, doing the replacing and reshuffling, if they must work with the knowledge that those they eliminate will immediately be branded, and may be deprived of food ration coupons'" (ibid., 73). In short, removing someone from a job could amount to starvation.

Political conniving and bureaucratic finagling dominate these vignettes, but the crucial question of expertise dominated. Claims of expertise could be easily dismissed as a cover for political appointments or the saving grace of a politically compromised bureaucrat. Nonetheless, the status of professional qualifications was just as much on the table as were the machinations of political parties. Interest groups lobbied crucial figures, hoping to secure for themselves a position in the new state despite (or because of) their ostensibly objective qualifications untainted by political affiliation or commitments. Notable figures in the emerging coalition were flooded with requests for patronage or public support as people scrambled to make the best of their precarious situation.

As the door to open politicking began to close in late 1947, any semblance of coalition politics fell away. Representative bodies, or what the Communist Party referred to as "bogus representative bodies" (*álérdekképviseleti szervek*)—such as producers' associations in viticulture, fruit, tobacco, and sugar beets—were abolished in September of 1948. Communists believed these organizations were led by kulaks and other reactionary elements, carrying far too much influence nationally and at the local level for the party to tolerate. Other government agencies established to represent the interests of the peasantry in 1945, such as the National Councils of Agricultural Affairs (Országos Földmívelésügyi Tanácsok), were abolished as well, with the Ministry of Agriculture taking on their advisory role. Interest representation was to be channeled into the new mass movement groups being established by the Communist Party, while experts—if they passed muster politically—were to be absorbed by the ministry. "Officials qualified for a pension must be sent into retirement (if it is not possible to dismiss them for disciplinary reasons)."[11]

The association for professionals in agriculture was treated less harshly than producer associations or agricultural councils. Initially the plan had been to keep the association, but with the founding of the new organizations designed to embrace the poorer members of the peasantry, the party's policy changed. Instead of simply dismantling the association, the Communist Party crafted an intermediary strategy in November 1948 to placate professionals by having them join the working peasantry in a larger umbrella organization. As the recommendation made clear, leaving professionals on their own, without connections with a mass organization, would put them beyond the reach of the party/state, which was keen to draw them into political indoctrination as soon as possible. These moves were to be tempered, without any signs of duress. "Initially a certain autonomy must be guaranteed to experts within the mass organizations, because only in this way is it possible for them to feel the new organization to be their own until they become absorbed in the work of local organizations."[12] Professionals would be encouraged to create a special section within the organization and allowed to elect their own leaders. Nonetheless, care was taken to ensure that in the local chapters around the country kulak and reactionary professionals not dominate for fear that "they block the work of our organizations and exercise their influence."[13] This relative autonomy was shortlived.

Show Trials

Another weapon the Communists' wielded in their assault on democracy was to fabricate trumped-up charges of espionage and treason against powerful forces allied against them, such as the Ministry of Agriculture trial mentioned at the beginning of this chapter. Hamar argues that after losing the elections in 1945, the Communist Party's hopes of dislodging the agrarian elite were dashed, so they followed a different strategy for infiltrating the ministry (see also Baczoni 2000). They began to keep track of apparently insignificant irregularities, illegalities, and inappropriate personal dealings, which were then woven into a grand narrative of larceny and treason in the summer and fall of 1948. "On the basis of evidence discovered in the course of house searches, as well as the confessions of the prisoners, it was ascertained that since the liberation, the Horthyite, fascist bureaucrats of the Ministry of the Agriculture *methodically conducted a whole series of damages of the greatest dimensions*" (*Szabad Nép*, 25 Jul. 1948, 3; emphasis in the original). Officials were charged with the misuse of funds and the destruction of state property, portrayed as a crime against the nation when resources were so scarce after the war. Exhibiting a callous disregard for the public's needs, ministerial officials spent monies on new office furniture, bought cars for ministerial employees, and ordered ceremonial plaques. "While the working class set off for villages to build schools week after week, sacrificing their Sundays and free time, [the accused] were flooding the Hungarian village with obscene works like 'Love, Engagement, Marriage' and other sorts of pornography from state funds" (*Szabad Nép*, 15 Sept. 1948, 1). A whole range of charges were brought against the staff of state farms whose difficult material circumstances after the war were held against them. For example, at the state farm in Mezőhegyes, horse herds had been severely depleted when force-marched westward with the retreating German army. Barns and fields were in shambles, and Russian soldiers commandeered the bulk of foodstuffs being produced. Nonetheless, ministerial employees were held responsible for the dire conditions, all problems being attributed to corruption and personal greed. At the same time that enemies were being exposed, accounts of the trial regularly assured the decent, hardworking bureaucrat that he would be protected from harm; indeed, the trial offered citizens a lesson in identifying unreliable elements in their midst.

A crucial element of the Communist Party's case against the ministerial elite was their alleged subversion of state policy and intransigence in the face of the new democratic order. They were tarred with the epithet "Non-Party Block" for refusing to take sides in the heated political contests of the day. According to the newspaper, the ability of the defendants to hide behind the façade of objective nonpartisanship made them more dangerous than fascists, whose politics were openly declared (*Szabad Nép*, 15 Sept. 1948, 1). In the guise of disinterested buraucrats, ". . . a significant portion of the weak-willed bureaucrats succumbed to the cunning fascist leaders" (*Szabad Nép*, 19 Sept. 1948, 3). More to the point, the top defendants were sympathetic to the conservative social policies of the Catholic Church. For example, Pernecky, a leading defendant, had penned two treatises two decades apart (1926, 1943) in which he extolled the virtues of "professional status hierarchy" (*hivatásrendiség*).[14] A select group of officials also met in the evenings at the ministry to discuss publications issued by Acto Catholica, for example, "Christianity and Democracy" or "Labor, Wage, Property," both issued in 1946. (The "pornographic" pamphlet on love and engagement was also published by Actio Catholica.)

The most credible claim among these charges concerned the land reform, that is, the strong doubts the ministerial elite voiced about the wisdom of dismantling manorial estates in favor of very small farms. They were not charged with direct obstruction; the Ministry of Agriculture did not participate in the allocation of lands, a task supervised by the National Council Supervising Land Reform (Országos Földbirtokrendező Tanács) and implemented by Land Claims Committees in villages (Donáth 1977, 51–55).[15] Numerous quotations attributed to the ministerial elite in the trial documents testify to their views. Perneczky was said to have voiced his opinion about the unfortunate consequences of land reform in the following way: "[the reform] didn't take technical issues of production into consideration, because the vast number of dwarf farms were not viable."[16] Solthy was chastised for publishing educational materials in 1946 on the behalf of the Ministry of Agriculture in which "they showed large estates to be the ideal unit for agricultural production."[17] The committee chair of the Food and Agricultural Organization of the United Nations in Hungary, Sibelka-Perlberg, was charged with having had materials compiled in which the analysis "deliberately misled foreigners, and represented Hungarian policy of landed property in a bad light . . . [claiming that] the land reform was not an economic, but a political move . . . because the international security of Soviet power required it . . ." (Hamar 1997, 43).[18] In short, their sins lay in not actively supporting the new agrarian regime. More grievous still was the temerity of ministerial officials who dared to question government policy, an act that by 1948 had risen to the level of a crime against the state. An ironic twist to the trial was its timing. All the defendants had been in jail for only a month when Communist Party Secretary Rákosi held his famous speech at Cegléd announcing the party's embrace of state farms and cooperative farming (Hamar 1997, 4), a move that allied Communist Party policies with the views of the allegedly fascist bureaucrats who also had preferred larger economies of scale.

Changing the (Intellectual) Guard[19]

Universities and technical colleges were also targeted for restructuring after World War II, with equally mixed results in terms of forcing staff changes and curricular innovations. As was the case with administrative reforms in government, debates within the agrarian community over the degree of specialization and problems of access to training were not new.[20] Basic questions about the relative value of practical training versus advanced studies at the university level were complicated by thinly veiled rivalries among historical institutions like Magyaróvár or Debrecen over claims to be the pinnacle of Hungarian agricultural research and training. Recognized specialties—the school of forestry in Sopron or the Agricultural Chemical and Paprika Experimental Station in Szeged—were initially exempted from these protracted battles. At the core of the major reorganization of higher education in agricultural sciences promoted by the Communist Party was a deliberate strategy to rein in errant teachers and schools too independent of the party's influence.

Immediately following the land reform, the Ministry of Agriculture issued a new regulation altering the contours of agrarian education. On July 1, 1945, all institutions engaged in vocational and professional training in agriculture—once under the aegis of the Religious and Public Education Ministry (Vallás- és Közoktatásügyi Miniszterium)—were moved to the Ministry of Agriculture "for controlling their unified supervision and management" (Walleshausen 1993, 175).[21] The educational ministry continued to be responsible for universities. In the meantime, the agricultural academies (*mezőgazdasági akademia*) in Debrecen and Keszthely were elevated to the status of technical colleges (*főiskola*), a status previously only awarded to Magyaróvár in 1942. Regulations also stipulated changes in the qualifications required to teach in technical schools, preventing anyone from teaching who had not already been named to the status of a faculty member of the university. In October 1945 the prime minister announced the establishment of the Hungarian University of Agrarian Sciences (Magyar Agrártudományi Egyetem), which would incorporate all the university-level institutions nationwide in horticulture, forestry, and veterinary medicine, with the school of agricultural sciences being divided up into four departments: Budapest, Debrecen, Keszthely, and Mosonmagyaróvár. In one sense, founding a truly centralized university for agriculture was a long awaited victory; it had been first proposed in 1844 and then again in 1920. The siting of the university in Budapest was puzzling, nonetheless. An editorial in the Peasant Party newspaper (*Parasztpárt*) in September 1946 expressed this concern: ". . . it was incomprehensible . . . why it was necessary to abolish the independence of institutions in various points of the country which for a hundred-two hundred years had functioned and flourished to everyone's satisfaction, and force them into a gigantic university" (ibid.).

During the summer of 1949 the Agricultural and Cooperative Department of the Communist Party prepared a report setting out the kinds of changes needed to improve teaching at the university and the costs involved. The tone was optimistic, boasting of a successful reorganization of the agricultural and horticultural divisions. They took particular pride in reporting having purged the large majority of kulaks and fascists from the student body.[22] The class composition of the student body now reflected the vision of a new professional cadre prepared "to keep pace with building socialism," though it would require providing remedial training for some students. "We have succeeded in raising the proportion of working class students from 1 to 9 percent and the proportion of poor peasants from 2.5 to 13 percent, as well as prepare 141 poor peasant students who had not finished high school for university studies in a two-year course . . . alongside the university. They will attend the university this school year."[23] Teacher training would become a regular component of the curriculum, in anticipation of having to train a new army of agricultural specialists. In line with the new political climate, "we have increased the strength of the people's democracy and our Party by putting in new department chairs."[24] The task of clearing out the enemies of progress in the provincial divisions of the university was more difficult. "The least democratic of the agricultural departments are those in the countryside. The majority of these students are declassé and kulaks. The proletarian element is only 1 percent while the 45 percent declassé and kulak elements have 54 percent of the middle stratum under their influence."[25] It is not surprising, then, to learn that all the provincial institutions, including the venerable school in Magyaróvár, were closed indefinitely.[26] But problems with undesirable class elements were not isolated to agricultural institutions, as demonstrated by the problems the party faced at the Economics University. "The large majority of older students [from previous cohorts] are reactionary, and in many cases fascists. About 15–20 percent are B-listed officials, military officers sympathetic to the West, etc. (From this perspective the old economics faculty is by the far the worst of all the universities)."[27] The party made plans to reject 1,300–1,800 of the 2,500–3,000 applicants anticipated in the fall term and also set up a committee to deal with the uproar that would surely follow their decision. Staffing problems also affected the Economics University, as there were few economists properly trained in Marxism-Leninism. The best qualified in the field occupied leading positions in the Communist Party, in state administration or in party education. New courses in political economy were up and running at the university but not necessarily taught by luminaries in the field (Péteri 1998, 200). Anticipating these problems, the prominent Communist Imre Nagy wrote in July of 1947 to Jenő Varga, a famous Hungarian economist living in Moscow, asking for two copies of foreign language editions (German, French, or English) of Soviet political economy textbooks. "If, despite this, it is not possible to acquire the books I've requested, then please provide me with the authority to discover via the apparatus from which publishing companies abroad, in England or France, it would be possible to acquire the Marxist political economy we need."[28] One wonders where Nagy found the textbooks he sought.

In a few short months, a disturbing trend in the number of students dropping out of the Agricultural University prompted the ministry to appoint another committee to examine conditions at the university. The report painted a far bleaker picture than the previous summary. The urgency to increase the number of specialists for the new economy had caused severe overcrowding and an extremely tense atmosphere (Balogh et al. 1978, 200). Lecture halls did not have enough seating, cafeterias could not serve the entire student body, and college housing was only sufficient for 700 students, leaving sixty without guarantee of a place to sleep. "The attire of a portion of the students (about 20 percent) is scanty, evident in part in the lack of winter clothing and footwear. In many cases students living in the college dorms have been forced to attend lectures in borrowed clothes."[29] Measures were taken to provide poor students with a clothing subsidy to get them through the winter months, but this was obviously only a temporary solution.

Research organizations—independent or those housed in ministries—were also targeted for serious restructuring. Prior to 1945 economic research was conducted at several different institutions. The most venerable, and long lived, was the Hungarian Economics Research Institute (Magyar Gazdaságkutató Intézet, or MGI), established in 1927 by the Budapest Chambers of Commerce and Industry to adapt the new insights of business cycle analysis to Hungary. "To this day their studies in monetary policy, investigations of the Hungarian national income, analysis of industrial investment and capital accumulation, examination of all sorts of economic problems in Hungarian agriculture, as well as periodical economic reports, are important sources of data and ideas for research in economic history" (Péteri 1998, 191–92). MGI was both a research institute keeping current with contemporary economic theorizing and a valued but always independent partner in the creation of national policy. After the war, the institute held on to its autonomy from political control, even when several members were appointed to high-level positions in the coalition government. This would end with the ascendancy of "socialist science," which by its very definition could never be detached from the party/state and the working classes (ibid., 189). MGI was dismantled in August of 1949, to be replaced with the Institute for Economic Science (Közgazdaságtudományi Intézet, or KTI). As a result of vetting former members for their party affiliation, Péteri calculates that approximately thirty of the thirty-four researchers employed at the institute were let go (ibid., 196). Only two of the researchers were members of the Communist Party; most were not allied with any party whatsoever, which constituted a major sin in the eyes of the party/state, drawing charges of passivity and indolence. Unfortunately, those who came to replace the unreliable researchers had to be offered remedial training. "The poor professional training of the staff [in 1950] is attested to by the fact that for the first half year of the KTI a course on the political economy of capitalism was organized for the institute's members" (ibid., 199). Péter Erdős, one of the temporarily assigned codirectors of the economics institute, complained in 1951 that he could not find people qualified to do independent research; the up-and-coming generation was going to require many years of experience before reaching that status (ibid., 200).[30] As Péteri points out, the loss of highly skilled statisticians and sophisticated economists from the playing field of national policy, precisely at the time when the economy was being centrally managed, was a problem. Indeed, it was devastating.

In the narrower fields of work science and agricultural economics, there were three institutes in place before the end of the war: one in industry, two in agriculture. The Work Science Institute (Munkatudományi Intézet) was established in 1940. A private firm staffed by engineers, its activities were comprised of devising output wage systems for industry, as well as studying the theoretical bases of establishing norms.[31] András Heller, a major advocate for manorial production (see chapter 1), founded the Agricultural Work Science Institute (Mezőgazdasági Munkatudományi Intézet) in 1943, housed at the Horticultural University in Budapest (Agárdy 1943). Mihály Kalocsay, whom I interviewed in 1997, remembered the institute being devoted to rationalizing tool use, not developing norms of any kind. (He also emphasized the strong German influence on this group, based in part on personnel from the institute taking trips to training farms in Germany.) The other work science research center in agriculture was the National Institute for the Study of Agricultural Organization and Production Costs (Országos Mezőgazdasági Üzemi és Termelési Költségvizsgáló Intézet; see chapter 2), sponsored by the Ministry of Agriculture. It was a successor to earlier institutes in manorial management; its research activities would be substantially curtailed by the summer of 1949.

Confusion reigned in the world of agricultural research, as a report penned in January of 1949 complained. "At present there are 121 agricultural research institutes and experimental stations. Among these only a few actually conducted scientific work, and those doing research work were divided into [groups working on] minor details. Where scientific and research work was pursued, it was done haphazardly."[32] The institutes were scattered among a variety of agencies—various departments of the Ministry of Agriculture, universities, and so on—dividing up their activities unnecessarily. Moreover, the large percentage of institutes' staff at all levels were incapable of conducting research.[33] The entire enterprise needed to be reorganized. Eleven new institutes were to be established, including institutes devoted to the study of agrochemistry, plant breeding, biology, animal husbandry, mechanization, soil science, plant cultivation, plant disease, horticulture, and forestry. The final institute listed in the report, the Agricultural Organizational and Economic Institute (Mezőgazdasági Szervezési és Gazdasági Intézet), had yet to be established; eventually it would become the administrative body charged with designing the work unit for cooperative farms.[34] The problem was solved in May, when the National Institute for the Study of Agricultural Organization and Production Costs was closed, to be replaced by the Agricultural Organizational Institute (Mezőgazdasági Szervezési Intézet, or MSzI).

Despite all the care devoted to restructuring agricultural research into a comprehensive set of institutions, important tasks still fell between the cracks. A good example was the basic work of collecting producers' prices in the agricultural sector. This posed problems for the still-active Hungarian Economics Research Institute, which in the absence of the proper data was having trouble figuring national income statistics. The head of the institute wrote a letter in July of 1949 to Erdei, who now was minister of agriculture, to complain,[35] wondering how they were to proceed.[36] The new institute (MSzI) had been charged with "exclusively research tasks," abandoning the once-crucial task of collecting data for other agencies.[37] This task fell to the Central Statistical Office.

A report reviewing the personnel office of the Agricultural Experimental Center issued on August 22, 1950, identified serious problems.

> The majority of researchers working in the field of agricultural scientific work are petit-bourgeois intellectuals, of whom a significant portion are old (especially the independent researchers). Their family connections, old milieu, and background hamper them, and this is evident across the board: in their work, their behavior, and their training along professional and political lines. They cling to their old, familiar methods, don't recognize the results and experiences of Soviet agricultural science, nor do they use them to support their own thinking. They judge their results on the basis of Western results, and so are satisfied with their work.[38]

The staff in the office were accused of focusing too narrowly on their own tasks and being insufficiently vigilant. They did not make the effort to explore researchers' past activities or old connections; "they don't recognize the enemy wrapped in the fog of professional knowledge."[39] They were also called to task for not insisting that researchers take part in political training. "Researchers do not participate in political training regularly, or at the proper level. As a result, they are uninformed (for example, 25 percent of the researchers do not participate at all)."[40]

While ridding universities and technical schools of reactionary elements was a top priority, it was not enough. Imre Nagy, whose experiences living in the Soviet Union informed his views, knew the task was different. In his speech to a meeting of the activists in FÉKOSZ (National Association of Agrarian Workers and Smallholders, or Földmunkások és Kisbirtokosok Országos Szövetsége) on the 19th of May, 1948, Nagy had a grander scheme in mind.

> Lenin frequently dealt with the problem of experts and demonstrated that a bad expert cannot be a good Communist. People don't follow a bad expert; a bad expert is not accorded public respect. What kind of Communist is someone who completely falls short? In order for our experts to take over the place of old experts and take control, it is necessary that they become better experts than the old experts were. They must expand their professional knowledge and keep pace with the development and results of modern agrarian science.[41]

Unfortunately, this task took a long time.

> With the socialist development of our agriculture, a burning necessity for properly trained, socialist-minded expert cadres has become an urgent need that is already felt. It is not possible to wait to fulfill this need until restructured technical institutions deliver appropriate experts from among educated young people. In the meantime we will train expert cadres in shorter courses from the ranks of adults.[42]

Accelerated courses of study were designed at the highest as well as lowest educational levels, training scientific researchers as well as state farm managers and college teachers.[43] (Unsurprisingly, these "quicky" classes proved inadequate to the task.) Universities and research institutes continued to be staffed by specialists with problematic backgrounds.

Postwar Economic Recovery

In the face of the massive destruction of the final year of the war, rebuilding the economy was a gargantuan task. Yet it also paved the way in principle for massive improvements in the infrastructure and manufacturing base of the nation. Unfortunately, these innovations were frustrated by a combination of international agreements mandating reparations, lack of funds (no Marshall Plan), and short-sighted plans to ramp up the economy. Of course, many of the centrally planned features of the economy had already been in place during the war. Thus the economic restructuring process was much swifter than administrative reform (Gyarmati 1981, 718).

As we have seen in previous chapters, advocates among political observers, social science researchers, and some bureaucrats argued relentlessly throughout the 1920s and 1930s that the state's role in managing the economy be expanded. In fact, the Hungarian government's policies toward the economy in the interwar period already entailed extensive state participation, either in direct control of nationally owned enterprises or through a variety of legal measures and regulations. This pattern preceded the Depression, but the role of the state in protecting Hungarian businesses and the health of the economy overall increased between 1931 and 1938 (Bojkó 1997, 138). Firms were regularly bailed out by the state, aided by provisions in the bankruptcy law that initially had been passed to facilitate the demise of unprofitable enterprises (ibid., 8). Monopolies were given special consideration, becoming the focus of state assistance as of 1931 (Berend and Ránki 1958, 26). The state set specific taxation policies or engaged in setting prices (often with the collusion of particular cartels) to dampen competition. The state also invested heavily in a number of private concerns, although these ties were often difficult to ferret out from official statistics.

> Firms established in Hungary before the First World War resisted liquidation, existing firms expanded, indeed, entirely new kinds of corporations started to appear, which appeared to be private businesses (e.g., joint-stock company, cooperative) but which in fact represented a unique combination of state capital and state intervention. Therefore the development of hidden state capitalism intensified, with the state acquiring stakes in private enterprises, either in the form of the majority of shares or actually the entire enterprise. (Bojkó 1997, 10)

The degree of monopolization of the economy grew in the 1930s. Approximately 60 percent of agricultural exports were handled by monopolies (Berend and Ránki 1958, 108n54). A number of crops were regulated by price as well as by trade. Ihrig provides a list of the crops under state regulation in 1935: "wheat, sugar beets, tobacco, linseed and flax, hemp, milk, firewood, wool, paprika, alcohol and potatoes. One must also consider that the price trends of wheat more or less influence other grains, indeed seed for fodder as well. Thus in the area of crop production the zone that is managed is larger than that of production that is not" (1935, 131).

Laws preventing the establishment of new firms eliminated competition, strengthening already existing monopolies (Berend and Ránki 1958, 109). The argument the minister of industry in 1935 made against new firms was simply "the danger of spreading limited capital too thin" (ibid.). The law eventually passed initiated a quota (or *numerus clausus*), forcing cartellization within industry. This was seen to be integral to the increasingly state managed economy, as an article in the newspaper *Stock Market of Budapest* (*Pesti Tőzsde*) in November 1935 made all too clear. "'There is no doubt that . . . registering the industrial *numerus clausus* represents a newer and forceful step toward a managed economy . . . it is obvious that this condition is not a temporary world phenomenon, but an overture toward a newer chapter of the economy'" (ibid., 109–10). In the banking sphere, increased concentration was evident. By 1938, 72 percent of all capital stock was in the hands of eight of the largest banks (ibid., 112). All these measures fed the interests of finance capital, industry, and, to a lesser degree, large-scale agriculture. As such, they demonstrated the power and efficacy of state intervention in the economy, giving further justification to those in favor of greater state management.

The state also was actively engaged in labor policy, passing regulations on the length of the work day, setting minimum wages, and, by the war years, setting maximum wage levels. Throughout the interwar period, and during World War II, the state consistently acted to moderate extremes in the price of agricultural labor, either by reigning in workers' demands in times of labor scarcity by setting an upper limit, or by forcing employers to pay a base minimum when the fortunes of workers had drastically worsened. The state also initiated public works projects in rural communities facing substantial shortages in winter supplies (ibid., 28; see also Steuer 1938). Regulations issued by the Ministry of Agriculture in 1926, 1928, and 1930 stipulated more carefully the exact role of national and county-level authorities in supervising the movement and contractual obligations of labor. Advocates of a better managed economy lamented the shortcomings of these agencies, most notably their inadequate methods of gathering information. During the war,

Reitzer complained about the consistent absence of proper numerical data, a failure impeding "the continuation of planned labor market policy" (Reitzer 1941, 996–97).

Another stage in the degree of state intervention was marked by the introduction of what came to be known as the Győr Program of 1938. Following Germany's lead in ignoring the ban on militarization set by the Treaty of Versailles ending the First World War, revisionists keen to enhance Hungary's power and reputation in the region began to call for serious plans to prepare for war in 1937. Crucial elements of the military began to lobby the government to invest in the armed forces. This initiated a period of militarization of the economy, during which approximately 10–20 percent of national income was to be routed to military investments (Berend and Ránki 1958, 277–78, 298). The role of Germany in the Hungarian economy also grew at this time. Approximately 50 percent of all exports ended up in Germany, while German capital interests held 12 percent of Hungarian industrial shares (ibid., 292). At the same time, thousands of Hungarian workers were employed in Germany. If we consider agrarian migrant workers alone, their numbers reached nearly 45,000 between 1937 and 1943, the peak years being 1938 and 1939 (Lencsés 1982, 177).

As elsewhere, the Second World War played a significant role in prompting further state intervention in economic affairs.[44]

> Beyond the direct equipping and provisioning of the army, [state intervention] spread to every area of economic life. The state became the largest consumer of industrial production, and it became necessary for the state to intervene in questions of production, such as establishing enterprises, the supply of raw materials and energy, regulating prices, the credit system, and the supply of labor power. (Bereznai 1943, 7)

By 1941, the state officially declared a fully planned economy (*tervgazdálkodás*). Problems of provisioning sparked the introduction of measures to ensure supplies made their way to soldiers. Citizens' needs were secondary.

After the war the dominance German interests had exercised in the Hungarian economy for nearly a decade was replaced by Soviet interests. In 1944, Germans dismantled factories in Hungary and shipped them west; as of 1945, Russians were shipping them east. Pillaging and looting accounted for some of the loss of economic assets after the cessation of fighting, but war reparations—paid in kind as well as in cash—took their toll on the industrial base of the nation. In addition to shipping existing equipment eastward, Hungarians were also required to manufacture machines and equipment to Soviet specification (Borhi 2004, 146).[45] As part of the peace treaty, assets formerly owned by Germans were appropriated by the Soviets as well (ibid.).[46] Once-active markets for Hungarian goods in Italy and Germany were now trained eastward, strengthening in trade the already substantial interests the Soviets had in the Hungarian economy. The Soviets established a number of joint ventures with Hungarians, following the signing of a treaty for economic cooperation on September 23, 1945. Clearly, by 1947, the Soviet Union—as shareholder, factory owner, landlord, and occupying army—controlled significant assets in the Hungarian economy.

The process of rebuilding the economy after the war intensified an already strongly centralized and managed economy bequeathed by the previous regime; demands for reparations only increased the Hungarian state's participation in the economy. Prior to the stabilization of the forint in August 1946, all of the productive capacity of the five most important industrial concerns was devoted to manufacturing to fulfill the conditions of reparations, dropping to only 60 percent as of September (Pető and Szakács 1985, 79). With time, what had been oversight of production by the state became outright ownership. As the Cold War intensified in 1947 and 1948, the vision of the Three-Year Plan once supported by the Communist Party was abandoned. Having swallowed the Social Democratic Party, and forced significant political enemies into exile, the Communist Party could now redirect its efforts to jump-start the economy from a more extensively state-led, state-controlled industrial base, while turning its back on promises to guarantee private property. The Marshall Plan was rejected, and in February 1948 new agreements with the Soviet Union were signed. The once-bright vision of a rejuvenated economy founded on investing in modern productive technologies was traded for an emphasis on increasing brute quantities. "Instead of reconstruction, there was only renovation, which displaced potential renewal. Therefore instead of introducing the quick technical and technological changes achieved during the war—of which only a small amount was perceptible in Hungary—they ... restored the earlier, obsolete equipment" (ibid., 122). The rush to the finish made it possible for the new party/state to declare a glorious and shortened end to the Three-Year Plan. Not surprisingly, a larger proportion of investments were made in industry than initially planned, leaving rural communities to bear the lion's share of the costs and burdens of reconstruction. The five most important heavy industry factories were nationalized, much to the dismay of their owners. The state now held a significant portion of assets in the economy.

> At the end of 1946 roughly 150 thousand workers, or about 43.2 percent of the employees in manufacturing and mining, were working in state firms. Among

them 75,000 worked in the five largest factories taken over by the state. The state role prevailing in the energy, raw, and basic materials sector exercised a significant influence over every branch of industry, indeed the entire economy of the country. (ibid., 81)

Finally, banks were nationalized in 1947, transferring the concentration of capital from the largest banks to control by the state.

> Nationalizing the banks—in addition to the large banks becoming state property and the overwhelming majority of banking houses falling under state control—resulted in nearly 60 percent of mines and manufacturing and roundly 80 percent of heavy industry becoming state property. Shortly thereafter, in March of 1948, the nationalization of factories employing more than one hundred workers followed, as a result of which by this time the socialist sector of the people's economy comprised nearly the entire mining industry and large-scale factory transportation and the large majority of manufacturing industry, approximately 80 percent. (Szabó 1986, 10)

So as Pető and Szakács explain, "nationalizations in March (1948) did not represent a basic change with respect to substantial growth in the role of the state in the economy, but rather created the conditions for the transformation of the institutional system, and a change in the relationship between the state and enterprises" (ibid., 99).

Conclusion

I have argued in this chapter that the common refrain about the Communists taking power in 1948 being a total economic and political about-face is misleading. There was no technopolitical rupture between administrative practices of the state and planned economy between the late 1930s and late 1940s in Hungary. Unprecedented changes in the organization of the state, economy, and educational system were envisioned by the Communist Party after the war, but emphasizing the party's aspirations for change to the neglect of views about administrative practice shared by the Communists and their conservative predecessors in government is inaccurate. Suggesting rapid and radical departures in practice also portrays the Communist Party as far more powerful than it was or could be. Adding an entirely parallel governing structure alongside state agencies—creating the ruling bodies of Communist Party administration—was a significant change, but as we shall see, it did not ensure the swift and effective adoption of Communist Party policy on the ground. Just as crucial to the restructuring as institutional intransigence was the absence of qualified personnel to put Communist Party ideas into practice. No textbooks on Marxist-Leninist political economy were available in 1947, and with the limited number of specialists already working in the government, no one could be found to teach Marxism-Leninism 101 at the Economics Faculty. Attempts to streamline the state and purge the bureaucracy of undesirables were hampered by countervailing pressures to install party favorites in key government offices. Continuities in state practice would also mean the Communists would have to compensate for the structural weaknesses of governing bodies they inherited from their predecessors. Better means of tracking the movement of labor, improved facilities promoting modernized business practices, and specialized training programs to prepare future administrators to manage accounts were sorely needed. If the party/state was going to succeed managing the nation's economy, crucial state capacities would have to be expanded. This was a tall order.

6

A New Matrix of Labor Value

> The balancing role of the market is missing, on which, in a bourgeois free economy, entrepreneurial descisions are judged—from the point of view of the capitalist entrepreneur—to be proper in rewarding profit or inappropriate for inflicting loss. Only one serious solution exists: the widest, scientifically organized, sustained examination of consumption and production, closely tied to the study of reducing operating costs and growing capacity.
>
> MAROSI 1947, 601

With the Communist Party firmly in control of the state apparatus, the gargantuan task of directing the entire economy was at hand. Unprecedented in scope, the state's role in economic decision making grew exponentially. Marosi described this change succinctly: "the most significant of the changes issuing from the introduction of a socialist planned economy is that business management and [national] economic considerations have become the same" (1947, 601). The managerial complexities were daunting; more was needed than simply increasing the size of the bureaucracy. Techniques designed to guide planning and specify strategies, and the skills necessary to use them effectively, became invaluable. The political significance of agricultural economics, business economics, and accounting skyrocketed, bolstered by their status as scientific practices that claimed to ensure precision and objectivity. Specialists in these fields were few and far between; moreover, virtually the entire community had been trained and practiced in "bourgeois" economics. Nonetheless, these specialists became crucial figures in policy development and training, focusing their energies on finally fulfilling their dreams of productive and efficient work places, though in a political regime very different than they would have imagined. The task became, then, to craft policy instruments that could achieve results effectively within the expanding scope of the plan.

The fledgling field of agrarian work science could boast a few doctoral theses, several scattered experiments, and a host of articles proposing strategies to improve productivity and modernize wage systems. No one, however, had actually designed a self-contained wage system on a national scale. Colleagues in industrial engineering had already been in the business of devising norm systems before the war, but these were solicited by individual enterprises for their own use. Manorial handbooks carried information on how much time and how many people were required to carry out various tasks at the farm, but the numbers were based on customary practices rather than scientific studies. The ambitions of agrarian work scientists would now be put to the test. Could they design a scientifically calibrated wage system for cooperative farms? And if they did so, would it work?

In the absence of a market, as Marosi explains in the opening quote, science would guide economic planners in keeping costs low while increasing the capacities of a planned economy. Scientific means of calculating costs, including the cost of labor, would render markets superfluous, even a distraction. An appeal to science as the basis of public policy was common in the Soviet bloc. After all, the ruling ideology of the Soviet Union, Marxism-Leninism, was inspired by the science of historical materialism. Unfortunately, the broad tenets of historical materialism offered Hungarians little help when it came to designing wage forms. One would be hard-pressed to find any reference whatsoever to Marx's labor theory of value in economic publications or government documents, while there were the extensive discussions of "bourgeois" techniques of estimating labor value. To be more accurate, the techniques themselves were never defined as "bourgeois," a designation only applied in the 1950s to specialists trained in the capitalist era. The techniques themselves were treated as innocent scientific tools to enhance productivity, a goal as important to socialist policymakers as to capitalists.

It is the purpose of this chapter to demonstrate how early socialist wage policy was crafted using the tools of interwar work science and with the expertise of trained practitioners. The new wage system was a novel configuration. Its skeleton was modeled on Soviet workdays (*trudodni*), a remuneration scheme for cooperative farmers adopted in the early years of collectivization in which workers' labor contributions could be assessed, while at the same time accommodating the seasonal vagaries of agricultural work, that is, crop yields depending on labor inputs that varied enormously during the year (Davies 1980, 131–33). Work norms were established by Soviet planners and ranked by skill level, arrayed on a simple scale from 0.5 to 2.5 (Gaponyenko 1950, 131). The flesh and blood of the system, however, was Hungarian: both the values of labor and the manner in which they were determined occurred locally, based on long standing practices in manorial farming and the technical insights of interwar work science. The innovative feature of the new wage system was found in its construction as a matrix, an internally consistent, fixed set of values (the relative proportion of skill required, physical difficulty, and significance to production). My argument is that a matrix of simple numbers was all that was needed to remunerate workers, rendering the market irrelevant. This technical wizardry would also render invisible the historically contingent cultural values of work and the investments in form creating a value matrix required. Only by tracing the technical details of the system's construction is it possible to ascertain what was locally produced, how, and why.

The chapter proceeds by showing why science mattered to Communist policy makers, describing the activities undertaken by the Work Science and Rationalization Society, and sketching policy debates that preceded the decision to collectivize agriculture. Then follows a brief overview of the ways the party/state attempted to force villagers to join collectives and how poorly these measures fared. The final section of the chapter describes exactly how the work unit system was created, and by whom.

Scientific Wages

The principles for a good wage system were simple: above and beyond ensuring the intensification of production, the system should be fair, consistent, and established to satisfy both workers and employers (Hegedüs 1947a, 20).[1] As the Communist Party gained greater control over policy, the issues of fairness and worker satisfaction would fall out of the picture, leaving only a trace in propaganda tracts illustrating the superiority of socialist wages over capitalist ones. Numerous articles were penned discussing the various options to be considered. Included among these were essays on wage policy, a central principle of which—considering the scarcity of capital and pressing need for economic recovery—was to design new wage systems that guaranteed a direct connection between wages and output.

A truly planned economy required technical norms. As Stalin explained, *"the technical norm is the great regulating force, which in production organizes the broad masses of workers to gather around the vangard elements of the working class"* (Mártonfi 1949, 1; italics in the original). Heeding Stalin's advice, the Hungarian party/state devoted enormous efforts in the first few years of socialism to labor studies at national firms, which included state farms and national enterprises. The project was extremely time consuming, as it entailed the careful analysis by newly trained personnel of every conceivable task in industrial and (state-run) agrarian production. A specific task would be identified, divided up into actions, and then subdivided again into the smallest element of each action, all of which would be timed by normsetters; if the work process was mechanized, both the manual tasks and work done by the machine had to be calculated separately (ibid., 64–65, 72–73).[2] Time and motion studies were commonplace in the late 1940s, but rarely had they been undertaken for every possible task in the workplace. In addition to timing every single unit of activity, attention also had to be paid to preparatory work required for the job. Once these times were recorded, they were multiplied by three factors: intensity, fatigue, and loss (*elfaradás és veszteség;* ibid., 74–75). In the course of studying workers' minute actions, work scientists were able to suggest more efficient use of time.

Science would furnish the tools to achieve technical norms "with mathematical precision" (Hegedüs 1947b, 36). Accordingly, workers' wages were not to be set by their trade or union affiliation, but on the basic notion of labor value determination, that is, "how much expertise the work in question requires, how difficult or dangerous the work is, and how long the worker in question has already spent in the trade or at the firm" (ibid., 37). In principle, as Hegedüs imagined, wages could be determined by formula, making the system "independent of the [person's] trade, the changing value of money, . . . and ingrained customs of the firm" (ibid.).

Until a more scientifically calibrated system could be designed and implemented, existing piece-rate wage systems (*akkord*) common in a number of industries would need to be seriously overhauled. And there were very pragmatic reasons for doing so: set the norm too high, and workers give up trying to reach that goal, eroding output; set the norm too low, and workers would forego extra effort, confident that their half-hearted measures would still bring a premium (Mártonfi 1946, 5). "Every business person knows what kind of serious problems arise when settling accounts of poorly determined akkord prices. A climate of constant friction, serious disagreements, and mutual distrust develops, all to the detriment of production" (ibid.). To convince the public that poorly calibrated norms were unacceptable, and to ensure continued increases in production, extensive efforts had to be expended to avoid this catastrophe. "*Basic and more effective innovations must be introduced* in the work output system" (Ihrig 1947, 2; emphasis in the original). An army of work scientists was needed.

Mobilizing Work Scientists for Socialism

Anticipating the possibility of greater interest in work science and rationalization, in 1947 a small coterie of work scientists with backgrounds in industrial work science established the Work Science and Rationalization Society; the Society would later become the Work Science and Rationalization Institute.[3] Their plan was to deploy the same techniques they had used before the war, building a clientele by advertising their services to factory owners and managers who were struggling to get their firms up and running.[4] The work of the society grew rapidly. By late 1947 the institute was well equipped with departments in rationalization, business management, energy management, and construction, and was in the process of creating specialized departments for home economics and heating. They had also established an education, documentation and propaganda department. The society sought out colleagues engaged in comparable work at the Agricultural Work Science Institute and the Free Trade Union of Employees in the Private Sector. The society also entered contracts with government agencies to oversee occupational training. "At the request of the Planning Office we established a working group of Psychotechnical Institutes, which will connect psychological technique with labor force administration in relation to selecting careers and directing apprentices and workers on their behalf."[5]

As the market for scientific wage setting grew, the society was well positioned to take advantage of the swell of interest in newly designed wage systems. By the 30th of November, 1947 they had completed rationalization and norm setting work for twenty-three firms, were in the process or near to finishing thirty-six more, and negotiating with another twenty-three. The kind of firms studied crossed a number of sectors: the steel and chemical industries, textiles, food supply, electricity, and construction. The kinds of services they offered, in addition to rationalization and norm setting, included price surveys and setting output indicators for both physical and white-collar workers. The society was also a boon to government ministries, which had been directed by the Party to explore the scientific techniques so praised by Lenin. As Kalocsay explained, ministries jumped at the chance to outsource this work to the institute. "The ministries gladly pounced on this institute. Finally there was a firm that dealt with this . . ."[6] As of November 1947, members of the institute had offered advice—"on theoretical and concrete questions"—to several of the most important public agencies crucial to the economic transition: the Supreme Economic Council, Industrial Production Council, Ministry of Industrial Affairs, Committee for Setting National Wages, and last but not least, the Planning Office.[7] The Industrial Production Council charged the institute with setting up a series of courses on rationalization to feed the demand for qualified personnel and to prepare cadres to manage the Three-Year Plan.[8] A new generation of analysts for industry passed through the institute's doors.[9] In Kalocsay's opinion, the project to rationalize the economy foundered, however, as the prospect of "completely norming the entire people's economy" was beyond anyone's capacities. Within a year, the institute was absorped by the National Wage Labor Committee (Országos Munkabér Bizottság), folding it into an arm of the government.[10]

Officials in the Ministry of Agriculture were just as keen as their compatriots in industry to avail themselves of the new services designing scientific norms. The Labor Relations Department and the Business Management Department of the Ministry of Agriculture solicited assessments of output norms from the ministry's existing network of state farms (research and experimental farms, stud farms, etc.). The norms were then evaluated by the National Norm Institute. A small sampling of the norms developed for various tasks include shaking and scraping sacks, loading sacks onto a railway car, digging tree stumps out of the ground, and (my favorite) a norm for making ice to be used by the state fisheries. Those designing the new norm systems took their work seriously, as did the National Norm Institute, as the following communication demonstrates. A letter from the Ministry sent to the National Enterprise of Silk Growers praised them for commendable norm testing, although it was pointed out that the values for the fatigue factor and loss bonus were supposed to be listed separately, leading otherwise to an artificially high fatigue factor.[11] Scientific measurement was paramount, after all.

If the energies expended to establish norms in industrial production may have been warranted—assuring a fairer wage—the calculation of norms for agriculture struck some as ludicrous. Kalocsay remarked that

> It was dreadful idiocy. . . . To introduce methods approximating industry into agriculture, where as a matter of fact a certain output wage [system] had already developed. . . . That is, in Horthy [interwar] Hungary . . . there were farms or mostly branches [of farms] where they used piece wages, they used regular, normal output wages . . . they were for industrial crops . . . but this was implemented in a relatively narrow area.[12]

Another expert in labor organization I interviewed, Latkovits, recalled his experiences working at a manorial estate before 1945. At one point a group of Germans visited the estate to conduct some experiments. "[We] nearly laughed out loud at these German work rationalizers. They began with different size shovels: a smaller one for a little worker, a bigger one for the bigger. The productivity of the smaller worker was smaller, bigger for the bigger worker. We laughed at this. We weren't forced to do this [scientific rationalization] . . . so we ridiculed them a bit. Especially when they showed up with a stopwatch, even the management had a good laugh about it . . ."[13] But it was now a different world.

Planning Agricultural Policy

In the first few years following the end of the war, new socialist states pursued different strategies toward agriculture. The character of collective farms and the rate of their introduction varied in the region, with Yugoslavia and Bulgaria leading the way in the early years (1944–1946) and Albania close behind (Wädekin and Jacobs 1982, 35). After the elections in 1947 securing the Communist Party's control of government, indeed even as late as early June 1948, the party had decided not to mount a vigorous collectivization campaign. The party's busy political agenda in late fall 1947—further nationalization in industry, formally nationalizing the banking sector, and progressively removing all other parties from the political scene—meant agriculture was not top on the list of priorities (Balogh et al. 1978, 156–60; Rainer 1996, 388; Pető and Szakács 1985, 85). But Soviet pressures to collectivize agriculture in its client states in Eastern Europe had increased in 1947 and 1948. The broad contours of collectivized production in the Soviet Union—joint ownership, labor brigades, shared costs and revenues—were familiar. The technicalities were less obvious but just as crucial to getting cooperative farms up and running. Finally, the tensions between the new socialist states and the Soviet Union, most notably the break with Yugoslavia, played a crucial role in finally convincing the Hungarian Communist Party to launch a swift collectivization drive.

The road to socialist agriculture lead through thick brambles; a jumble of abstract theoretical debates over historical development were intertwined with questions over the degree of latitude Eastern European socialist states could exercise in setting their own national strategies, all refracted through the lens of Soviet political history. It was simple to draw the conclusion from Soviet history that peasants could not be left to their own devices, their atavistic attitudes towards property and production constituting insurmountable barriers to socialist agriculture that the state would be forced to smash. What was less clear to the leaders of new socialist states was how this specific historical experience from the east would translate into policies at home. Disagreements over basic issues concerning property relations, class struggle, and mobilizing the peasantry were complicated in Hungary by clashing personal and political agendas within the party, as well as the ever-looming gaze of the Soviets. How could they commit themselves to the class struggle without endangering the country's food supplies in the bargain? A reasonable compromise appeared to lie in more discriminating policies of class differentiation, building alliances with poorer peasants while punishing wealthier farmers.

In Hungary, the national policy plan being crafted by the party in late

1947, early 1948 touched only superficially on agriculture. Even though Imre Nagy was recognized as the specialist in agrarian issues among the ruling elite of the Communist Party, he had fallen out of favor and was consigned to a minor role in policy making. Nagy did not draft any portion of the guiding principles touching on agriculture, only being asked to comment on the draft once it had been written. Concerned about the direction of party policy, and the role of agriculture in particular, he submitted a critique to the party leadership in December 1947, followed by a more extensive essay written for the journal *Social Review (Társalmi Szemle)*. It was never published. Nagy's intervention broke with party protocol, challenging the hierarchy of decision making within the Communist Party and offending colleagues whose vision his criticism targeted, notably Ernő Gerő, who directed industrial development (Rainer 1996, 377–400).[14] Challenging party policy pitted Nagy against the rest of the Central Committee, a battle of wills that lasted more than a year and finally resulted in Nagy's being censured. The debate took the classic form of Marxist-Leninist rhetorical style—dueling by selective quotation from Lenin, Stalin, and Engels—but touched on the most basic questions about what kind of economy Hungary had and would have in the near future.

Two different versions of socialist development dominated policy discussions. Gerő, the industrial sector's advocate, promoted a vision of rapid nationalization of the economy and the introduction of central planning, accompanied by a punitive drive against wealthier peasants and expansion of the state sector of agricultural production (Rainer 1996, 379–80). Nagy was insistent that the party/state avoid exaggerating the danger of wealthier peasants, or kulaks as they were called in Communist parlance. He feared that harsh measures against kulaks, as well as those who might fall under the designation unfairly, would endanger the Communist Party's potential alliance with rural communities. He also worried about the status of agricultural production under conditions of political duress. Clearly, his long years in Moscow witnessing the turmoil in the 1930s weighed heavily on his mind when he heard plans being discussed to speed up collectivization.

Lenin pronounced in his famous tract "The State and Revolution" that the small peasantry "spontaneously" bred capitalism. According to this logic, the dangers of class enemies could be extended (at least in principle) to the *possibilities* of small peasants becoming kulaks, even though they were not yet so. It also meant that one could not simply allow the peasantry to evolve slowly into a comfortable class of private producers. In contrast to his more conservative comrades in the party, Nagy was not convinced that the character of farming in the countryside had become firmly capitalist. Opposing a harsh campaign against wealthier villagers, Nagy proposed strengthening the middle peasantry as an alternative strategy, a position for which he was roundly criticized.[15]

The subtleties of differentiating middle peasants from kulaks, of distinguishing between improvements in life style and burgeoning capitalist villages, were summarily dismissed in the criticism mounted against Nagy. Stalin's dictum would prevail: "there is only one kind of capitalist farm, since it rests on the private ownership of the means of production."[16] In the diatribe issued against him, Nagy was said to have underestimated the strength of capitalism in the countryside, being "deaf and blind" to these perils and willing to "makes concessions" to capitalism.[17] Furthermore, the critique continued, "as the spokesperson for the 'interests' of the privately farming, petit bourgeois peasantry, Comrade Imre Nagy actually has acted as the protector of capitalism, kulaks, and the private ownership of the means of production against the interests of the working class and socialism and so has deviated from the path of Marxism-Leninism."[18] Nagy's views were described as "diametrically opposed to the party's village policy." In his own defense, Nagy pointed to similar debates being conducted in the fraternal parties of neighboring countries of Romania and Bulgaria.[19] Nagy eventually capitulated, exercising self-criticism in fall of 1949. He was removed from the Central Committee regardless. Since the debate around village politics was caught up in struggles within the Central Committee for position and power, it lasted far longer than the time it took to finalize the decision about collectivizing agriculture.

The Soviet leaders were also unsatisfied with the inadequate discussion of agricultural policy in the Hungarian Communist Party's guiding statement of principles, as a report prepared for Suslov, the secretary of the Soviet Ministry of Foreign Affairs, in June made very clear. The Hungarians were faulted for paying more attention to technical questions than to the social issues at hand, that is, not addressing the party's relationship to different segments of the peasantry (Izsák and Kun 1994, 268). In contrast to the Hungarian plan that had described the land reform as an achievement eliminating landless peasants, the Soviets insisted that the actual consequence of reform was to strengthen kulak farms. As a consequence, a much more direct intervention by the party/state was required to halt the inevitable move toward a fully entrenched capitalist farm sector. To deal with abberations in Eastern Europe, the Soviets called a meeting of the Cominform (Communist Information Bureau) in June 1948 in Bucharest; it was attended by representatives of all the countries under Soviet control. At the meeting, Rákosi, First Party Secretary of the Hungarian Communist Party, advocated for a position tempering the strong take on collectivization the Soviets were formulating, drawing on the support of Polish comrades who shared his concern about private farming. "[The Poles and Hungarians] would have preferred a more 'restrained' text, in which the notion of 'collectivization' would have been replaced by 'cooperation' and 'joint production,' and in which there had been discussion of spontaneity [voluntary participation] and a gradual approach" (Rainer 1996, 394). As Rainer notes, Rákosi's hesitation indicated that he was fully aware that a strong-armed push to collectivization would meet stiff resistance at home (ibid.). The Soviets would have none of it, and the pressure to conform built.

More was at stake than agricultural production. At the heart of these battles was the growing demand by the Soviets that Eastern European regimes stop experimenting with alternative paths to socialism (Rainer 1996, 393). The Poles and Hungarians were chastised for straying from the path, charged with mouthing the proper slogans but backing off from the Marxist-Leninist line when it came to policy at home. The Soviets insisted that "the policy platform properly emphasizes that it follows the ideology of Marxism-Leninism and employs and develops the teachings of Marx, Engels, Lenin, and Stalin. But it does so stipulating that this would be 'in accord with Hungarian conditions'. . . . This stipulation, under the pretence of adapting to Hungarian conditions, legalizes the opportunistic revision of Marxism-Leninism" (Izsák and Kun 1994, 267). The Hungarians bowed to pressure from Moscow; Tito did not. "The Cominform statement called the Yugoslav Party 'a narodnik kulák party,' because it refused to intensify the rural [class] war during the transition to socialism" (Rainer 1996, 394).[20]

In response to the Soviets' criticism, in June 1948 the Hungarian Communist Party platform had to be revised, repudiating earlier policy statements in which they had declared "'the Party protects peasant private property'" (ibid., 392). An ominous editorial appeared in *Szabad Nép* on July 4 warning of Bukharinist tendencies[21] promoted by enemies within the party, clearly a thinly veiled reference to Imre Nagy (ibid., 395). In August, Rákosi announced in a speech at the annual wheat festival in Kecskemét that peasants' fortunes would only improve if they embraced cooperative efforts, forsaking the selfishness of private production. The push to collectivize had begun. To forestall any further confusion, in September Rákosi specifically asked Nagy that "'under no circumstances [was he] to commit the Party to any kind of peasant private property'" (ibid., 397) in his public statements. With the decision to collectivize private producers foisted upon them by the Soviets, the Hungarian Communist Party began its eventually unsuccessful campaign to round up the peasantry into large units of collective production.

Fiscal Coercion and Pressures to Collectivize

The party/state distinguished two major groups within the agrarian community: working peasants and wealthy peasants or kulaks. The basis of the division was an assumption that kulaks owned more property and lived off the labor of others, whereas "working peasants" farmed their own land. Due to legal constraints on employing people on private contracts, it was virtually impossible to farm without working oneself, so this distinction did not map well onto actual farming practices during the 1950s. Also, who was designated a kulak was subject to wide variation, depending on the discretion and often mean-spiritedness of local officials. At the start of the collectivization campaign, the party/state was confident in being able to convince working peasants of the superiority of collective production; punitive measures against kulaks were intended to force them out of agriculture altogether. Neither of these expectations was realistic.

The social terrain of agricultural production in the 1950s was dominated by private farming, contrary to the usual depiction of the Stalinist era as a time ravaged by collectivization. In 1948, 80 percent of the land (13 million kh of 16 million kh or 7,410,000 and 9,120,000 ha, respectively) was being farmed privately (Erdmann 1987, 379). By contrast, the total acreage in cooperative farms in 1949 was 65,670 kh (37,342 ha), of which 49,519 kh (28,021 ha) was arable (ibid., 386). The size of cooperative farms increased in the course of the next seven years, but they were still far smaller than private farming.[22] Private farms were consistently more productive, even though cooperative farms were cushioned from failure by government support. Yields on privately owned land were higher and the costs lower. The incomes of private farm families exceeded that of cooperative farm members, who also earned less than workers at state farms (Donáth 1977, 152, 155). The often-touted superiority of collective production was not demonstrated in practice. "These results [at cooperative farms] were especially not convincing for smallholders since they saw that even though the large farms got fertilizer, machines, and mechanized labor, their results did not reach, or barely passed the results of, small farms despite the state's support" (ibid., 152).

The state imposed substantial burdens on all farms throughout the 1950s, in various forms of land tax, fees, food requisitions, and compulsory sales of produce to the state at artificially low prices.[23] Private farmers bore the lion's share, but not simply because they constituted the majority of producers. The dismal performance of the collective sector had to be made up for, putting greater pressures on an already steeply progressive rate structure by targeting the surplus of larger private farms. Just as important to party officials, however, was the desire to impoverish wealthy farmers and drive the peasantry en masse into collective production. While thousands of wealthier peasants abandoned agriculture entirely, poorer farmers by and large clung stubbornly to their property, farming under severe constraints.

All property owners, cooperative farms included, were assessed a land tax, as well as required to submit a wide range of products and perform services for the state, the level of obligations being determined on a progressive scale according to the size of arable acreage in the household.[24] Table 6.1 summarizes rate increases by the size of the farm. Forcing peasants to surrender produce above and beyond the crops delivered to fulfill tax obligations had been necessary in the years following the war but became increasingly important as the party/state ramped up for industrialization after 1948.[25] Cooperative farms were also obligated to submit produce to the state, but the rates they were subject to were equivalent to the levels set for the poorest farmers.[26] Land tax could be paid in wheat, as was the case for additional crop requisitions demanded of every household. Peasants were also required to turn over milk, eggs, poultry, and lard.[27] Wine, grapes, potatoes, onions, and hay were

TABLE 6.1. Requisition multipliers: Changes between 1948/49 and 1950/51

Property category based on arable land (kh)	Property category based on arable land (ha)	Change in multiplier
up to 5 acres	up to 2.85	1.0–1.3
up to 6 acres	up to 3.42	1.1–1.4
up to 7 acres	up to 3.99	1.2–1.5
up to 8 acres	up to 4.56	1.3–1.6
up to 9 acres	up to 5.13	1.4–1.8
up to 10 acres	up to 5.70	1.6–2.1
up to 11 acres	up to 6.27	1.8–2.3
up to 12 acres	up to 6.84	2.0–2.5
up to 14 acres	up to 7.98	2.2–2.8
up to 16 acres	up to 9.12	2.4–3.0
up to 18 acres	up to 10.26	2.6–3.2
up to 20 acres	up to 11.40	2.8–3.5
up to 25 acres	up to 14.25	3.0–4.1
up to 30 acres	up to 17.10	3.2–4.6
up to 35 acres	up to 19.95	3.4–5.1
up to 40 acres	up to 22.80	3.6–5.5
up to 50 acres	up to 28.50	3.8–6.0
up to 80 acres	up to 45.60	4.0–6.4
up to 100 acres	up to 57.00	4.2–6.9
up to 120 acres	up to 68.40	4.2–7.1

Notes: Increase in requisitions required of families based on size of their farm, showing the crackdown on wealthier peasants.

Source: Erdmann 1987

added to the list of targeted produce later. Farmers were required to enter contracts with state agencies to sell a set amount of their produce at a price set by the state, far below market price. These were the so-called C coupon contracts, which were officially defined as voluntary but were anything but (Erdmann 1987, 386). Fees for services, to be paid in kind, included the cost of any work done by machine stations on private lands, as well as fees for threshing and milling wheat. Any surplus of a family's harvest that remained could be sold in the open market, where peasants could fetch a high price, but it did not make up for the losses imposed by selling to the state at low prices. In 1948–49, when the requisition and tax rates were first established, "peasants surrendered 24 percent of cereals and 13 percent of all the other important crops without any compensation" (ibid., 381). Most people were able to fulfill these demands in 1949: small and medium-sized farms fulfilled their obligations by 82 percent, kulaks by 68 percent (ibid., 381). With time, rates of taxation and requisitions increased dramatically for kulak families, in line with the party's policy of crushing a potential threat to socialist production. "The peasantry's tax burden increased to the greatest extent in 1951 . . . it rose nearly 70%" (Pető and Szakács 1985, 183). The obligations at small and middle farms doubled at the same time (Estók et al. 2003, 292).[28] "In the wake of the ruthless collection of requisition obligations—the so-called 'attic sweepings'—it turned out that on the one hand, out of 1.2 million peasant families 800,000 didn't have enough grain for bread and seeds, while on the other tens of thousands of cooperative farm members were left without the per capita ration" (Estók et al. 2003, 293).

In addition to imposing heavy taxation rates, the state issued regulations requiring that villagers follow directives on farming practices, such as clearing dry corn stalks from the fields by a specific date or turning the soil promptly in preparation for the spring sowing. Neglecting these statutes could result in severe penalties. In the first few years, wealthier peasants were also charged with a "development fee" to contribute to the expansion of socialist agriculture. State regulations extended to wage contracts. "We must coordinate the wage question with those measures of our village policy we are taking to sharpen the class war."[29] Two regulations were issued in 1949 stipulating working conditions for employees and sharecropping rates. In the past, regulations on labor contracts had specified the lowest and uppermost wage levels, but in the revised version specific wages were mandated, according to ten categories that differentiated between a wide range of agricultural tasks. Each category stipulated the hourly wage, and two different values for monthly labor, depending on whether it was high season or low. Government regulation 2.950/1949 shared with previous labor regulations extensive paragraphs on all aspects of the labor contract, for example, the length of the work day, amount of free time to be given to workers, overtime pay, and payment in kind. This included the bread ration, which was the responsibility of the employer to provide. In March 1949, a regulation was even passed outlawing the right of kulaks to fire employees who had worked for them since the first of January, since the widespread layoffs would cause problems for the unemployed and for the level of agricultural production (Erdmann 1987, 386). In the face of stiff competition from the private sector, the regulations on wage rates and in-kind compensation at state and cooperative farms were calibrated to give them an advantage, as the following passage in a March 1949 report describing the theoretical principles behind the ruling for the private sector makes clear.

> In particular we took into consideration the Organizing Committee's ruling that we must establish increased demands against kulaks in the area of in-kind benefits.... [W]e paid attention to the wage level of state farms in terms of the wages paid by kulaks. We made sure that kulak wages exceeded somewhat the state farms' wage so that [state farms] would be able to significantly increase the wage naturally with in-kind payments.[30]

A crucial feature of the 1949 regulation, overriding the conditions set out in the 1948 regulation, was that private employers were not allowed to pay workers with piece rates or akkord wages. The state wanted to secure a monopoly for itself on the use of those powerful instruments.

In the first years of cooperative farming, the party/state recognized three different kinds of farms, distinguished by the way they worked together and how they shared property. Type I was the least collective form of farming, often referred to as an association (*társulat*) rather than a cooperative. This organization permitted a group of landowning peasants to commit themselves to working together on the lands owned by each family: ploughing, tending to crops, and harvesting. In the second kind, type II, the level of cooperation was greater, as was the range of collective goods. Peasants continued to own their lands privately, but it was amalgamated into one large area. All the tools and farm animals were owned jointly, and the work throughout the season was done together as a group. The third kind of farm, type III cooperative farm (collectively producing cooperative group, or *közösen termelő szövetkezeti csoport*) was organized along the lines that are most familiar in discussions of cooperative farming: all the land, tools, and farm animals were owned collectively, and the work was shared by all members and their families (Erdei 1956, 18).[31] The fortunes of cooperative farms—the number of members and size of acreage—waxed and waned throughout the 1950s, in relation to the level of punitive requisitioning: taxes, fees, and other related charges. By September of 1949, there were 389 cooperative farms nationwide, with 5,581 members. This represented 0.25 percent of the agrarian work force, which numbered 2,196,185 (Donáth 1977, 97). At the same time, there were 171 state farms, employing 2,801 workers.[32] The distribution of the farms regionally shows large variations, with far greater numbers clustering in the Alföld in the eastern part of the country than in Transdanubia in the west or the northern counties.

Private farmers were also negatively impacted by a series of measures designed by the party/state to modernize agriculture: consolidating holdings and expanding planning to all enterprises. When a cooperative farm was established, members would surrender their land to the group. It was very common for families to own land parcels in a number of different fields surrounding the village. Since these parcels were acquired over generations, as one's patrimony by marriage and inheritance, peasants were very reluctant to part with individual plots that had been in the family for generations. The extreme parcellization of private holdings was inimical to the economies of scale envisioned for modern agriculture, particularly once mechanization was seriously underway. Small cooperative farms could also benefit from consolidation, but the scale of consolidation was out of proportion to the needs of 1950s farms. If as a private farmer one was unfortunate enough to find one's property surrounded by lands dedicated to a new cooperative farm, then these lands were expropriated, to be replaced with lands elsewhere in the vicinity. "In place of the land that had been [initially] surrendered to the farm, the lands given back in exchange were allocated, from the producer's point of view, in less advantageous and farther away fields. Often the allotted property was also of less value" (Donáth 1977, 139). Between July of 1949 and March of 1953, nearly 40,000 farms owned by wealthier peasants were appropriated, equaling approximately half a million hectares (39,527 and 553,527, respectively). Poorer families were not spared either; farms of less than fifteen hectares were also appropriated if they were in the way. In the same time period, 68,033 smaller farms, equaling 201,617 hectares, were lost to their owners (ibid., 140).

The ultimate goal of the party/state was to move everyone working in agriculture into type III farms, and eventually to eliminate these "remnants of capitalism" with the full transition to state farms.[33] Types I and II were there to assist those who still hesitated to relinquish private farming. But, in Rákosi's eyes, warming to cooperative production would be swift, as he confidently predicted in 1951. "The superiority of cooperative production even at this simple level will be demonstrated so that.... in the majority of cases after the first harvest members of the I. type of farm will move in the direction of the higher types of cooperative" (ibid., 18). Rákosi's optimism was misplaced; working in less developed cooperative farms was not transformative. Only the severity of forced deliveries and taxation were effective in coercing peasants into type III farms. By June of 1953 there were 5,200 cooperative farms, of which 3,700 were type III, 1,500 of type I and II. By June of 1953 approximately 200 specialized producer groups (*termelői szakcsoport*) had been established, whose activities were primarily devoted to beekeeping and growing rice, vegetables, and fruits (ibid., 20). Erdei noted an increase over the course of 1953–54 in the variety of different producer associations alongside types I and II farms.[34] With the return of the conservative wing of the Communist Party in 1955, these trends were reversed. "Of the 507 cooperatives established in 1955, a total of 127 were of the simpler type" (ibid.). These designations would become obsolete in the second, and far more brutal phase of collectivization in 1959–1961, at which time virtually all farmers would join type III cooperatives.

In the 1950s new cooperative farm members (type III) were overwhelmingly from the poorest families. In 1949, approximately 70 percent of the cooperative farm members had been totally landless before the land reform, coming from the ranks of manorial servants and day laborers.[35] Virtually all the rest had owned less than 4 kh (2.28 ha; Donáth 1977, 142). The dominance of the rural poor—or the rural proletariat, a more precise description for our purposes—has been well documented; the trend characterizes collectivization drives in other Eastern European socialist states as well (Iordachi and Bauerkämper 2014; Kligman and Verdery 2011; Wädekin and Jacobs 1982) The reasons for this are not surprising. Agrarian proletarians—manorial servants, day laborers, migrant workers—had never been able to run a viable family farm, having lived their lives working on contract for others. Once the land reform redistributed land, their hopes ran high, but they were defeated by the lack of basic tools, owning no barns or stables and very few livestock. Their dependence on the mercy of wealthier landowners to gain access to tools and draught animals dragged on. In the first years after the war, labor inflation was common, especially in areas where livestock and draught power had been decimated the most during the war. This meant that poorer villagers were required to work for several days, and even up to a week, to reciprocate having borrowed tools for a day from their wealthier neighbors, giving them less time to work on their own crops and seriously reducing their yields. So joining cooperative farms meant they would have access to tools, more land, and perhaps livestock. The rural proletariat also shared another important characteristic. They had extensive experience working with nonfamily members in relatively large groups, in strong contrast to peasant families whose work was conducted almost entirely with small groups of kith and kin. It is important to note, therefore, that a small number of villagers benefitted from collectivization, even when the fortunes of cooperative farms waned. The chance to earn a living by running a farm gave them a sense of pride they had long been denied. As we shall see in chapter 8, differing class attitudes towards cooperative farming in village communities would erupt into serious conflict over property rights and labor when cooperative farms were allowed to disband in the summer of 1953.

WORK UNITS ARE BORN

The new work unit system for type III farms was announced with fanfare in the newspaper for the agrarian community, *Szabad Föld*, on the sixth of March, 1949. The table was unveiled at a national meeting of all cooperative farm leaders called by the Ministry of Agriculture for the twenty-second and twenty-third of February, in order to come to agreement, in concert with the ministry's specialists, on how to "measure work" for farm members. As the article proudly declared, full cooperative democracy blossomed at the meeting. "[The leaders of cooperative groups] know and approve [the fact] that the government will not pass even one important provision of the law or significant measure *without having first discussed it with those affected*" (ibid.; italics in the original). In the course of the meeting, five subcommittees, including cooperative leaders and Ministry of Agriculture experts, were charged with reviewing every single number, after which it was decided that the numbers generally should be increased (ibid.). The first published work unit book saw the light of day on March 7, 1949.

In late January the head of the Organization Institute within the Ministry of Agriculture, Berend, was informed that he was to prepare a work unit table immediately.[36] He proceeded with his crew to put the initial plan together, asking for representatives from relevant groups to participate: the Cooperative Farm Department of the Ministry of Agriculture, National Association of Working Peasants and Agrarian Workers (DÉFOSZ), state farms, and the National Committee to Set Wages. As Berend explained in his report, they consulted the norm tables from the Soviet Union and Bulgaria. In his presentation to the Agricultural and Cooperative Policy Department on the 14th of February, he offered two possible variations: "One of the variations was prepared more boldly, while the other was rounded off."[37] The two plans only addressed work in the fields; animal husbandry had not yet been tackled. The two variations each had their disadvantages. The danger arising from plan

A, in which the differences between tasks were more starkly expressed, was that it would pose problems for managing labor, since workers capable of performing both the easier and harder tasks would opt for those that would earn them more work units. The danger of plan B, in which the difficulty between similar tasks was downplayed, was that it would encourage attempts to distribute work units equally among all the workers, a practice to be avoided at all costs.[38] It was decided that the bolder version, plan A, which had consistently higher values than the second option, would be preferred. This was the version sent out to cooperative farm leaders to review before the national meeting. Although Berend and his colleagues reviewed the Soviet system in the initial stages of developing work units, the actual values for tasks were established using output figures from manorial estates as a guide.[39] In short, this meant that the norms ended up being higher than those used in the Soviet Union.

In contrast to Berend's dry bureaucratic report, Ferenc Kalocsay relayed a far more colorful account of how the value of work units was established.

> There was this László Balla, Uncle Laci, an elderly farmer, he worked in the Ministry of Agriculture, at that time in the Labor Affairs Department. . . . This crowd was given the task of making the first work unit book. Uncle Laci was an old experienced farmer, and, if I'm right, had worked in Reichenbach's entourage [of work scientists] before he would have ended up at the Ministry of Agriculture. He just simply sat down together with two other colleagues and along with a glass of good brandy from Szatmár, they went through all agricultural tasks, crops, etc. "How much do they do at your place, this much, then that's about . . ."[40]

Kalocsay underscored the difference between the approaches taken to cooperatives and those established for state farms. "When these guys went out to state farms they performed lots of measurements. It was organized centrally. . . . For cooperative farms, what happened was that there was this kind of 'scraggly beard' (*kaporszakállas*) department. That's what they called them. They were old agricultural experts, who on the basis of their own experiences estimated norms, they worked along these lines."[41]

The initial values specialists had devised were then subjected to analysis in the field, a process that continued for several years. I spoke with one of the field specialists, Mihály Kuzmiak, who proudly declared to me that he had designed the work unit system. Kuzmiak shared with the "grey beards" a background in manorial farming. He had a degree from the Academy of Agriculture in Debrecen and had worked for two years at a manorial estate in Veszprém County. He also brought invaluable experiences from what he referred to jokingly as his educational sojourn in the Soviet Union, that is, when he worked at a *sovhoz* while a prisoner of war.

> ... to a certain degree I got involved in all of this because of my experiences abroad ... [that's why] I got the job of developing the Hungarian version of the work unit system from the Ministry of Agriculture. I ended up at the [National] Institute for the Study of [Agricultural Organization and] Production Costs and, well, they were looking for someone who could solve this issue. The choice fell to me [because] of my Russian experiences ... and historical background ... which in part were data that came from practical life, and in part from literature. ... [earlier] I had worked at a manorial estate ... I was interested in the management and organization of work ... what can people accomplish, how to do that ... my notes from that time were very useful, when I ended up [doing this project] ... my acquaintances called me the King of Work Units. ... The theory on which I built was ... unquestionably the calibration and distribution system accepted at kolhozes ... but I built it entirely on Hungarian conditions, domestic possibilities.[42]

In his account, he returned to the region where he had once been employed as a steward to consult former estate workers about their views of the relative values of tasks. He also traveled to other regions to augment the materials he had gathered in Szatmár. "Every number that ended up in the work unit book ... there wasn't a number in it that I hadn't discussed out there with cooperative farm members, not bosses, ... the simple cooperative farm members."[43]

At this point, the story of designing Hungarian wage units suggests a strong continuity with earlier practices in manorial estate management and work science. Yet a more careful assessment of the new work unit system reveals significant departures from earlier wage scales, in several important ways: degree of freedom, shifting metrics, increased precision, and degree of formalization. The dictates issued by the party/state substantially reduced the degrees of freedom exercised in allocating work units. In the work books used on manorial estates, the tables describing the requirements for various tasks were presented as average values for the average worker, with the caveat that a number of factors—weather, quality of the soil, the organization of work, oversight, and premium awards—would have a crucial influence on results. Furthermore, it was important that the proper tools were used to achieve these results.[44] The new work unit system was designed to standardize wage scales within cooperatives across the country. In its original formulation, cooperative farms could vary the values within the scale by 30 percent, that is, allowing 15 percent below and 15 percent above, to allow for local conditions such as character of the soil, type of terrain, and so forth. A column was included in the work unit book to record what had been locally decided. Farm members were cautioned, however, that variations could only be introduced to the level of output—they could alter the amount of acreage worked or the number of quintals gathered in the course of the day. Fiddling with the numerical value assigned to a task was strictly forbidden, since the values of work units reflected the relative value of tasks within the overall labor system (Földmívelésügyi Minisztérium 1949, 7). In short, the work unit system constituted a matrix of value, designed by officials to set the value of labor scientifically rather than through customary means of market mechanisms.

The second difference, conceptually of far greater importance, was to change the metrics by which labor would be evaluated. In the work books used on manorial estates, activities were listed in terms of the output to be expected during a ten-hour day, the standard also used by work scientists before the war. When a task was usually completed by more than one person, then the output for 1 kh (0.57 ha) was set indicating the number of people needed to finish in the course of a ten-hour day. It bears emphasis that a work day more loosely defined was also the basis for labor exchanges in villages among kin and neighbors and also for day labor. Units of longer duration—such as migrant labor and manorial contracts—were also defined temporally.[45] Even sharecropping—be it corn, tobacco, or harvesting wheat—was understood in terms of a temporal unit, such as the growing season or the length of the wheat harvest. In a standard ten-hour day at a manorial estate, two men and two women were required to harvest one acre of wheat, whereas loading a shock of wheat onto a cart took three minutes, reflecting the primacy of temporal metrics (Reichenbach 1944, 458–59).

In the new work unit system, ten-hour days were maintained as the period during which a specific output was required, but finer gradations in task performance and a new structure of value were added. The same caveats about average working capacity and a standard ten-hour day held. The numerical values of work units, however, had now been calibrated by considering the specific requirements to pursue a task: significance, skill, and physical difficulty. Easy and minor tasks like sweeping the yard were accorded far less value than harvesting wheat, which was both difficult and extremely important. Even workers assigned to the same task, such as sowing wheat, were differentiated according to the relative skill and difficulty their portion of the job required. For example, in the case of a sowing machine (with fourteen seed drills) drawn by a team of two oxen: the man steering the machine earned 1.4 units, the man leading the oxen 0.8, and the man sitting at the back watching the seed drilling earned 1 work unit, or 0.22, 0.4, and 0.28 work units for each acre, respectively (Földmívelésügyi Minisztérium Termelőszövetkezeti

Főosztálya 1949, 63). Each task was assigned a work unit value and positioned within a general hierarchy of tasks represented by the final matrix. The scale defining value in the Soviet kolhoz initially had 7 points (0.5–2.0), with increments of 0.25 units. Hungarians chose a finer gradation, preferring increments of 0.2 between the lowest value of 0.6 to 2.0 as the highest (Földművelésügyi Minisztérium 1950, 8).

To convey the sharp difference between the two systems, I offer a few examples comparing the manorial handbook and the new work unit system, but note in passing that the relative values of the tasks themselves were fairly constant. Whereas the manorial work book suggested that in the course of a ten-hour day, 40–50 shocks of grain could be transported by one man with a team of horses approximately 1.5 kilometers, in the work unit book the same task was assigned the numerical value of 1.2 work units for transporting 50 shocks of grain 1.5 kilometers, or alternatively (as shown in the adjacent column) transporting ten shocks of wheat would earn the worker 0.24 work units. The task of hoeing corn for the first time in the season required four to six workers to complete an acre, according to the manorial handbook. In the work unit book, this same task—now distinguished in value for light, medium, and heavy soils—required one person over a ten-hour day to hoe 400, 300, or 250 square fathoms, respectively, earning the worker one work unit, which translated into 0.25, 0.33, or 0.4 work units for each square fathom.

In principle, since the value of tasks had not altered significantly, one could imagine the new system being implemented simply by farm members doing things as they had been done before: spending a day hauling grains or hoeing corn. The state demanded more: greater precision in measurement and attention to keeping records. The complexity of calculations and the level of attention required to be precise are illustrated in a brief passage drawn from a brochure on the work unit system describing the way to figure the value of harrowing a field before spring planting.

> In summary, the work unit to be recorded for 7.5 kh (4.3 ha) and above is 1.24. We calculated with the help of the sixth column that for one acre one gets 1.16 work units, so if one exceeded 1.5 *hold*, then 0.24 work units should be recorded. Don't use the work units in the sixth column (the so-called crutch numbers) as a multiplier, because they are only figured accurately up to only two decimal points, and so the multiplication doesn't provide accurate results. (Földművelésügyi Minisztérium 1950, 10)

Kalocsay ridiculed the entire endeavor.

Latkovics quipped that Russians were lazier and less productive, forcing Hungarians to raise the norms to meet their own standards of diligence:

"with their norms we would have been laughed at."[46] These prejudices were commonplace, bolstering nationalist sentiments against an occupying power, but Hungarian attitudes towards the Russians' work ethic were of far less consequence than different ecologies.[47] Constructing an alternative system was necessary for the simple reason that the crop profiles were very different in the two countries, the Soviet Union being dominated primarily by grains, whereas row crops and other more labor-intensive garden crops played a substantial role in Hungarian agriculture.

> Hungarian practice took off with this mania in pursuit of exacting precision. In Hungary there were people who were used to this precision, German precision, who said, "What's that, the Russians say that hoeing corn ranges from 400 to 1,200 square fathoms? How do these 'Ivans' manage this, what a backward group, when the outer limits are so amazingly great. . . . Then we should figure out a large limit that has been influenced by the soil, incline, weediness, rain, etc. Let's build in these criteria and then it will be substantially more accurate." And so it was, but it was impossible to use. Who in the hell traipses out every morning [to check] whether the ground is moist or not, etc. . . . that's why it failed.[48]

Precision required oversight. But in fact who decided how much work was completed? Who guaranteed that the work performed met the standards of quality the farm wanted to achieve? These tasks were delegated to brigade leaders, who finally decided what the daily work's value was by assigning a work unit. Whether cooperative farms would appoint brigade leaders, or even organize brigades, remained an open question.

Cooperative farm members accumulated work units all throughout the year, but only at the end of the year were the numeric values translated into goods and cash. The bylaws stipulated how income was to be handled. Outstanding debts, tax obligations, and rent owed to members who contributed land to the farm were dispatched first, then monies were set aside for operating funds and future investments, and the remaining surplus—crops and cash—was divided according to the numbers of work units performed by each member. Any advances of produce and money received during the year would be subtracted before each member was given his share of the year's proceeds. At the annual meeting, the total numbers of work units performed during the year was announced. Using this number to calculate how much each work unit was worth in kind and cash was the first order of business. The number of work units recorded for each person would be figured out when it came to figuring out how much the person was still owed by the farm by subtracting any advances received earlier in the year.

The new work unit system introduced an unprecedented degree of formalism to figuring wages. Newspapers and magazines explained endlessly that the work unit system would introduce a fair measure of people's contribution, just as scientific norms would in industry. But this could only be guaranteed if farm members kept good records of everyone's work on a daily basis and tracked all the expenses and outlays in the course of the year. Great effort was expended explaining how the work unit system should be used: in brochures, public lectures, and evening classes. At the 1949 meeting of cooperative farm members, questions were raised about how this whole system would work. "Mihály Lippai worried about the decimal point. According to him peasants don't understand the decimal point and so there will be confusion." Answering this concern was István Fekete, the head dairy hand at Városföld farm in Kecskemét: "With regard to the decimal point, although many cooperative farm members have completed only two years of school, *they will now learn* what a zero and a decimal point is for" (*Szabad Föld*, 6 Mar. 1949; italics in the original). Truer words were never spoken.

Conclusion

Under pressure from the Soviets to collectivize agriculture, Hungarian party officials scrambled to set up government policies to guide this enormous undertaking. New legal codes were written stipulating the degrees of cooperation various forms of collective farms were required to follow and the parameters of wage systems for private and public labor contracts in agriculture. Significant increases in state requisitions—grains and other crops—were imposed on villagers, with the hope that fiscal pressures would force them to leave the private sector. Bureaucrats in the Ministry of Agriculture looked to see what kinds of wages were being used in Soviet collective farms and what had been adopted by Bulgarians. The actual construction of the work unit, however, was done in-house by old estate stewards and work scientists. Drawing substantially on long-standing temporal metrics used at manorial estates and personal experience in large-scale agriculture, the bourgeois specialists crafted an entirely new matrix of labor value, weighing all tasks against each other in terms of three crucial variables. Within months the ministry published a book 120 pages long listing the values for cultivating a number of different crops. Unlike their comrades in the Soviet Union and Bulgaria, however, the ministry also included a slightly altered work unit system to reward people working in animal husbandry. At least six amended versions of the book were issued that year. And more were to come. The party/state's onslaught of requisitions and fines did not achieve its goal of mass collectivization. Only a small percentage of villagers—primarily the rural proletariat—chose to follow the new path, but their fortunes only slightly improved. Their compatriots still working in private farming were more successful farmers, but their coffers and barns were raided by state agents. As the chapter to follow explains, this resistance to collectivization would bring even greater violence to the lives of villagers, in particularly brutal forms of class warfare.

Early cooperative farms in Eastern European socialist countries shared many of the features of the Soviet *kolhoz*: a moderated system of norms for assessing farm members' labor contributions and a division of labor based on brigades. The dark shadow Soviet collectivization has cast over the history of collectivization in Eastern Europe, combined with the claims that socialist economic policies constituted a radical change from previous governments ("technopolitical rupture-talk"), however, has meant that the manner in which cooperative policies were designed and implemented was not worth further investigating. Surely, the entire enterprise was Soviet-inspired (Berend 1996, 39; Kovrig 1979, 258–59; Wädekin and Jacobs 1982, 258). To counter this assumption I have offered evidence demonstrating the strong influence of interwar German agrarian economics and work science on Hungarian wage policies and practices in the early phase of collectivization, with particular focus on the specialists who performed the work and the techniques of commensuration and valuation they deployed. The empirical materials also strengthen the argument I am making about techniques of commodification that are adopted in the socialist period, as the wage forms were designed to tie the worker's output directly to forms of remuneration. This was true in the absence of both a labor market and a monetary wage. Finally, I hope to have illustrated how formalizing practices like mathematical equations and arithmetic operations can only be fully understood if the indexical meanings of the formulae are explored by examining the local, historically contingent character of their use.

7
Administering Coercion

The ambitions of the new socialist state were grand: scientifically calibrated wages, plans for every economic enterprise big and small, aided by an eager cadre of officials willing to enforce the rules. The organizational and technical challenges to the state bureaucracy were dwarfed only by the doubt and distrust of the nation at large. Assessing the value of labor in a socialist economy entailed far more than measuring bits and pieces of time; it also extended to evaluating whose labor was worthwhile and whose was not. Moral imperatives claimed by the party carried little weight when the changes demanded seemed unnecessary or even silly, such as measuring the time it took to lift a sack of grain or keeping careful records of each and every action performed all day long. Directives from on high filtered down through a maze of offices staffed by people whose political commitment, expertise, and authority varied tremendously, with the result that actually implementing the vision of the new socialist economy was frustrated at every turn. Unwilling to be patient, the party unleashed vicious henchmen whose cruelties fostered havoc. Confusion reigned, fear spread, and anger grew.

The dominant motif in discussions of collectivization is the loss of property, the party/state's appropriation of private land and livestock for collective farming. I propose a different approach, that is, to examine the social imperatives entailed in working collectively. This trains our attention on the activities people were engaged in, whether they were members of cooperative farms or bureaucrats in party/state agencies, to make cooperative farms work. What sorts of changes were required to alter the way people worked together and the manner in which they valued labor? What kind of "investments in form" would be required to reach this goal? Time spent, energy expended, resources mobilized, regulations enforced. How would these efforts contribute to a more lasting infrastructure that would enable the party/state to manage labor and make it more productive?

Infrastructures are not built in a day, as they say. Neither are they constructed everywhere in the same way or at the same pace. How policies on collectivization and class warfare were implemented in Hungary varied a great deal from county to county and sometimes even village to village. The social history of a region and its geographic location influenced how it was treated by party officials; in the halls of the party/state, counties acquired reputations. The history of class relations in various regions would prompt different degrees of supervision by party leaders. For example, in Győr-Moson-Sopron County—a county reputed to be inhabited by very wealthy, very Catholic peasants, who were expected to be hostile to socialist policy—officials were held to a much higher standard, being punished for not setting up a party cell in every community by late 1949. In contrast, Zala County—a poor region characterized by mining and a barely viable subsistence agriculture, whose inhabitants were assumed to be friendlier to the socialist vision—bureaucrats would calmly report in 1953 that they had neglected to establish party cells in village communities and perhaps should consider doing so now. Counties on the western border were admonished to keep on the lookout for enemy activity, including the corrosive effects of Titoist infiltrations from across the southern border, whereas counties in the east were not called to fight this battle in the Cold War. The political ambitions and personal agendas of people in charge influenced how strictly policies were enforced, a stance that could be abandoned when alliances among central figures in Budapest or local officials shifted. These were not the most felicitous conditions for implementing the grand plan for collectivized production envisioned by the Communist Party. Rarely, of course, may we expect smooth sailing when attempting to change in people's lives and welfare so drastically. It is precisely for this reason that paying attention to what people did, rather than what they planned to do, will give us greater analytic purchase on the process underway.

The chapter begins by describing briefly the economic and bureaucratic difficulties faced in the first few years of the People's Republic of Hungary. It proceeds to describe how cooperative farmers learned (or not) to use work units, followed by a discussion of labor competitions that were an analogous means of tracking and rewarding labor. I then turn to analyzing the increasingly violent ways the party/state attempted to dictate how people would work in agriculture from 1948 to 1953. The chapter ends by examining the ways the party/state bureaucracy assisted or impeded its goals in the first few years of its existence.[1]

Problems to Confront

In the first years of Communist Party rule, economic conditions worsened. Investments in industry and infrastructure were substantially increased, but the monies were not used efficiently, spread across a range of projects all competing for funds. Whole new towns, like Sztálinváros, were constructed de novo; gargantuan industrial plants absorbed monies and people. National plans were not sufficiently detailed to guide implementation, stipulating only the amount of money to be spent (Pető and Szakács 1985, 189–90), or they were entirely unrealistic, such as the assumption that collectivization would take only a few years to complete. Meanwhile, the political climate darkened. "The entire [Communist Party Second Congress of 1951] was permeated with the spirit that—alongside the hysterical emphasis from the Stalinist perspective on the increasingly deepening class war—in the near future one could expect the imperialist forces to start a world war" (Balogh 1986, 20). Rumors of foreign invasions and state seizures of all private property hung in the air like ghostly spirits, not easily dispelled by party officials. The personality cult surrounding the Communist Party leadership, and their beloved patron and friend, the wise and venerated Stalin, flourished. "Our working peasantry owes our beloved father, Comrade Mátyás Rákosi, deep gratitude for directing inexhaustible care towards villagers" (*Győr-Sopronmegyei Hirlap*, 12 Dec. 1951). The new bywords in every government report, editorial, and party circular became discipline, vigilance, and consolidation (*fegyelem, éberség, megszilárdítás*).

State offices were saddled with enormous responsibilities and grew exponentially in response. Of course, some of the expansion in administration was driven by coalition-era politicking (see chapter 5). Those complications did not evaporate as soon as the Communist Party took power, as a number of non-Communist Party members stayed in place in the interests of maintaining the semblance of a multiparty coalition, the Hungarian Independence Popular Front (Magyar Függetlenségi Népfront). Confronting an unwieldy governmental apparatus, repeated calls were made to rationalize administration—trim staff, eliminate unnecessary procedures, streamline administrative functions, improve efficiency—but these attempts were at odds with the pressing demands of planning an economy and waging class warfare on all fronts. To give a sense of the proportion of increase, in 1946 5,168 people were employed by the Ministry of Agriculture and its subordinate offices. By 1955, this number had grown to 18,370 (Valuch 1996., 13). Furthermore, the rapid escalation of duties required of state agencies was taking place while ministries and other government bureaus were being reconfigured. Unfortunately,

FIGURE 7.1. Time is money! Don't waste it.

this created an unstable organizational environment. Romány, minister of agriculture in the 1970s, recalled:

> The modification of the governing (and with that the oversight) structure of agriculture was striking: prepared frequently in secret, nearly every six months. For example, at times state farms operated under the direction of an independent ministerial organzation, at other times they were assigned to the Ministry of Agriculture and other ministerles. Their county headquarters, trusts, regional management centers were literally closed down from one day to the next, reorganized, moved to a new city. (Romány 1996, 429)

As a result, lines of authority within and between agencies of the government were often unclear. Staffing was plagued by a lack of adequate personnel, as was true in other domains of public life (see chapter 5). For those in office, stability of tenure was not guaranteed, either because there was frequent reorganization of government offices or because people fell victim to purges within the party.

The rapid expansion of the central bureaucracy caused serious problems for lower-level bureaucrats. "The problem is that a Ministry of Agriculture with more than 1,000 people oversees a county department of 114 people, which oversees a district department of 15–18 people and the district department supervises communities in which, in the best of cases, one person works in the apparatus along agricultural lines."[2] Local officials consistently complained about the amount of paperwork they were required to plow through, the burden of which seriously impeded their ability to implement directives or innovate politically at the local level. These were not empty complaints. In 1952 First Party Secretary Rákosi specifically complained to Ferenc Erdei, minister of agriculture, about the inflation in regulations, a concern sparked by two separate reviews of county-level paperwork that had shown how serious the issue had become.[3] Using the example of burdens imposed on the Csongrád County Council between the first of January and the 29th of April, Rákosi wrote "Thirteen hundred and eleven [notices] arrived from the Ministry of Agriculture. From the Ministry of Transportation, 389, the Ministry of Public Education, 380, from the Ministry of Health, 356, the Ministry of Collections, 27. In other words from your ministry almost as many as the four ministries combined."[4] Erdei acted on Rákosi's advice, while reminding him that the Ministry of Agriculture's mandate was far more encompassing than that of other ministries. "The Ministry of Agriculture has 18 departments and the council has 8 groups of agricultural departments, each of which govern an area comparable to a sector of industry."[5] Nonetheless, Erdei was able to report in mid-July that the ministry had reduced the flood of circulars, although less than the 30 percent he had promised to achieve.[6]

Learning Work Units

The introduction of the work unit did not proceed effortlessly. A review of party documents and contemporary newspaper articles provides extensive evidence of the difficulties surrounding cooperative farm members' introduction to and adoption of work units. Misunderstandings and disagreements were common. To contend with the problems, party/state officials mounted a series of winter seminars for cooperative farm members. In the calm of a winter's eve, party lecturers tried their best to explain the work unit system, emphasizing the mechanics of rewarding effort justly.

FIGURE 7.2. "Free two week evening classes. Agricultural course. 'How do we increase yields?'"

In hopes of making the system easier to understand, the agitprop cadre—those charged with taking Marxism-Leninism to the masses (*népnevelő*)—offered simple, concrete examples of the way that work units would be assigned on the basis of some task. Unfortunately, in some cases officials gave inaccurate advice, such as promising that the state would guarantee a fixed monetary amount for each work unit.[7] A recurring problem was the less than enthusiastic performance of officials themselves, which brought stern reprimands from their superiors. "It is characteristic of Bolsheviks that they don't carry out 50 percent of the set tasks, but 100 percent. That's why our job is to overfulfill the plan . . . our working peasants like to learn, one just needs to deal with them."[8] In Zala County, classes explaining work units were being held in December of 1952, three years into their use, not a good sign. It is interesting to note that the success rate county officials reported back to their superiors verged on 100 percent. The phrasing used in the communiques was nearly identical: "The agitprop cadre dispatched to the farm identified mistakes, which the members of the farm realized were their own, and they resolved to put the work unit into practice in the coming years. This will eliminate 'egalitarianism.' The members understand that [using the work unit] is the only way to have a just division [of goods]."[9] All the evidence was to the contrary.

In the first few years of cooperative farming, it was very common for work units to be equally distributed among all members of the farm. In these cases, farm management was accused of the worst possible sin in socialism: "egalitarianism" (*egyenlősdi*). This was a central heresy. Socialism, as Stalin had long argued, rejected bourgeois egalitarianism; the industrious were to be rewarded and the lazy punished. But as one informant explained, members were heard to exclaim, "Everyone has a stomach."[10] Taking time to distinguish between everyone's contribution to the task seemed superfluous. "The opinion of this group, which they have insisted upon, is that everybody worked alike, so why should they bother calculating work units?"[11] In some instances, farms simply recorded how many days were worked (labor day, or *munkanap*), a recording system comparable to presocialist agrarian exchanges and so bereft of careful scientific calibration. While not the fine-grained tool officials hoped farm members would use, the work day was also the base temporal unit for work units, with the important caveat that the work had to take an entire work day to perform.[12] Time and again officials underscored the significance of differentiating between tasks. "Naturally, a work day is not the same day when someone picks weeds as when one hoes."[13] The issue, however, was not just difficulty but also skill and expertise. At the first county meeting of workers from cooperative farms and machine stations in Hajdú-Bihar, held in June 1950, the audience was reminded of Rákosi's recent insight. "There are artists among bakers, but the land also has artists, who can conjure much more from the land than others."[14]

The dangers of wage equity feared by the party and proscribed as egalitarianism were overblown. The far more common problem was tailoring the work unit system to suit local social prejudices. In a few cases, farm workers wished to pay women less, in accordance with earlier practices of differential day-labor wages.[15] Of course, this would have only compounded gender inequalities, since the work unit system already incorporated an implicit gendering of tasks. More frequently, management would vary the number of work units apportioned for different tasks, depending on the alliances, allegiances, or kin status of the cooperative farm member. Cliques internal to cooperative farms were common; allegiances among extended kin groups or between specific class segments were forged against others. In Romania, brigade leaders who were accused of distorting a member's contribution were said to have stolen with a pen (Kligman and Verdery 2011, 183).

More basic, however, was the division between those who worked and those who did not. A large number of cooperative farm members neglected their obligations or simply refused to work on collective projects, preferring to devote their time and energies to working the land they were allotted for family subsistence. Refusing on principle to participate in socialist farming was common. In other cases, farm members hired someone to do their share of work at the farm and then took on projects for wealthier peasants farming privately, who would pay them more. This was unacceptable, József Berend proclaimed at a meeting of party leaders in Zala early in 1949; avoiding collective work in this way meant that cooperative farm members had themselves become exploiters, a clear sign of encroaching capitalism.[16] It was difficult to sustain the members' commitment, much less enthusiasm, however, when they were regularly reminded how much worse their situation had become since joining the farm.

> At the Kinizsi cooperative in Hajduház there are very many men with large families who didn't get even a penny; in fact, they still have debts. . . . They allege that their children remain unclothed and can't even go to school. They don't know what to do with their children; they starve more than when they weren't even in the cooperative. Many working peasants work at the state machine shop in the vicinity, which increases the farm members' bitterness. They pay attention to the income of working peasants, who are constantly earning 100 forints a week. The comrade who had been dispatched [to Hajduház] explained in detail the principle of the members' inappropriate view and pointed out that the value of the cooperative's shared wealth, which is second in the district, exceeds more than about 300,000 forints. After this the membership of the cooperative calmed down but asked that if possible some sort of clothing credit be extended to them so that they could buy a small overcoat for themselves and their children.[17]

At times like this, finding the right people to manage a cooperative farm was crucial. "It is a major deficiency that some comrades only joined the cooperative farm because they are communists and want to set an example. But they don't see large-scale farming clearly, and keep one eye on private farming."[18] Of course, if discipline were not strict, people could start to expect compensation no matter what their contribution had been. As one cooperative farm president complained at a county-level meeting in Zala, "the president before me was considered a kind man but he fulfilled the membership's every request. . . . I told the comrades that I will not be a good person, because I'll give a bonus, money, to those who deserve it because of their work."[19] At a farm in Győr-Moson-Sopron County, a substantial minority of the membership thought of themselves as wage workers, like those at state farms, and expected to be paid every week, or at least every month.[20]

A continuing problem throughout this period was the inclination of brigade leaders and other farm officials to neglect the recording of work units. Work unit books were fudged, numbers being drawn from the air when managers needed to explain their annual budget review. As a general rule, accounting was not taken seriously, considered a burden rather than a necessity. This problem of neglecting accounting procedures would continue at least until 1956; indeed, as the work unit system became more intelligible to farm members, problems with accounting—keeping books, and keeping them accurate—became the primary focus of county-level interventions in cooperative farm affairs (see chapter 8).

Additional means of encouraging membership to do their fair share included exhortations in newspapers and radio programs and visits to family homes by party activists, often made by industrial workers enjoined to enlighten their less advanced brethren in the socialist homeland. Party officials also organized visits by working peasants to well-run cooperative farms in their vicinity, in the hopes that seeing a successful enterprise would demonstrate the feasibility of cooperative farming. Hungarians who had visited state and cooperative farms in the Soviet Union were guests at community meetings to describe their experiences in the great socialist paradise.

Not just cooperative farm members were sent to the Soviet Union; privately farming peasants were also included in the group. The reason for this was simple, articulated in a memo setting out the goals for a visit to the Soviet Union:

> Above and beyond becoming familiar with socialist agriculture of the Soviet Union, they should become fond of kolhoz life, so that after their return individually working peasants themselves would join the cooperative farm. They

FIGURE 7.3. "Hungarian emissaries of the working peasantry in the Soviet Union."

would convince individually farming peasants by their example and instruction that the sole proper road to progress is farming collectively in a cooperative. In the course of this study trip, every single member of the delegation should acquire a deeper love and more gratitude for the Soviet Union and Comrade Stalin.[21]

Some returning travelers were greeted at the train station with a brass band and boasted that recounting their experiences encouraged villagers to join the cooperative farm. Others had a far colder reception, finding the preparations for community meetings inadequate or nonexistent; frequently even the local party secretary and other party members did not attend the event. No matter how interesting their travelogues may have been, they were not believed by villagers, a clear sign to the party that class enemies were poisoning the atmosphere. When shown films of Soviet collectives, Romanian peasants were awed by the bounty and size of farms but saw little connection to their own lives (Kligman and Verdery 2011, 241). In Hungary indifference and sometimes outright hostility was common. "I was traveling in the tanya region [eastern Hungary] and having conversations at [people's] homes, but nowhere could we achieve any result. They say that they believe that in the Soviet Union they produce more and better, but this doesn't interest them. Just leave them be, or maybe they should just shoot them in the head."[22]

Jump-Starting Socialism

Staged competitions at work are well known from Soviet history as Stakhanovism, named for the worker who was the party's poster boy for enthusiastic rate busting. But competitions were a far more general phenomenon. In Hungary, competitions were held with a variety of goals in mind, for example, reducing waste, improving record keeping, encouraging innovation, preparing a business plan, and even recruiting new members to the party. Competitions were designed to increase the intensity of work, and so reach higher productivity, by teaching workers to think about their labor in terms of finer and finer gradations: dissecting the components of a task and dividing up the day into smaller and smaller fragments of time. This would guarantee that effort and output would be drawn more tightly together; it would also activate the personal desire for gain that planners assumed was crucial for piece-rate systems to work. "A wage system must be developed on the basis of which the worker can figure out his income while he works."[23] Moreover, introducing mechanisms to methodically increase output—the practice of measuring and rewarding output exceeding plan targets—was integral to a planned economy. In Hungarian Stalinism, planning was not simply the means to set state budgets but a device to direct the actions of every citizen, most effectively achieved through staged competitions. "In the interest of further development and consolidation of the competition movement, it is extremely important that we break down the factories' plan and bring the plan down to the work bench, to every worker."[24] Having fused state economic policy with business management, the party/state had saddled itself with a gargantuan task requiring unprecedented scales of fine-tuning and oversight. The danger of introducing imbalances in production did give pause. "We must prevent the possibility that various degrees of overfulfilling the original plan's tasks in various branches of industry in the end leads to the distortion of the basic features of our plan."[25]

Competitions became the means by which workers were disciplined in the new socialist workplace. Justifying the sped-up pace, party leaders argued that challenging each other to improve the way land was tended at the farm was not coercive, at least not in the same way that unemployment had forced people to work under horrific conditions in capitalism.[26] Goals for competitions were set to ensure that specific tasks being demanded of cooperative farm members were seen to, and to pace tasks in the course of the year. Targets were often incremental steps in a labor process, as a call for nationwide competition in the spring of 1950 attests. For example, in one competition, farm members promised to increase yields by 20 percent over those in 1949. In a handwritten proclamation, they explained that to reach their goal, they would complete the following tasks: spread manure on one-fourth of the farm's plowland; not leave manure piled up for more than three days, and once it had been distributed, leave it no longer than five hours before it was plowed under; weed all row crops perfectly; hoe corn four times; harvest wheat by machine and finish threshing within two weeks; and complete every job ahead of the targeted dates in the farm's business plan (*üzemterv*).[27] Setting a goal of overfulfilling the yearly milk quota by 100 percent or the poultry quota by 20 percent was a familiar tactic, identical to the government's ratcheting up of taxation levels and requisition demands (*Zala Néplap*, 26 Mar. 1952).

The charge of instilling new work habits would fall to the new generation of agrarian professionals trained by the socialist state. As the resolution of a national meeting of cooperative farm party secretaries and agronomists in 1952 declared: "Become fighters for the new, the better, for modern methods, with which it is possible to achieve greater results. Teach cooperative farm members the basic regulations and consistently demand their implementation ... encourage the adoption of deep ploughing, ploughing up the stubble

FIGURE 7.4. The heroes of the labor competition for new production methods

in time, proper treatment of manure, sowing in time . . ."²⁸ Until then, a short-term solution was to offer quick classes for cooperative farm presidents and local party officials, in the hopes they would succeed in teaching villagers the new techniques. In addition to coursework, competitions also demanded establishing specific metrics to judge what would be considered a reasonable measure of having fulfilled output, and means for comparison. "At the 'December 21st' cooperative, the two groups are neck and neck in tending plants so it is extremely difficult to judge. I worked out the data of both groups at the site. The work methods of each are the same, and they are fighting a hard battle with each other in competition for the contest flag."²⁹ In the case of competitions involving machine stations, calculations had to be performed

FIGURE 7.5. Be among the first in the competition to submit requisitioned crops!

in relation to how much fuel had been saved or overconsumed, and also in relation to the judicious use of parts.[30] And finally, what exactly was to be measured also had to be ironed out; people had taken to measuring the tractor's output, rather than that of the tractor driver. "This seriously interferes with the monumentum of the labor competition."[31]

Competitions were a valuable instrument for calibrating the plan mechanism. They were also considered excellent tools to enlighten the political consciousness of workers. The process of individuating workers on the shopfloor was a crucial feature mobilizing workers on the path to socialism: "the best tool for mobilizing the masses is individual competition."[32] The intensive individuation of persons in the Stalinist era in Hungary is commonly overlooked, overshadowed by the pageantry of enforced collectivity: mass rallies, holiday parades, and the public reading of newspapers organized daily in many factories. Heightened individual responsibility nestled firmly within collective production.[33] Bulletin boards mounted in factories or in front of the town hall posted winners of specific competitions, with the percentage of overfulfillment prominently displayed alongside their names; local newspapers listed the accomplishments of outstanding workers, often accompanied by photographs. These honor rolls offered public affirmation of the righteousness of competitions for the state, as well as being a means to shame those not displayed.

In principle, developing the system of effort and output in industry was a matter of refining already existing structures, such as piece rate wages.[34] In agriculture this kind of incentive structure was virtually nonexistent. Measures to increase productivity required not only new wage systems but also fundamental changes in how work itself was organized. Intensifying the pace of work could only occur if more effective means of achieving results were put in place. In agriculture, this entailed designing tasks to be performed in work gangs or brigades that were calibrated to *individual* results. Working together at a cooperative farm was *not* the same as working with one's family or a band of migrant workers, since now every single person's contribution had to be evaluated separately from everyone else's. "It is extremely important that in contrast to private farming and formerly manorial agriculture, when everyone was forced to work in the job assigned to him, now one can work in a sphere of work appropriate to one's abilities and so in this way production also increases."[35] A mass of individuated workers unfettered by the drag of an antiquated collectivity was the goal. Competitions were ideally suited to instilling this new sensibility.

Competitions in agriculture were planned for every sort of agricultural community: state farms, all three types of cooperative farms, machine stations, and private farmers, although the emphasis fell on public and jointly owned properties.[36] "The purpose of a competition in agriculture is the quantitative increase of our agricultural production, improvement of the quality of our produce, the reduction of production costs primarily in the social sector of agriculture and among individually farming working peasants in such a way that small-scale private farms be advanced not in the direction of capitalist but of cooperative development."[37] Competitions were intended to be launched at every possible level of organization: "not only should cooperative farms pair off in competition, but brigades within groups, work teams within brigades and individual group members should also invite each other to compete."[38]

Many of the problems haunting the introduction of work units hampered the success of competitions as well: confusion, indifference, and bureaucratic mishandling. The point of participating in competitions was lost on many farm members. Explanations steeped in Cold War rhetoric could only have muddied the waters. A proclamation from the National Peace Council published in the Győr-Sopron County newspaper exclaimed: "We struggle most effectively for peace if we fortify our free country . . . into such an iron wall that every vicious attacker will rebound with a bloody head. Every brick, every newer furrow . . . is a blow to warmongers and contributes to the great work of protecting our peace" (*Győr-Sopron Megyei Hirlap*, 7 Dec. 1950, 5).

But what was the virtue in finishing tasks before the deadline set in the cooperative farm's business plan, or value of promising to cut the dried stems and leaves off the potato plants after harvesting? Answers were not forthcoming, except in the convoluted language idealizing planning and promising a future of bounty. Often competitions were treated by local officials as routine exercises to be held with little regard to the level of productivity or the quality of work performed. In fact, it was very common for authorities to report having mounted a competition that had not taken place. Improvements in organizing labor brigades looked impressive on the books but were hard to find in practice. This was especially the case with pressures to individuate output. "One of the serious impediments is group accounts. Many times they don't know how to measure individual output.... It is also a disadvantage of the competition movement that in many places they still work according to time wages, or the workers are not experienced in the work, they are less able to fulfill the norm, weekly income is small, and so the competitive spirit is missing."[39] In the final analysis, leverage to compel participation was limited. This was true even at state farms and machine stations. At a meeting discussing a round of competitions tied to changes in wage policy, "the county person responsible for the [machine stations] (in Vas County) sarcastically noted: 'The committee can do what it wants, but it still can't launch a competition at the machine station, because the tractor drivers will not make any more pledges.'"[40]

The immediate response of party officials was to berate those charged with leading propaganda efforts. If people suspected that competitions were fixed, then they had not been properly politicized. Insufficient efforts had been made to promote a new spirit of rivalry in the citizenry. County reports frequently mentioned the absence of any public display in local communities. Bulletin boards keeping track of the competition figures or honor rolls listing the winners of competitions were neglected, as were the possibilities of using the public address system to broadcast results. The efforts by ministerial personnel and movement operatives (officials in the so-called mass organizations that were controlled by the party/state) were also found wanting. In March of 1950, a party resolution called on officials in both the ministry and the mass organizations to work out the glitches in their relationship over competitions and develop better means of cooperating with each other. A similar demand was issued in February of the following year, demonstrating that the problems had not gone away. The Union of Working Youth (DISZ, Dolgozó Ifjúság Szövetsége) and the National Union of Working Peasants and Agrarian Workers (DÉFOSZ, Dolgozó Parasztok és Földmunkások Országos Szövetsége) were criticized for neither organizing nor promoting competitions. "There are still state farms and machine stations where the hooray atmosphere dominates—a competition at any cost—after which the competitive spirit dies down."[41] Officials responsible for managing agricultural competitions needed to be assigned at every level of government, from lowly council to the inner sanctum of the Ministry of Agriculture leadership.[42] The ministry's propaganda office was charged with gathering information daily on the situation with competitions across the country and funnelling that to the press. Unfortunately, the problems were not confined to keeping good records. Machine stations were prevented from competing for lack of proper tools and replacement parts. Construction of stables at cooperative farms was hampered by a scarcity of experts, while those experts employed at the ministry did not offer any professional support to encourage the development of ideas and suggestions initiated by workers themselves. To top it all off, the Ministry totally neglected the private sector when it came to organizing competitions.[43]

Detractors spoke up against competitions as pernicious and counterproductive. In a letter written to the editor of *Free Land* (the cooperative farmers' newspaper) in the summer of 1949, a self-described "old commie" made his views known. "I know what a competition is like. It is harmful to the community, harmful for the individual. People are not the same, one is slower, the other more agile. For this reason alone competition is bad, because the slow one will never be able to do what the agile one can."[44] Of course, this was precisely the point of competitions. He continued: "If this happens everywhere in the country, then one can be certain that the people of the country will starve with collective production." His prediction would have come true, if not for the aggressive appropriation of villagers' food stores in the increasingly oppressive crackdown on peasants in 1952.

Class Warfare Moves to the Countryside

Class warfare in the countryside is very difficult.[45]

Class warfare in Hungary during the 1950s took a variety of forms: some rather ordinary, others more extreme.[46] Measures designed to make private farming impossible for wealthier peasants were couched within broader tax policies and requisition orders, which expropriated substantial portions of every farm's surplus (see chapter 6). Primitive accumulation in Stalinism fed on the agrarian sector voraciously, no matter what one's class background.

"In 1952 compulsory requisitions became the primary tool of state procurement. The proportion of state supplies from collections increased from a 22 percent share in 1950 to 46.8 percent in 1951 and 73.3 percent in 1952" (Pető and Szakács 1985, 184). The blunt edge of state violence—seizing property, physical battery, and jail sentences—was meted out to far more villagers than some hypothetical roster of class enemies would have warranted. Coercion was an equal-opportunity tool.

Private farmers were required to enter contracts with state agencies to sell a portion of their produce at a price set by the state, which was far below market price. As a party official remarked at the time, "Among individual peasants generally the view is that 'the free market does not concern us because there isn't anything to sell.'"[47] Table 7.1 shows the discrepancy in prices paid. Even though most peasants continued to farm privately, they were not to be left behind in the march to the future. All the measures taken to ensure that cooperative farms were adopting more advanced methods of cultivating plants and tending to animals were also expected of private farmers. Officials realized that private farmers initially would be at a disadvantage setting up a business plan, as they did not employ agronomists or bookkeepers. So the Ministry of Agriculture initiated a propaganda campaign in the fall of 1949 to inform private producers of their responsibilities.

> Propaganda should enlighten everyone on the planned fulfillment of ploughing and sowing, extending to every phase of fall agricultural work, such as the proper preparation of the soil, sorting and treating the sowing seeds, mechanized sowing, the proper use of artificial fertilizers, caring for the crops, the advantages of contractual production, etc. In the interest of imparting the results of work well done, the press office of the Ministry of Agriculture should publish appropriate posters and illustrative brochures.[48]

TABLE 7.1. The wholesale and average prices of produce from peasant families and cooperative farm members, 1949–1952

Form	Percentage of overall sales (%)				Prices (1938 = 100)			
	1949	1950	1951	1952	1949	1950	1951	1952
Tax (in kind)	6.76	6.98	9.00	7.62	0	0	0	0
Requisitions	10.96	16.40	33.52	62.69	383	385	413	398
State purchases	47.41	45.42	27.75	4.03	563	533	635	676
Free market	34.87	31.20	29.73	25.66	703	961	1901	2041
Average price					554	605	879	800

Notes: Wholesale and average prices of goods submitted to the state or sold; proportion of produce that was appropriated as taxes and obligatory sales to the state and what was left to sell on the open market.

Source: Erdmann 1987

Plan indicators, ostensibly methods recommended for sound farming, became a powerful tool in the class struggle. As the Party was wont to emphasize, cultivating plants "is a question of political work and class warfare."[49]

Demanding these tasks placed an onerous burden on private farmers. A whole new repertoire of images used to brand class enemies and paint their heinous crimes was regularly deployed, but it was individuals who suffered for it.[50] Nearly 400,000 peasants were jailed or fined in the early 1950s for offenses that might otherwise seem extremely minor (Donáth 1977, 149).[51] Articles regularly appeared in local newspapers naming the offenders, giving their street address, and then providing details of their crimes. For example, in November of 1950 we can read in the *Győr-Sopronmegyei Hirlap* about five kulaks living in Sopron, including a widow, who were accused of sabotage for not sowing as much acreage as the plan had stipulated (4 Nov. 1950, 5). At the other end of the country, another scheming kulak was accused of sabotage when he chose to farm as before. He had neglected the wisdom of state authorities and spaced his seedlings according to tradition. "Kulak X. Y. spaced his sugar beets 70 centimeters apart, and tobacco a meter apart, for the whole community to see, since he is an 'expert'" (*Hajdú-Bihar Néplap*, 14 May 1950, 1). A county document from Hajdú-Bihar lists, among others, the following infractions and punishments over six months in 1952: A kulak who did not finish his spring planting by the 31st of March, was sentenced to a year in prison and a fine of 3,000 forints, while an elderly man (seventy-three years old) who had not cleared his land of weeds and plowed the soil at a stipulated depth was sent to prison for a year and a half.[52] The cruelty of these measures was incredible. Surprisingly, being a party member in good standing was not even a guarantee against being treated as a class enemy. In a personal letter to Erdei from Pál Szabó, the populist writer, we find the following anecdote about conditions in his home village.

> Among others on the shame board I found X. Y., who in the past had been a good opposition small peasant. Because of me, he was a member of the Peasant Party from the start of the liberation [a front for the Communist Party]. Later he finished party school here in the center, after which he became the district secretary, and now he is on the kulak list. . . . Since he was unable to fulfill his delivery obligations, especially eggs, the district court sentenced him to four months.[53]

Only Szabó's intervention at his appeal hearing secured his release.

The party's hope to purge the countryside of kulaks was coming true, though with unanticipated results. To escape from the heavy burdens they faced, thousands of families reduced the size of their holdings substantially,

A KULÁKOK A BÉKE ELLENSÉGEI

Dolgozó népünk éberen őrködik elért eredményeink felett és a szabotáló, feketéző, cselédnyúzó, titóista ügynökeikre, a kulákokra kemény ököllel sujt le. Mint például legutóbb a megye egyik legelvetemültebb, mindenre képes kulákjára,

155 holdas kulákra, aki a multban rossz élelmezés mellett napkeltétől éjszakáig dolgoztatott hat cseléddel. Most pedig úgy igyekezett szabotálni, hogy nem kapálta meg 2000 négyszögöl burgonyáját, amivel **50 mása burgonyát** vont el a népgazdaságunktól. Az elvetett 5 hold napraforgóját sem művelte meg, így ki kell szántani. A **4.000 darabon** felüli tojásbeadási kötelezettségéből csak **200 darabot** teljesített. Elkésve kezdett hozzá **32 hold gabonájának** learatásához s ezzel **10 százalékon** felüli szemveszteséget okozott

De békebizottságaink éversége leleplezte ezt a megátalkodott gonosztevőt és a nép bírósága még aznap meghozta felette a megtorló ítéletet, az öt és félévi börtönt, 3.000 forint pénzbüntetést és 15 holdon felüli vagyonának elkobzását.

Békeharcosok, legyetek éberek!
Leplezzétek le minden ellenséges elem aknamunkáját!
Őrködjetek a nép állama felett, jó munkátokkal adjatok választ az ilyen aljas mesterkedéseknek!

Fk. Schweier Ferenc — Uj-Somogy Nyomda, Kaposvár 2705 — Nyf. Mojzer I.   1951.

FIGURE 7.6. "Kulaks are the enemy of peace! Working people vigilantly protect our accomplishments and strike down kulaks with a heavy fist, kulaks who commit sabotage, participate in the black market, exploit their servants, and are agents of Tito." In the upper picture, the kulak hoards a basket of eggs, while listening on the radio to the Voice of America, London, and Belgrade. He is charged with not farming properly and not fulfilling requisition orders. The charges include not having hoed 2,000 square acres of potatoes, the result of which was depriving the people's economy of fifty tons of potatoes, and being late to harvest wheat, thus causing more than 10 percent loss of grain. In the lower picture, he languishes in a jail surrounded by rats and cobwebs.

or simply gave up farming, "gifting" their land to the state. Wealthy peasants had no other options, as they were forbidden from joining cooperative farms. Poorer peasants had the option of joining, turning the title of their land and their livestock over to the group. A year-long campaign in 1951–1952 strongly urging private farmers to join a cooperative farm increased the percentage of farm members from 5.7 percent to 15.2 percent and their proportion of plough-land from 9 percent to 14.5 percent (Balogh 1986, 31).[54] "We would characterize the relation to collective production at the time more accurately if we say that [private farmers] backed in. After all, their thoughts were still focused on their abandoned farms. They concentrated their attention and labor on the household plots that remained" (Donáth 1977, 141). Increasing the membership of cooperative farms did little to improve their performance, which continued to lag behind that of private farmers. As of early 1953, one million hectares had been ceded to the state, the majority of which was abandoned during 1950 and 1951. All of the land left behind lay fallow. Young people fled to the cities, leaving the elderly at home to tend the family garden. "Between 1949 and 1953 the wage labor force engaged in agriculture dropped by a quarter million.... nearly every one in the quarter million were of working age and experienced in production and farming" (ibid., 139–41). Augmenting one's income by taking on extra jobs had been necessary for many since 1949, even for cooperative farm members. "Between 1949 and 1952 the number of peasant families earning a wage, i.e., a two-income household, grew from 360,000 to 420,000. In 1952 among the families with less than 15 kh (8.5 ha) of land close to 35 percent of the family members worked either in industry, commerce, or transport" (Pető and Szakács 1985, 203). Meanwhile, state farms faced chronic shortage of workers, hiring prisoners and soldiers at peak season to bring in the crops. The consequence of these harsh policies was to endanger the nation's food supplies. Between 1949 and 1953, the population in Hungary had grown 6 percent, while the acreage devoted to growing wheat dropped 21 percent (ibid., 204). In 1951, foods began to be rationed, initially only flour and sugar, but then bread, milk, meat, lard, and soap were added to the roster (Balogh 1986, 328–29). How much one was to receive as a daily ration was calibrated, as calorie money had been before, according to one's occupation. Now, however, not only workers doing physical labor, but technical, economic, and administrative leaders, were also provided a higher ration, as were those doing high-level intellectual work and prize winners in labor competitions (ibid.). Intellectuals and government officials were no longer last in line for food allocations, as they were in the early years of rebuilding the economy in the 1940s.

A Problem with Definitions

A recurring problem in the class struggle was the confusion over classifications: where to draw the line between peasants and kulaks, between legitimate private farming and nasty capitalists who would spell the destruction of socialism (see also Kligman and Verdery 2011, 326–41). Soviet leaders expected Hungarians to quash any incipient capitalist class immediately. The language of kulak, taken from Soviet history, and class enemy (or class alien, which captures the Hungarian more accurately) did not sit easily in the large majority of Hungarian rural communities. Families who were to become stigmatized as kulaks were often respected members of the community, pillars of village politics and economic leaders. The difference between a comfortable peasant household and an impoverished, struggling family was obvious, but their social allegiances far outweighed the differences peasants saw between themselves and urban residents, commercial interests, and, most glaringly, the landed aristocracy. "No one pities those people whom the village considered scoundrels, i.e., middlemen, pubowners, or such, if they end up on the kulak list. After all, the village has always had and still has a healthy sense of justice."[55] There were some regions in which the history of intravillage class animosities was more pronounced (e.g., the southeast corner, known as the Stormy Corner, *Viharsarok*), where terms such as "greasy peasant" and "fat-necked peasant" were local parlance, but this was the exception. In his report on a series of meetings held with activists in several northern counties in late May 1950, András Hegedüs noted with dismay that a lot of confusion plagued the definition of kulaks. "They don't see kulaks as a class category, the most dangerous stratum of the bourgeoisie. Rather they think that this is some kind of derogatory label that they can use for anyone they want. Phrases like 'kulak epithet' or 'we classify them as a kulak' are very widespread."[56]

Deciding upon strict definitions was difficult. Initially the biggest problem was caused by regulations being issued in the course of several months that were at odds, and public pronouncements that were unclear or contradictory. The figures stipulating the property holdings conferring kulak status released from August to November of 1948 were set at two different levels: 15 kh [8.55 ha] (150 gold crowns) and 25 kh [14.25 ha] (250 gold crowns) (Erdmann 1987, 384).[57] Moreover, comparable plots of land in different regions of the country produced different yields, due to the quality of land, the terrain, and the weather suited to specific crops. Combining the size of the farm with the land's value recorded in tax registers helped to sort these differences out, but only if attention was paid to these differences. In his letter to Erdei about the conditions in eastern Hungary, the writer Szabó expressed deep concern about problems with land valuation and requisition quotas. In his region, the soil was very poor, so the size of farms was misleading.

> The obligations of peasant farms were progressively raised so high that peasants with bad land were not able to fulfill their obligations. But when their

cases leave the village and go through the district committee or the county and end up at the court, naturally officials look exclusively at just the number of *hold*-s, and they treat them like other kulaks. The anguish they experience is nearly inhumane.[58]

Meetings held for county officials and party workers devoted much time and energy to raising questions about who was a class enemy and debating the way policy needed to be implemented. Lengthy debates took place pondering the fine distinction between economically restricting kulaks, the official policy of the party, and liquidating kulaks, which was discouraged as too radical, at least publicly. Warning his colleagues against distorting policy, one official then proceeded to muddy the waters further: "we need a policy with which we squeeze kulaks. We don't want to liquidate them, just drown them in their own blood."[59]

The frustration of senior party officials was obvious; even their own delegates sent to the countryside to clarify matters could not come to agreement. "In many places it's disturbing that in the midst of their propaganda efforts work comrades will quarrel over a person, whether he is a kulak or not, whether he can join the [cooperative farm]."[60] But senior party officials themselves contributed to the confusion. In a 1949 speech meant to clarify the regulations, Rákosi cited Lenin's definition as a guide: those who occasionally produce a surplus and sometimes hire workers. But then he added, kulaks are those who "had a position at church, who didn't give his daughter's hand in marriage to a middle peasant, who, in addition to farming, speculated or was a middleman or kept a farmhand year round" (Erdmann 1987, 385).

Over time, as pressures increased and the fortunes of wealthier peasants worsened, distinguishing between restrictions and liquidation seemed even harder to fathom.

> In many places our party functionaries don't understand why a kulak, whose current financial condition doesn't exceed that of a working peasant, is still a kulak. Bewilderment is also expressed about how one can differentiate between one kulak and another. How can kulaks in some places be in management positions, why are they able to work in industrial firms or at state farms, while at the same time they can't be members of the cooperative farm? ... The question is posed this way: "our kulaks have become impoverished, indebted, many are dropping out. Where is the border between restricting or liquidating kulaks?"[61]

It became increasingly obvious that the status of class enemy was arbitrary and open to interpretation. No matter how much time was spent parsing these urgent policy questions, the actual process of identifying kulaks and class

FIGURE 7.7. "Exploiters, you are kicked out of the cooperative!"

enemies of any background was highly subjective, dependent on the whims of people who were responsible for the final decisions made. The integrity or treachery of those in charge—local administrators and party officials, activists and agitators—determined one's fate. Official regulations would be disregarded in one community and obstinately enforced in another. The status of class enemy was also a mobile category, moving quickly from house to house as local politics dictated, so that the names of poor peasants would be prominently displayed on the village Shame Board while wealthier villagers caroused with local officials in the pub. Authorities could be bought off, wealthier peasants slipping money under the table to avoid having to fulfill their requisition allocations. In other cases, bribing took place more publicly. "When the accused was summoned to the police station in Nyiradony and questioned on the 4th of August, 1952, the accused suggested to the presiding policeman that if he tore up the record and not charged him, he would give the station a calf. They let him go. In the evening he took a calf up to the police station, gave it to the officer in command, who tied it to the fence."[62]

Winning Hearts and Minds

Purging society of unreliable elements and dangerous adversaries was the first step in the Communist Party's plan to build socialism. But rallying everyone to be vigilant extended far beyond the immediate task of banishing class enemies. Vigilance, along with discipline and consolidation, would be expected of everyone in the factory or the farm, a state of readiness that would heighten productivity and strengthen labor organization. The party/state would only be successful, however, if it inculcated an emotional commitment to socialist victory. This was particularly true in rural communities thought to be the breeding ground of capitalist predilections. Thus winning the hearts and minds of working peasants was an unrelenting exercise in psychological strategizing. "We absolutely must deal with the psychology of our peasantry."[63] Local officials were encouraged to take the emotional temperature of the community, to gauge the mood of villagers toward each other as well as toward the party/state. In fact, weekly reports submitted by district officials were entitled "Mood" or "Atmosphere Reports" (*hangulatjelentés*). Class warfare entailed more than just identifying kulaks. The priesthood (referred to as clerical or black reactionaries, *fekete reakció*) and other reactionary elements, be they conservative peasants or former gendarmes, had to be challenged and marginalized. Figuring the mood of the village and accurately judging the true danger of the enemy required serious and sustained effort. It also required a full understanding of the social determinants of temperament and reliability, qualities of personhood that were directly related to one's class position.

Stalinist documents portray the kulak and his fellow travelers in antisocialist crime as unreliable and wicked. Kulaks were not simply mean-spirited or stupid. They were constitutionally incapable of recognizing the truth, their reason clouded with improper instincts about work, about political rights, and about democracy. The incorrigible class alien stood in stark contrast to the formerly downtrodden peasant, portrayed as naturally supportive of the party/state's missions and tactics. Poor peasants were inherently good, trustworthy, and open to the historical truths of justice the Communist Party offered the country's citizens. Despite these stark dichotomies, some properly "class-positioned" villagers could also turn out to be working against the interests of the state, either under the sway of pernicious elements or out of a deluded sense of independence. In such cases, they had to be punished as harshly as if they were class enemies.

Fine-tuned skills were needed to decipher intentions, far beyond the appearance of actions or spoken words. Time and again lowly village officials were chastised for not learning the signs of sabotage. Outward signs of unreliability were clearly marked by class background, but these were insufficient to identify who in the community was actually propelled by nefarious purposes. Central authorities were at pains to explain to local authorities that those villagers accused of sabotage who denied culpability could easily be masking their real feelings or simply be unaware of their true sentiments. One could not trust the declared intentions of the enemy, since dissimulation was his métier. Who complied with socialist policy had to be examined with care. Astute observers could discern the wily ways of the enemy and were quick to explain to their comrades the true nature of the crime. One could not be swayed by public pronouncements, such as a kulak's declaration that cooperative farms were superior to private farming. Local officials had to be shown that a kulak could never have believed this, but that saying so was the means by which the kulak planned to wreck the farm from within.[64]

Sabotage assumed planning and forethought and potentially also entailed enticing others to participate in one's heinous deeds.

> [Kulak X. Y.] announced that in three days he would obtain another engine, because the current one, the kulak says, is not a paying proposition in terms of the national economy. The comrades were duped by the kulak's demagoguery and didn't realize that by doing this the kulak took the threshing machine and the engine out of work for two full days. Only when the supervisory committees explained it to the comrades did they understand the kulak's method of sabotage.[65]

A broken machine was no longer simply the result of wear and tear; it too was sabotage.

> The party committee has determined that the backsliding and the careless operation that has taken place on kulak machines is sabotage. Bearings are burning out every day. Pistons are breaking, or as in Szőcs, where only one machine worked in the entire community, the head of the piston broke ... as a result the motor was rendered completely useless.... Kulak X. Y.'s machine was put into operation without being repaired and after a few days of bad threshing the machine completely broke.[66]

So whatever happened, the kulak was commiting sabotage, whether he took time to repair a machine or didn't. Alternative explanations, such as problems with providing services to the right places at the right time, were not poor management on the part of the party/state but the result of deliberate obstructionism.

Bureaucratic activists had to be encouraged to question their own judgment, since clever enemies could devise less obvious but equally effective methods of outwitting the party/state. Shenanigans surrounding the use of machines were shrewd but more obvious than other sneaky and underhanded techniques. Rather than confronting cooperative farm members head on, enemies sought out more vulnerable parties. "Kulaks tell wives that their husband's skin is hanging off his bones . . . so he shouldn't be allowed to work at the cooperative farm. Wives then fight with their husbands until they leave the farm."[67] When conditions were already strained, ne'er-do-wells stoked the fires. "Wheat harvesters have no meat or lard and are demanding a just solution. These kind of complaints are frequently arriving at the city council. The mood is being forcefully aggravated by hostile elements."[68]

The vastly changing social universe being ushered in by the party/state required one actively question one's previous moral judgments, as well as one's perceptions and intuitions. Just as formerly well respected members of the community could turn out to be extremely unreliable and dangerous elements, so too other phenomena could not be taken at face value, could not be assumed to be transparent and self-evident. One may think all is well, but that is the wrong assumption. Depending on the confidence of workers in the party/state and in the future was not a wise strategy, as kulaks were always up to no good. The greatest danger was to fall into a false sense of calm. One document records criticism of local officials who took pride in class peace in their communities. If calm reigns, the document explains, then the party official is neglecting his primary obligation of maintaining a fear of the enemy and constant vigilance.[69] "The enemy's agitation has been stronger than ours."[70] Peace is illusory; struggle is constant. Vigilance became a byword and recurring refrain. This was even more the case for those living near international borders. "We [in Zala] are Tito's neighbors and so we must be more vigilant and wary against the enemy than in other counties."[71]

Party officials' success in winning the hearts of villagers depended on managing villagers' emotions, as well as their own. A spirited campaign against the enemy was demonstrated by working peasants coming to share the party's views. If not, then party officials had not succeeded at their job. "The kulaks uncovered in Mihály were not made sufficiently detestable to the working peasantry."[72] Rákosi promoted the effectiveness of internal emotional disciplining. At a national meeting, he explained, "the best disciplining is developing an atmosphere in which loafing, shirking, and insubordination are shameful and odious."⁷³ Public shaming was a central technique of the party's enlightenment campaign. "You must warn members who stir up these ambitions that if they continue to do this, they will be stigmatized."⁷⁴ Posting names publicly was considered an especially effective tool, as seen in quick results. "By inspecting and punishing kulaks, and then publicizing their infractions, the rate of work among those who had fallen behind accelerated."⁷⁵ Neglecting the fine tool of shame and public humiliation was noted and brought to the attention of authorities overseeing local party activities. "It is characteristic of the work of the village council in Kaba that the Shame Board in the village's main square has been empty for weeks even though the names of very many of the laggards could be posted on the board. Having conducted a faulty inspection meant that not one single kulak was reported by the 14th of May for having neglected to tend his crops."⁷⁶

Identifying the enemy would become a crucial component of bureaucratic expertise. Another crucial skill would be the ability to convince villagers of the veracity of the party's claims. Several different techniques were used to persuade villagers, ranging from what was seen to be a simple task of conveying information to more intimidating inducements. In contrast to the party's highly suspicious treatment of class enemies' motives and beliefs,

FIGURE 7.8. Public shaming of slackers in the plowing competition. The caption reads, "It is your patriotic responsibility and in your own interest to finish [spring] sowing. Great shame if your name appears among those falling behind."

their treatment of villagers deemed to be more receptive to their views was extremely simple. One merely told villagers what had happened and expected the truth of the statements to be accepted at face value. When someone had to be relieved of his duties, then the reasons for his dismissal were to be clearly communicated to the cooperative farm members and the working peasantry. For example, an accountant was discovered to be an enemy interloper. He had already been kicked out of his job at another farm because he had neglected to pay the state taxes due on farm revenues; he had even gone so far as to gather tithes for the church. In his new job, he had not been distributing monies according to work unit standards, showing that his belligerent attitude toward state policy continued unabated.[77] Villagers had to be taught to fear the enemy, especially if they were on the front lines of the war against sabotage. Occasionally efforts to demand military readiness of everyone bordered on the absurd. "The night watchman hadn't been informed of his task. He doesn't recognize the class struggle and underestimates the enemy."[78] Night watchmen (*éjjeli őr*) were usually older men whose greatest skill was finding the most comfortable hay stack to snooze in all night. Demanding vigilance of the old codgers would be funny except for all the horrific stories people would tell about falling victim to an overly zealous party official.

Goading party/state officials to ratchet up the level of fear, anxiety, shame, and suspicion could end up harming the party's goals. Techniques of psychological motivation were difficult to learn, much less master, as they required a discriminating sensibility not shared by all party/state officials. High-ranking party/state leaders could not rely with any certainty on officials to do their job right.

> Certain of our party organs and authorities have improperly applied the intensified war begun to restrict kulaks, and their work must be ceaselessly supervised. In Tét they made the mistake of having kulaks arrested who had contravened rules, but the working poor and middle peasantry of the village did not know the exact reason for the arrest, at most only afterwards. Rather than turning the poor and middle peasantry against the kulak, they slipped into [an abuse] of administrative procedures.[79]

Overweening comrades endangered the socialist project, as when party officials proceeded to appropriate the assets (land and animals) kulaks had not voluntarily surrendered to the state. Overzealous actions also tended to reinforce kulak propaganda about the party's goals. "If these abuses continue, then we will not be building socialism in the village, but hindering it."[80] The dangers of party misbehavior also required disciplining authorities in charge of village affairs. Party/state officials had to be chastised for misunderstanding Marxist-Leninist theory and distorting Communist Party policy in village affairs. Even if functionaries knew Leninist slogans, they were not putting them into practice. "The reason for distortions is in significant respect the ideological illiteracy of functionaries."[81] One had to learn how to treat poor, middle, and kulak peasants differently. "Left-wing deviations also isolate the small and middle peasantry from the cooperative farm groups."[82] Officials were faced with a difficult task, and not all rose to the occasion. "It's easier in the village to bring restrictive or punitive measures against a few or more kulaks than to convince the small and middle peasant of the advantages of cooperative farming and turn them against kulaks, the enemies of cooperative farming."[83] As one official in Hajdú-Bihar noted,

> We're impatient with workers. We don't know how to treat them patiently. In the Soviet Union socialist agriculture developed over 15 years. It's true that here we must develop it quicker, because on the basis of their example we are progressing on a well-worn path. But one mustn't refer to small and middle peasants as kulak stooges who due to their backwardness themselves do not join the cooperative farm. We must heed the voluntary principle. One shouldn't create a mood of hurrah and hoopla.[84]

In the final analysis, disciplining party officials both in techniques of class war and in party policy would be required to ensure that problems would not develop.

Bureaucratic Tangles and Faulty Mechanics of Coercion

The execution of party policy depended on a well-functioning bureaucratic apparatus: agencies pursuing clearly delineated tasks and agreeing on lines of authority. It also relied on government and party institutions being staffed by people willing and able to do their job effectively. Neither of these was firmly in place in the early 1950s in Hungary. Bureaucratic growing pains accounted for some of the difficulties, as discussed earlier, while clashing personalities explained others. These are garden-variety political problems. There were, however, certain administrative problems that could be traced directly to the character of one-party rule in a Marxist-Leninist regime. The simple issue of finding qualified staff, common in all modern government bureaucracies, was complicated by the dilemma the Communist Party wrestled with when weighing expertise against political conviction (see earlier discussion in chapter 5). These problems were exacerbated in the countryside, where professional expertise and a commitment to Communist Party ideals were much harder to find. Unfortunately, the entire edifice of centralized planning required a vibrant presence in every community, as Communists knew well. "The party is a motor, which hums if wound up, and if it is missing there is nothing to provide an impetus to work."[85] Engines of progress were few and far between in rural communities.

In the 1950s, class warfare and collectivization were stymied by the difficulty of extending the party's influence to the countryside.[86] Coercive techniques were only as effective as the apparatus carrying out policies. The poor state of local governance bodies was a constant refrain in period documents. Functioning party committees in villages and at cooperative farms were scarce. Even if a party committee had been set up at the village council, it did not mean that the local cooperative farm had taken steps to form one. Often party committees existed solely on paper. As one official at the county meeting of machine stations in Zala County declared, the local party committee was worthless (*"egyenlő a nullával"*).[87] Elsewhere, the very idea that outsiders would dictate local affairs was rejected. Across the country in Hajdú-Bihar County, county interference offended members of the cooperative farm, who bristled at the notion that they should be taught how to cultivate crops. One member of the party committee's directorate went so far as to tell government officials to stay away. "[We] don't need to be given advice from county experts, or have agents from the district talk about [labor] competitions at the cooperative farm. [We are] fully aware of what needs to be done."[88] Under these circumstances, bureaucrats and activists charged with educating the folk (*népnevelés*) were often less than enthusiastic about their job. "Site visits must be utilized better. It is not permissible for the worker sent out [as a visiting delegate] to sit in the cooperative farm office for an hour doing nothing. This sort of thing has happened."[89] Ardent supporters of the party wrote letters to the editor of *Free Land* complaining about the dereliction of duty they witnessed: "X. Y. from Börcs [Győr-Moson-Sopron County] writes that on the occasion of the [yearly village] festival, people shouted songs making fun of the five-year plan sitting right at the open window in the cooperative farm hall. No one there stepped up to take them to task."[90]

Positions of authority—in the government and in the party—were occupied by all sorts of folks. Concerns about wealthier peasants holding positions of power—at cooperative farms or the town council—were matched by disappointment in the quality of cadres taking jobs. If the local party was predominantly constituted by less respected segments of the community— "lumpen" is the term used in party documents—then their influence was seriously compromised.[91] "Our party organizations underestimate the Party's strength. They don't dare fight openly against the local reactionary priest, because they think that the priest has greater authority.... The large majority of our party organizations await dispositions from above."[92] High-handed and disrespectful actions taken by a local party leader could alienate everyone in the community, as well as infuriate their superiors. "The party secretary [at the cooperative farm in Ujrónafő in Győr-Moson-Sopron County] is impatient and dictatorial with the people. He dispenses orders the way a boss would have done in the old days. He doesn't accept criticism. The members say that he characterizes the slightest criticism as destructive and calls the critic seditious. In the beginning he even threatened to bring in the [secret police]."[93] Local officials in charge of requisitioning were held in the greatest disdain, considered thieves and scoundrels.

Repeated efforts were made by central authorities to improve the reputation of party members by conducting a careful review of their status in the community. This would rid party organs of unfortunate hangers-on and opportunists who had joined for the wrong reasons. Unfortunately, damage control had to be exercised at all levels of the government and party apparatus, not just in the countryside. Great care was taken to prevent the Communist Party's image from being tarnished by prominent figures' public drunkenness and lewd behavior, especially when party delegations visited the Soviet Union. Party members' private lives were to be exemplary models of propriety, a notion of morality bearing a close resemblace to bourgeois norms of heteronormative family life. Individuals of questionable aptitude or those with moral failings who came to the attention of important party officials were subject to review, and if their behavior was deemed inappropriate, measures were taken to lessen the harm they could do. In cases of prominent party members, the consequences could be mild, bringing a rebuke and perhaps reassignment to a less influential position; at other times, harsh measures were taken.

The need to instill the principle of individual responsibility (*egyéni felelősség*), that those in charge were to be held accountable for their public agency, was a recurring struggle, a constant refrain in criticisms of government at all levels, from ministries and party committees to town councils and cooperative farms. No doubt some of the hesitation officials demonstrated was motivated by their indifference to the task at hand, as was the case with village party secretaries and cooperative farm presidents, but in other instances officials' caution derived from the fear of being punished for taking responsibility for decisions made. This could lead to paralysis, when officials were simply unwilling to act even within their own sphere of authority.

Personal vendettas easily surfaced in performance reviews and interagency negotiations, as is the case elsewhere. The complicating factor in dealing with every type of personnel problem within the socialist administration, however, was the ever-present role of the secret police in dealing with conflict. This very real threat far outweighed the usual consequences of common office, or interagency, politics elsewhere. To this day, the Stalinist era in Hungary is remembered predominantly in terms of the arbitrary and wanton use of force by the party and secret police. One could not anticipate with any certainty the outcome of a charge of incompetence or allegation of enemy sympathies, so leveling charges against one's adversaries could result in their losing their job, being sent to jail, or execution.[94] This was a problem for powerful officials as well as lowly bureaucrats; virtually all of the Communist Party leaders in socialist politics during the 1960s and 1970s had been jailed at one time or another in the 1950s.

Secrecy enveloped the machinations of the ÁVH (State Protection Authority, Államvédelmi Hatóság), but it also permeated everyday government work. A curious feature of socialist economic planning agencies and ministerial bodies was the need to keep much basic information secret, even those documents that were needed to develop policy. "Preparing the 1952 plan was generally made more difficult because every plan target arrived stamped as top secret. As a consequence workers in planning got the plans from the administrator of confidential documents at around 9 in the morning, and then at 4:15 p.m.—right in the busiest time at work—they were returned to the administrative office . . ."[95] István Friss, head of the Central Statistical Office and founding head of the Economics Research Institute, complained about the impossibility of training new economists when the statistical information required to strengthen their analytic capabilities was kept secret.[96] Effective administration was also hampered by the often chaotic state of planning; the dimensions of plans were constantly changing. "In Hungary in 1952, the then-current five-year plan was changed 472 times, and the yearly plan for 1952 113 times. In 1953 the Council of Ministers and the Planning Office changed the plan 225 times, and plans as passed to enterprises were changed 71 times" (Swain 1992, 71).

Another issue hindering the workings of the party/state was the discrepant agendas of government agencies, which were not easily reconciled. Nor did they suit the Communist Party's self-image of ruling over a monolithic and cohesive party/state apparatus. Tensions over conflicting missions and decision-making procedures erupted, as, for example, when the demands of the Food Ministry to procure meat conflicted with the Ministry of Agriculture's mandate to maintain herds. "The department of livestock raising has tried to bridge the difficulties with the meat supply by providing help in weeding out sheep. However, fulfilling our sheep breeding plan, and ensuring wool production, is a top-ranked national economic interest."[97] It was simply impossible to satisfy the demands of both institutions, requiring in this case the intervention of Imre Nagy, minister of requisitions, to sort things out. A frightening example of the possible consequences of such a clash at the local level puts the furious exchange of letters between bureaucrats into perspective. Collecting the greatest amount of produce was the requisition agent's task, while the agents representing the food industry had little use for rotten produce or emaciated pigs. Refusing to accept goods could have horrendous consequences, even though it was precisely what the food industry would demand. "In every issue of the Procurement Bulletin one can read extensive correspondence . . . telling of fines, which were imposed for not fulfilling the enterprise procurement plan or for abuses. In fact, on April 23, 1950, the workers' court condemned to death two managers of the Meat Sales Enterprise for a crime committed against public supply. They carried out the sentence that very day" (Szabó and Virágh 1984, 175).

Divergent agendas were not the only problem. Over the course of time, the Ministry of Agriculture found itself in the peculiar position of being both an executive body of the party/state and an agency promoting the interests of farmers themselves. This double role caused problems when representing the interests of agrarian communities clashed with the political and economic agenda of other factions. Ernő Gerő, minister of state, was intent on funneling state funds into the industrial sector with little regard for agriculture, putting him regularly at odds with Erdei.[98] In one case, a clash arose over a regulation stipulating that cooperative farms prepare an estimate of receipts and expenditures between May 30 and July 10 to guide budgeting over the course of the summer and fall. Described as a measure to consolidate cooperative farms, its purpose was to ensure that advances of grains cooperative farms gave members over the summer represented the proximate value of work units at the end of the fiscal year. In other words, the regulation was designed to reign in the common tendency at farms to forget to set aside operation funds and to neglect their obligations to the state in favor of distributing the year's proceeds directly to the membership. The Ministry of Agriculture submitted a proposal sketching the problems with the new regulation and possible ways of circumventing them. A study was also conducted to assess the effects of the new policy. The Ministry of Agriculture feared the prospect of a serious degradation of people's commitment to working at the farm, once they knew the level of their outstanding debts and could figure out what they might earn by the end of the year. In short, once poorly performing farms confronted the cold, hard facts straight on, they would demoralize the membership, feeding the cycle of low productivity and dependence on the state to keep them afloat, precisely in the midst of the busiest time of the agricultural calendar. An ensuing debate on the relative health of the cooperative farm sector considered whether insisting the farms honor their obligations to the state and credit agencies first would leave members without anything to show for their efforts. The push to tighten requirements could seriously affect village life. "We use forecasts to the utmost to mobilize all the reserves of the cooperatives, to assist in fulfilling their plans and increasing production. Despite this a significant number of farms show a shortage of feed and in fewer cases wheat shortage has also arisen."[99] The dilemmas surrounding planning and the extraction of surplus in 1952 exacerbated an already difficult relationship between Erdei and Gerő. The level of animus is obvious in the tone of a letter Gerő sent Erdei in the midst of this disagreement in June 1952.

> It is completely incomprehensible to me how the Ministry of Agriculture time and again becomes the center and origin of essentially antistate tendencies. The "investigation" [into cooperative farm budgeting] shows this at least, just as in the past more than once we ran into similar initiatives coming from the Ministry of Agriculture. Of course we don't want the Ministry of Agriculture to embellish the situation of cooperative farms, that it portray a more attractive picture than what's real. We can reasonably expect, however, that after such a careful and thorough investigation and deliberation that it give its opinion and take a position. This kind of one-sided stand and its promotion seriously harms our people's economy, our state, and, in the final analysis, the cooperative movement as well.[100]

Gerő's vitriol is intense, essentially accusing the Minister of Agriculture of being a traitor.

Conclusion

Punitive measures taken by the party/state were not successful in driving peasants into cooperative farms, resulting instead in widespread pain and suffering. The strategy of class warfare the party followed between 1949 and 1953 garnered little support in the countryside, and even county officials in the party/state were hard pressed to execute orders when the very definition of kulak was unclear. Attempts to entice villagers to the party's cause by making emotionally charged pleas fell on deaf ears. Calls to militancy and vigilance introduced a period of widespread distrust in the countryside; neighbors shunned neighbors, and families turned inward to protect themselves. Class warfare was defeated in part by the poorly equipped or ill-suited soldiers sent to battle: most rural party secretaries and officials charged with carrying the ideological banner made only halfhearted attempts to alter people's behavior, while others seized the opportunity the sanctions the party/state provided them to exercise terror. Meanwhile government agencies clashed over tasks and party leaders squabbled about lines of authority and national priorities.

In the early 1950s, the collectivization of agriculture faltered. Most landowners refused to join cooperative farms, until growing tax burdens and punishments meted out by the authorities became too much to bear. A small number of the villagers welcomed the opportunity to join cooperative farms, virtually all of whom had been members of the agrarian proletariat—migrant workers, residents of manorial estates, day laborers—who lacked the tools and skills needed to farm successfully on their own. But they too resisted the party/state's interference in their affairs, preferring to share equally in the farm's proceeds rather than adopt the new wage system being advocated by officials in Budapest. Both groups suffered profound confusion when presented with new regulations dictating the depth of a furrow or stipulating that records be kept on how long they had worked. While the overwhelming rejection of collective production in the early 1950s has long been recognized in the literature on Hungarian collectivization, less attention has been paid to the simple difficulties of mastering the steep learning curve socialist production represented. People were perplexed and baffled about the changes being sought, as much as they were dismissive of outside interference in village affairs.

Communist Party rule in Hungary during the 1950s has long been described as chaotic and anarchic. While true in many respects, this image obscures features of the party/state that were less visible but just as important. Large segments of the governing apparatus spent their days engaged in mundane work: writing reports, attending meetings, dealing with correspondence, and filing paperwork. Sequestered in grey buildings, this bureaucratic machinery was hidden from view. As we will see in the next chapter, the routinization of bureaucratic work would increase as the party/state rethought agricultural policy.

8

Fighting over Numbers

Stalin died on March 5, and by June 1953 Rákosi had been ousted from the position of prime minister. Citing the disastrous state of the economy, Soviet leaders sought to change the political dynamics of the leadership and, after lengthy discussions in Moscow, chose Imre Nagy as the new prime minister. The socialist project in Hungary entered what has been called the "New Phase," ushering in crucial changes in policy and releasing thousands of political prisoners from labor camps. The changes in agricultural policy were immediate, as was the emotional and political impact of freeing the victims of unjust incarceration. The agenda of the new government gave greater freedom to farmers: peasants were released from the heavy burden of outstanding tax arrears weighing on their shoulders, and cooperative farm members were given the choice to return to private farming. Greater leniency was shown toward previously castigated class enemies, especially experts and specialists who were trained and had worked in capitalism. More effective means of administrative oversight had to be devised, putting to rest tactics based on the simple use of brute force. New forms of discipline had to be designed: in the final analysis, just as coercive, but far less violent in their daily execution. Doing this would require a major overhaul of government agencies, streamlining the bureaucracy and improving the qualifications of the staff. Turning the country around in the face of severe economic and political hazards was a difficult and near-impossible task.

Throughout Eastern Europe and the Soviet Union, the post-Stalin era is referred to as the Thaw, evoking images of a cruel Russian winter melting away and leaving fresh soil behind. The analysis here will take a different tack. The usual focus on policy reform overshadows the slow march toward a more entrenched and effective apparatus of rule taking hold. Policies were tweaked and twisted to accommodate the new political line Nagy and his allies introduced, but the weight of normalizing procedures in the socialist apparatus grew apace. A number of respected research institutes were rejuvenated, providing a firmer empirical foundation to national policy and a means to exercise criticism under the party/state's imprimatur. But the commitment to centralized economic management continued. A more self-confident apparatus also led to open questioning of the value of the Soviet model, giving officials a chance to craft wage policies that could be more effective at increasing productivity and rewarding hard work in the context of Hungarian cooperative farming.

The story of cooperative farming under Nagy's leadership in 1953–55 has usually been told in relation to the widespread dismantling of cooperative farms. The principle of voluntary participation in cooperative farming that Nagy had advocated in 1947–49 (see chapter 6) had finally been adopted as national policy. Lost in this picture is the struggle that ensued over the collective assets of cooperative farms as they broke apart, a rift exposing very different conceptions of the value invested in cooperative farming based on longstanding class attitudes about property and labor. Farm members who had donated land and livestock when they were forced to join the cooperative expected to be recompensed for these goods, claims that were deemed illegitimate by other members who saw labor as the sole source of wealth accumulated at the farm. As a consequence, who had worked at the farm and how much became crucial issues to resolve. The debate over the source of value and its measurement that had preoccupied work scientists, and then party/state officials, had now become a battle to be fought at the village level, one often waged by wielding the once-maligned work unit as a powerful weapon in litigation adjudicated by county authorities.

As a matter of fact, the question of who was working at cooperative farms and with whom had been a vexing issue throughout the early 1950s. The slow entrenchment of the governing apparatus in the post-Stalin era was paralleled by villagers having to contend openly over the division and organization of labor. The party/state's insistence that cooperative farm members had to work in brigades rather than with family members or on their own was rarely respected. What it actually meant to work in a brigade—who participated, how much and for how long, and how plots of land were apportioned to groups—was just as confusing to county authorities as the directives on class warfare had been. Concessions on the part of farm leadership to work groups composed of family members or the individual apportionment of plots of land under the guise of brigade organization were frequent. All this attests to the difficulty of defining what collective production actually meant. This conceptual dilemma was as pressing as the question of the source of value and just as fundamental to the definition and success of cooperative production. And it too, like the struggle over the value of cooperative property, was a problem pitting farm members and neighbors against each other. Of course, most landowners continued to farm privately, and state agencies continued to promote cooperative production. The important point is to stress that the usual depiction of a dyadic struggle between rural communities and the centralized state apparatus in the early 1950s must be discarded in favor of a far more nuanced view that recognizes the conflicts occurring over labor and material goods within and across villages as well.

This chapter covers the period from Nagy's ascension to the position of prime minister in 1953 to the eve of the Hungarian revolution of 1956. The first section of the chapter reviews the criticisms Moscow leveled at the Hungarian Communist Party leadership when Krushchev took power. The criticisms led to a revised estimation of professional expertise and changes in bureaucratic decision making, and an initial attempt at revising the work unit system. Some agencies in the government, such as the Audit Committee, took on a larger role in overseeing budgets and expenditures at cooperative farms, shedding light on additional forms of expertise such as bookkeeping required in the new party/state. The next section of the chapter discusses the intense class warfare over land and labor that erupted between poorer and wealthier farm members as cooperatives fell apart and addresses less public clashes over the way work was organized at cooperative farms. The chapter also addresses attempts on the part of farm members to block efforts imposing a more collective spirit of collaboration at the cooperative. The chapter ends by describing efforts to rethink the work unit system as a whole.

Moscow Calls

The Central Committee of the Communist Party met on June 27 and 28 to redirect policy, and exercise self-criticism for the dismal state of the country. Following the script the Soviets had dictated during the Hungarians' visit to Moscow in mid-June, four major issues were identified as problematic: (1) industrialization had been undertaken too rapidly, especially with to regard to heavy industry, neglecting the actual conditions of the country and the living standards of the people; (2) agriculture had been slighted, and moves to collectivize were unnecessarily swift; (3) the Communist Party had ignored the people's needs and, as a result of poor economic policy, lowered the workers' living standards; and (4) the Party had sanctioned the use of harsh measures, pursued too many legal and petty offences, and treated the population arbitrarily (Balogh 1986, 497–500). Party leaders were chastised for not ruling "collectively": an inner circle of powerful leaders had dominated decision making. Those in power ignored the signs of dissatisfaction among the populace, especially in relation to the use of force and punishment, resulting clearly from ideological and theoretical backwardness (ibid., 501–3). The Soviets also looked askance at the expansion of ministerial positions. "It's odd that in Hungary there are more [government] ministers than in the Soviet Union" (Estók et al. 2003, 296).[1]

The Dilemma Surrounding Expertise

From the inception of the new party/state, the role of expertise in research, policy making, and socialist management was a tricky problem (see chapters 5 and 6 for discussions relating to earlier problems with experts being positioned in the government and university). Creating a new society committed to the principles of scientific socialism required the active participation of highly skilled experts and emotionally committed cadres. Ideally these qualities would inhabit one soul; commonly they did not. While bourgeois expert and class enemy may have become synonymous in early socialist politicking, the nature of expertise was a far more complicated problem than simply tossing out the old guard and bringing in the new. First and foremost, there was no new. Virtually none of the leading party figures in Hungary, or any of their subordinates, had any training in historical materialism, either as a set of theoretical principles or as a guide to policy making. Those who had greater familiarity, such as Imre Nagy or György Lukács—both of whom had studied and lived in the Soviet Union for decades—were treated gingerly by the leading faction of the party, occasionally courted, often criticized and silenced. More to the point, just how and in what ways bourgeois experts could assist in building a Marxist-Leninist party/state was unclear. In principle, scientific socialism was a politically committed creed. Whether one owed one's allegiance to the working class, however, was not necessarily a good criterion for deciding if a professional's credentials were in order. Public pronouncements proudly declaring that decrepit bureaucrats mired in reactionary politics had been dismissed clashed with private deliberations acknowledging the need for expertise at the highest levels of the party. In a meeting of the Central Committee's Agricultural and Cooperative Department, this policy was explicit: "The guiding principle in relation to dismissal among agricultural experts is that since we are in such a sorry shape, we should only lay off open enemies. With experts we must examine whether they are individually honest, whether they work well, and don't work against the Party's policy."[2]

Comparable sentiments were voiced by the head of the party's publishing house, Spark (*Szikra*), in 1952. "We don't need to get rid of every unreliable individual from Szikra [Press], only those individuals who are useless because they are untrustworthy."[3]

Two countervailing forces were at odds: ideological inflexibility, often matched with excessive zeal, and the pressing need for qualified staff to implement party/state policies. The clash of these tendencies made the fate of experts extremely unpredictable. This dilemma haunted every domain and every level of the party/state and became more acute as the economic fortunes of the country worsened over time. Not surprisingly, the problem of expertise was particularly important in research and policy making. Recurring campaigns were waged against idealism (i.e., Western genetics) in biology[4] and the aberrations of social science. In the new world of Stalinist epistemologies, positivism was ridiculed as a pseudoscience and "glorifying facts" as the "pathological symptom" of its "methodological dead end," in contrast to the radical notion of class-aligned relativism (Péteri 1998, 190). It was difficult, however, to be assured of epistemological conversions, as specialists became good at quoting the right passages from the enshrined literature, all the while holding on to their old ideas.[5] Soviet advisors weighed in on the issue of expertise as well, although their concerns focused elsewhere. In their view, research institutes were not being well managed. Greater attention had to be devoted to providing the ministry advice on planning based on a thorough review of the Soviet and Hungarian literature.[6] The disconnect between abstract ideas and practical experience was a deficiency the Soviets found especially worrying.[7] In 1952, the agricultural economics department at Gödöllő came under serious criticism. Nazarcev, head of the Soviet delegation specializing in plant protection, identified serious deficiencies, notably the completely abstract nature of lectures and the absence of any analysis of the practical experience of vanguard agricultural enterprises incorporated into the curriculum.[8]

Problems of expertise also haunted the state farm sector. The public face of agricultural modernization depicted the advantages of large-scale production, enhanced by the innovations adopted from the far superior socialist agriculture of the Soviet Union. In a speech delivered at a meeting of party secretaries and agronomists from cooperative farms in Zala County, one official claimed that large-scale production under socialism was the ideal place for the creative juices of experts to flow, in stark contrast to the hidebound world of manorial estates. "Why was there need for so few experts in the previous system? It's because the large landowner or renter only went after profit. They only employed a few experts, and they wanted to turn them into their own instrument in the interests of exploiting and oppressing workers. Honorable professionals could not have felt comfortable in such a situation. . . . In the manorial estate system, workers despised the expert whom the landowner or renter respected."[9] Small farms under socialism also set boundaries to the creative powers of agricultural specialists. State and cooperative farms were the proper workplace for progressive experts. The speaker continued: "Soviet dialectical materialist science gives agronomists a free hand."[10] In fact, dialectical materialism was revolutionary. "Soviet dialectical materialist science arms the expert with a theory that assumes that it is possible, indeed necessary, to change nature."[11]

The private face of collective agriculture managed by the state portrayed a different story. The predominance of manorial estate personnel at state farms and machine stations was constantly being criticized by government and party officials. A report detailing the state of affairs in 1950 at the two most prestigious state farms, Bábolna and Mezőhegyes, found the two farms performing far below expectations, no doubt a result of sabotage by class enemies. "The leading experts are under the influence of clerical reactionaries; the hand of the enemy is evident if one surveys work organization."[12] In the absence of adequate personnel with proper political commitments, the final solution was to continue employing those who held less extreme views and train a new generation alongside them to prepare for the future. This did not always bring the desired result. Serious tensions marked relationships between "long-established experts" and new cadres, fed by a very strong anti-intellectual bias among many Communist Party members. Similar prejudices were faced by agronomists at state farms and cooperatives who had to fight hard to be respected by the locals. Pleas to raise the salaries of agricultural experts often fell on deaf ears. As a general rule, villagers were also inclined to dismiss the value of schooling beyond an elementary education, especially when it came to knowledge in farming. Agronomists familiar with agrarian economics, defined here as "the economic and fiscal employees of the cooperative farm," were being sent on errands not related to their work, for example, buying up potatoes or overseeing the weekly market. In actual fact, agricultural economics experts were expected to teach courses, review regional planning strategies, and keep management up to date on policies, in addition to their daily tasks of overseeing bookkeeping and the fiscal affairs of the cooperative farm. To convince villagers of the value of having an agronomist on hand, officials in the county agricultural office had to issue a circular explaining just what the agronomist's responsibilites actually were, ending with a caution from Erdei that neglecting this announcement would bring swift punishment.[13] Villagers' lack of respect for experts, especially those who had worked in capitalism, was validated by the tendency of some former manorial stewards to be slow in discarding old habits acquired at manorial estates: a harsh tone, dismissive comments, impatience, and intolerance. Ironically, it was precisely this high-handed managerial tone that would be adopted by many party and government officials to demonstrate their elevated status under the new regime.

Negotiating the line between party vigilance and economic performance could cause serious problems for conscientious managers when they ran up against overly zealous party officials. Facing a problem with local precinct officials, the manager of a state farm in Tolna County wrote directly to the Minister of State Farms, András Hegedüs, in April 1953.[14] At the end of February, the party committee of Paks (in Tolna County) issued a directive requiring the state farm fire all class enemies within two weeks. This caused a whole cascade of problems for the manager. Dismissing conscientious and qualified personnel would have serious negative consequences for production. A number of well-qualified and dedicated specialists would fall under this categorization, a point the state farm manager illustrated with extensive descriptions of their talents. For example, "there's the head agronomist, X. Y., who has worked at the farm since 1949 as the managing agronomist. He has always performed his work excellently and conscientiously. The forementioned comes from a peasant family, his parents were manorial workers. . . . The only objection against him is that he was a manorial steward in the past."[15] The state farm manager's first line of defense was to argue that he was not authorized to fire specialists. "The answer I got to this from Comrade X. Y. was that alongside every professional I should put an honest worker, who in two weeks can master the expert's work. Then the expert should be dismissed."[16] Adding to his misery, the state farm manager was unable to rely on any local political support to fight his cause because the local party secretary had been expelled for immoral conduct. Because he resisted the party official's directive, the farm manager was reprimanded, and he feared, as he explained in his letter to Hegedüs, that he would soon be kicked out of the party too if he didn't meet the next deadline set by the precinct officials. To his relief, Hegedüs came to his aid by ordering a review of the facts. Scolding local officials for their inappropriate behavior, Hegedüs explained: "One can't do this kind of work rashly and by fits and starts."[17]

Problems also arose when budgetary constraints conflicted with perplexing legal formalities. One anecdote is particularly memorable. Disciplinary measures had been initiated against a farm manager accused of being in cahoots with a kulak. Even though it had been six weeks since the decision had been taken to fire him, the farm manager had still not received the manager's dismissal notice from the ministry, leaving the former employee on the farm's books even though the accused was already in prison. An official in the secretariat's office of the ministry refused to act until he had received the court's verdict. This frustrated the county official to no end, since he was unable to pay the new employee who had stepped into the position. "It seems to me that the ministry has the right to determine someone's guilt, at which time it is entitled to investigate the case. It shouldn't be placing further difficulties in the farm's way. We have already dispatched a new farm manager, but the wage fund only makes it possible to pay one farm manager [at a time]."[18] In addition to showing how difficult it was to find qualified personnel at the time, this example also demonstrates the degree to which bureaucratic practices were becoming increasingly formalized and rule bound, as the governing apparatus solidified.

The lack of respect shown specialists led to a serious "brain drain" in the agricultural sector, exacerbating the already dire situation of recruiting competent personnel. This needed to be rectified. When acquainting the nation with the changes he would initiate in policy in a July 4, 1953, radio speech, Nagy devoted special attention to this problem.

> Unfortunately, it still happens—though the government has resolved to alter this radically—that intellectual work, and generally the intelligentsia, especially the old intelligentsia, isn't accorded proper respect. Often they are surrounded by a climate of distrust, the end result of which is neglect, at the same time when in every area of our life—economic, cultural, scientific—we have a shortage of experienced, qualified intellectuals. The socialist development of the people's economy has created unusually broad possibilities for their activities. Despite this—on the basis of distrust or in the course of groundless purges—more than once well-intentioned intellectuals have been treated in a way unworthy of the People's Democracy. They were deprived of the opportunity to employ their knowledge in their speciality for the good of the country. The government will forcibly put an end to these kinds of improper and impermissible procedures, and it is [the government's] firm resolve to redress these grievances. (Rainer 1996, 535–36)

Nagy's admonitions about treating experts with respect, and using their knowledge to build socialism, were heartfelt, but changes were slow in coming.

Institutional changes required a shift in priorities. Agencies across the board began with a thorough review of policies in place, after which evaluations of personnel needs and budget requirements were decided. Within a year, many prestigious members of the old guard were rescued from obscurity, or from imprisonment, to reform party/state policies, including among them a number of prominent figures in agrarian work science during the interwar period. For example, in 1949 the prominent agrarian work scientist Reichenbach (now known as Nagypataki) had been forcibly retired from the university, but he got by working as a consultant at the Ministry's Agricultural Institute and the Stockbreeding Institute. After 1953 he joined the new Economics Research Institute.[19] Ihrig, who at the end of the war was director of the Institute for Agricultural Policy (Mezőgazdaságpolitikai Intézet) and the Dunavölgy Agronomics Research Institute (Dunavölgyi Agrártudományos Kutatóintézet), lost his job when both of these institutions were abolished. He found piece work at the National Agricultural Library as a translator of technical literature. In 1954 he was called back to professional scientific work at the Institute for Agricultural Economics and then to the Economics Institute of the Academy. István Heller, a strong proponent of manorial agriculture and founder of the Agricultural Work Science Institute in 1943, was also sent into retirement in 1949, but by 1955 he managed the Work Studies Department of the State Farm Research Institute.[20] Only Witthen, who was a steering committee member of OMGE and had been employed at the MGI before the war, had managed to keep his job at the Ministry of Agriculture throughout the entire period. Table 8.1 illustrates the professional qualifications of employees in leadership positions at the Ministry of Agriculture on May 14, 1954. Professional expertise in fields related to agriculture was rare among the high-level cadres. This may not have been important, since the day-to-day work of crafting policy was handled by mid-level managers, a good proportion of whom had been trained or had worked in agriculture. Along with their prominent colleagues who had been rehabilitated, they would contribute substantially to revising agricultural policies.

Owing to the poor state of the economy, there was little money to devote to organizational restructuring, and ever-present concerns about rationalizing government (i.e., limiting the expansion of government agencies and offices) substantially limited the latitude agencies enjoyed. Moreover, there was little incentive for officials to adopt new behaviors. Accordingly, the ability of agencies to follow the new course effectively was severely constrained. As a case in point, I quote from a letter sent to Imre Nagy from the president and secretary of the Academy of Sciences in the fall of 1953. In the face of ever-increasing demands to provide advice and conduct research in support of policy changes, the academy found itself overwhelmed with work.

> Various state organs—the National Planning Office, the Ministry of Finance, the National Workforce Committee, etc.—approach questions of scientific research with the attitude and methods characterizing state administration or

TABLE 8.1. Information on leading and mid-level cadres in the Ministry of Agriculture[1] (May 14, 1954)

LEADING CADRES	Age	Original occupation	Party membership, date when joined	Professional or vocational training	School	Communist Party school attendance[2]
ministerial deputy director	30	agriculturalist	52 MDP	stockbreeder	postgraduate studies at the Agricultural College	2 years in corr.
ministerial deputy	40	worker	45 MKP	auto mechanic	3 years of middle school	1 year
ministerial deputy	38	worker	43 SzDP, 45 MKP	construction worker		1 year
ministerial deputy	43	worker	45 MKP	steel pourer	Agricultural College corr.	1 year
ministerial deputy	32	gardener	46 MKP	gardener	2 years of gardening courses	1 year
ministerial deputy	34	mechanical engineer	44 MKP	mechanical engineer	Polytechnical University	1 year
director general	45	upholsterer	28 MKP	upholsterer	commercial high school	—
director general	36	teacher	46 SzDP	teacher at agricultural school	Agricultural University	2 years
MID-LEVEL CADRES						
assistant undersecretary.	38	worker	36 MKP	licensed farmer	Agricultural College	—
chief of department	40	lawyer	45 SzDP	lawyer	University of Law	—
chief of department	28	locksmith	45 MDP	locksmith	6 years of elementary school	1 year
chief of department	42	industrial worker	45 MKP	industrial worker	6 years of elementary school	5 months
chief of department	30	civil servant	46 SzDP	economist	Economics University	—
assistant undersecretary	37	blacksmith-locksmith	36 MKP	blacksmith-locksmith	6 years of elementary school	1 year
chief financial officer	34	industrial worker	45 MKP	tool and die maker	4 years of middle school	—
department head	40	civil servant	45 MKP	practiced farmer	Law School	—
department head	28	agronomist	51 MDP	agrarian economist	Agricultural University	—
director	41	plant breeder	46 SzDP	licensed farmer	Agricultural University	—
vice director	53	estate steward	48 MKP	licensed farmer	Agricultural College	—
director	48	stockbreeder inspector	47 MKP	stockbreeder	2 years of vocational school	—
director	34	veterinarian	—	veterinarian	Polytechnical University	—
director	53	agricultural inspector	—	licensed farmer	Agricultural College	—

director	50	agrarian worker	45 MKP	agrarian worker	4 years of elementary school	3 months
director	37	veterinarian	45 SzDP	veterinarian	Agricultural University	
director	31	poultry breeder	45 SzDP	poultry breeder	3 years vocational school	3 months
vice director	69	teacher		veterinarian	Agricultural University	
vice director	53	teacher	45 MKP	licensed agriculturalist	Agricultural University	
vice director	32	agronomist		licensed agriculturalist	Agricultural University	
delegate director	52	agrarian inspector		licensed agriculturalist	Economics University	
director	30	civil servant	50 MDP	licensed agriculturalist	Agricultural College	
director	40	teacher		licensed gardener	Horticultural College	
delegate director	51	estate steward		licensed farmer	Agricultural College	
director	42	agricultural worker	45 MKP	agricultural worker	6 years of elementary school	5 months
department head	41	civil servant	48 MKP	accountant	commercial high school degree	
director	35	forestry engineer	47 SzDP	licensed forestry engineer	Polytechnical University	
director	30	agriculturalist	46 MKP	agriculturalist	agricultural high school certificate	1 year
director	63	civil servant	19 MKP	machine technician	high school graduation certificate	
director	32	gardening inspector	51 MDP	licensed gardener	Horticultural College	
vice director	32	agronomist	45 MKP	licensed agriculturalist	Agricultural University	3 months
vice director	41	mechanical engineer	45 MKP	licensed mechanical engineer	Polytechnical University	
director	27	engineer		licensed mechanical engineer	Polytechnical University	
department head	34	agricultural worker	45 MKP	agricultural worker	7 years of elementary school	6 months
vice director	39	estate steward		licensed farmer	Agricultural University	
vice director	39	estate steward		licensed farmer	Agricultural University	
director	41	estate steward	42 MKP	licensed farmer	Agricultural University	
director	42	estate steward	51 MDP	licensed farmer	Agricultural University	

(*continued*)

TABLE 8.1. (Continued)

LEADING CADRES	Age	Original occupation	Party membership, date when joined	Professional or vocational training	School	Communist Party school attendance
director	37	agrarian teacher	—	licensed farmer	Agricultural College	—
director	48	estate steward	—	licensed farmer	Economics University	—
director	34	estate steward	45 MKP	licensed farmer	agricultural high school certificate	—
director	44	agricultural worker	39 SzDP	agricultural worker	6 years of elementary school	—
director	43	estate steward	—	licensed farmer	Agricultural College	—
director	34	agronomist	46 MKP	licensed farmer	Agricultural University	—
director	30	estate steward	45 SzDP	licensed farmer	2 years of vocational school	—
director	56	industrial worker	15 SzDP	industrial worker	5 years of elementary school	5 months
director	32	gardener	51 MDP	gardener	horticultural high school	—
director	27	gardener	52 MDP	licensed gardener	Agráregyetem	—
vice director	32	forestry engineer	45 MKP	forestry engineer	Polytechnical University	—
director	38	forestry engineer	48 MKP	forestry engineer	Polytechnical University	—
director	42	cabinetmaker	31 MKP	cabinetmaker	4 years of high school	—
director	42	cabinetmaker	45 MKP	cabinetmaker	6 years of elementary school	1 year
director	36	forester	46 MKP	forester	4 years of high school	—

Notes: This table lists the position of leading and mid-level cadres at the Ministry of Agriculture in May 1954. Included is information on their age, original occupation, party membership and date joined, formal education and where acquired, and whether the person attended Communist Party school. Individuals without training in any field of agriculture tend to cluster at the highest levels of the ministry, but most of the personnel have a background in various fields related to agriculture.

MKP   Hungarian Communist Party (Magyar Kommunista Párt) (1944–1948)
MDP   Hungarian Workers' Party (Magyar Dolgozók Pártja) (1948–1956)
SzDP Social Democratic Party (Szociáldemokrata Párt) (1944–1948); as of 1948, allied with MDP

[1] András Hegedüs, MDP-MSZMP 278 f., 74 csop, 31 ő.e., pp. 4–7.
[2] Correspondence school

other spheres of the people's economy. They do not understand sufficiently either the nature of scientific research or the developmental necessities of scientific research and so [don't understand] the proper and sound methods of evaluating requirements. In many cases, in our opinion, the result is inappropriate, mechanical, and therefore damaging. . . . We think that as a consequence of [staff] reductions . . . the Academy finds itself in a situation in which it cannot meet its objectives under the current conditions. In the final analysis, this will lead to a decline in the standards of scientific research.[21]

Reconstituting research budgets was not the only change that accompanied the New Course. Far more significant, as Péteri has demonstrated, was the rejection of Stalinist scholasticism for "naive empiricism," that is, paying attention to facts on the ground as the basis of national policy. This would bring an end to statisticians at the Central Statistical Office being chastised for producing "wrong data," or numbers that did not accord with the party's public proclamation (Péteri 1993, 149). Péteri emphasizes that this epistemological shift was not a turn away from Marxism but a return, as István Friss, a prominent Communist and soon head of the new Economics Research Institute, conceived of it, to the strengths of a scientifically grounded Marxist economics (Péteri 1997, 300). Describing the research program he initiated at the new research institute for economics, Friss explained:

> *We strove conscientiously to gather facts, possibly all the facts relating to the various phenomena, and to study these [facts] exhaustively considering all their possible connections in order to be able to come to more and more exact inferences concerning the inherent connections, regularities, movements, and conditions of development of the phenomena and processes.* We did our best to consider everything that had been written about the phenomena under study (or about phenomena related to them) by researchers (especially Marxist researchers) before us. But *we have never regarded anyone's statements as sacred, [especially not] if they weren't confirmed by carefully made factual observations.* (Péteri 1997, 300; italics in the original)

The insistence on turning to the analysis of "actually existing socialism" was fueled by more than a simple respect for scientific legitimacy; many believed that having created policy in the absence of empirical analysis was a crucial reason for the failure of the regime's economic vision. In a speech given to the Academy of Sciences in June of 1954, Nagy said as much. "[Nagy] went so far as to identify the backwardness of economic research as the very root of 'mistakes committed in the field of economic policy'" (Péteri 1993, 153).

A significant consequence of the new approach to research was the founding of the Institute of Economics (Közgazdaságtudományi Intézet) in 1955 under Friss's supervision. This replaced the failed Institute for Economic Science (Közgazdaság-tudományi Intézet, or KTI), which once had been touted as the avant-garde of socialist economics.[22] KTI had never found a full-time director and was permanently understaffed. The new institute, on the other hand, built on the strengths of two valuable groups: formerly outcast experts and a new generation of young economists who had been trained at the university since 1949. Friss could now proudly proclaim on the 26th of September 1955, "The fact is . . . that for the first time organized Marxist economic research is being conducted in our country, which portends well."[23] The institute would be at the center of reform economics and in 1956 would contribute increasingly critical voices to the public debates over economics. It would also be the site where the work unit system was once again submitted to serious scrutiny, a point I will return to later in the chapter. Needless to say, the institute's commitment to critical scrutiny of party policies, and its tendency to attract former political prisoners, never endeared it to the conservative establishment within the party.[24]

Correcting Work Units

By 1951 it became increasingly evident that the work unit system had serious problems. The system was poorly structured: "the members' share was based on monetary accounting and in many instances the allocation in kind was unfair, not in proportion to the work completed."[25] In response, the Council of Ministers drafted a regulation altering the way that income was distributed at cooperative farms. The Council of Ministers' first recommendation was to adopt "that well tested system, which in the cooperatives of the Soviet Union they have employed for many long years in the interest of strengthening cooperatives, remunerating good work fairly, and so constantly increasing the members' welfare."[26] The Organization Institute (see chapter 6) in the Ministry of Agriculture was charged in mid-June with reviewing the work unit system. By fall, a series of proposals were submitted to guide revision.

The report submitted to the party described the current work unit system as baffling, incomplete, repetitive, and too rigid.[27] Serious discrepancies in the value of tasks between branches of production led to workers rejecting some kinds of work outright.[28] Values for machine-aided tasks or new agrotechnical innovations were missing from the work unit book. Allowing cooperatives to adjust norms up or down 15 percent to accommodate local conditions had not given farms enough leeway, evident in the common practice of farms changing the output of specific tasks by 40–50 percent above or below the stipulated figure in the work unit book.[29] And as anticipated, the complicated decimal system calibrating the values of tasks became an impediment. "They can't count with work units. They aren't clear about decimals, so that many of them set a higher value on 0.16 than on 0.5. We have taken steps to clarify this, but it would be necessary to deal with this within the framework of winter courses."[30] Most problematic of all was that the system was not conducive to advancing the principle of socialist distribution. In its current form, the connection between effort and reward was obscured: "they are not inclined to evaluate various tasks on the basis of the energy invested and the expertise needed for this."[31]

Simplification was in order, and the answer was close to hand. First and foremost, the scale used in the original work unit system in 1949 was to be abandoned, replaced by the Soviet practice of a seven-stage scale calibrated by 0.25 units, from 0.75 to 2.25. The introduction to the work unit book was now to include a clear explanation of the basic principle of socialist distribution: a metric rewarding effort calibrated according to physical difficulty and skill. New procedures for altering work unit values were initiated, although they still placed significant barriers on local autonomy. (The new regulation permitted changes to be made in the course of writing up the coming year's business plan, to be approved by the farm membership and precinct officials; except under exceptional circumstances, no changes could be made in the course of the agricultural season.) The average daily output values of tasks could not be increased, except in order to rectify discrepancies in relative values identified in the structure. Anticipating the need to augment the work unit book to keep pace with agrotechnical developments, ongoing field research and data analysis were to be incorporated during revisions.

The inherent problems of the work unit system were also a topic of discussion in the corridors of the university, carefully described in the 1952 lecture notes on agrarian economics from the Economics University.[32] The Hungarian work unit system was systematically compared with the Soviet wage form, and in several respects found wanting. Surprisingly, a crucial difference between the Soviet and Hungarian system was the degree of freedom allowed farm members in Soviet *kolhoz* (cooperative farms). "The application of our work unit system differs from the work unit system of the kolhoz. In the Soviet Union the norms suggested by the government and their work unit values are only standard indices, which the cooperative assembly adopts to local conditions. [In Hungary] it is only possible to adopt norms to local conditions; the value of work units is established for the entire country by the work unit book published by the Ministry of Agriculture."[33] A more glaring problem was the discrepancy between the quantity and quality of work performed. The Soviet system made allowances for rewarding the quality of work, above and beyond the amount completed. "*A significant step in the development of the work unit,*" the report concluded, "*is the introduction of the supplementary work units.*"[34] In the current situation, workers could be rewarded for doing sloppy work, while their more industrious colleagues would be punished for taking more time to do their job. The only way to ensure that quality and quantity be connected was to build it in to the incentive structure. So, for example, if the brigade "that, as a result of good work, has harvested more than the average, or more than stipulated in the plan, or in animal husbandry they reached higher yields, *then they obtain a 1 percent supplementary work unit for each percentage of overfulfillment.*"[35] The additional payment would then be distributed to the members of the brigade proportionately according to the number of work units each person had fulfilled. The reverse was also true. Workers would have work units subtracted from the sum they had accumulated to date if they did not meet plan goals (although the deduction could not exceed 25 percent of the total number of work units). This solution was expected to be a panacea for all sorts of problems besetting wage policy. "It prevents '*egyenlősdi*,' makes the use of work units more economical, and in this way increases the value of the work unit. It drives overproduction of the plan and it focuses the individual interest of cooperative members on increasing work productivity to a large degree."[36] Introducing this kind of bonus system was suggested as a solution to this problem, though with the bonuses rewarded in kind rather than in money. In principle, this might have been a good idea, but since there were virtually no records kept at cooperative farms on the amount of work completed, it was nigh impossible to figure out whether in fact workers had exceeded average yields in any particular field.

LEARNING THE BOOKS

In the first years of cooperative farming, there was widespread indifference to bookkeeping and written accounts. Keeping records was considered a waste of time by farmers. "All the writing and administrative work is considered an unnecessary burden."[37] County officials impressed upon cooperative presidents the importance of bookkeeping. "Good bookkeeping is an indispensable condition of proper management."[38] It was common in county records to see references to problems with cooperative farm managers forcing office personnel to work in the fields, rather than permitting them to compile information on farm accounts. In other cases, cooperative farm members refused to pay administrative personnel for their work in the office.[39] There was no sense that such records benefited the farm or its members. When accountants were forced to participate in the harvest, they lagged behind with their books. Without proper assistance they found themselves in a pickle. "[Having to work in the fields] discourages bookkeepers from bookkeeping, especially in situations when they have to enter a more difficult entry, they don't find a solution and simply don't enter it."[40] Fed up with poor working conditions and tired of farm members' expectations that they work without pay, bookkeepers (very often women) regularly threatened to quit their job. Domestic politics also played a role here. With their wives being treated rudely and not being paid on time, husbands complained; it didn't help that their wives weren't getting their housework done.[41]

Party and government officials had to contend with the frustrating experience that cooperative management—presidents, bookkeepers, and even pursers—were often unable to settle accounts.[42] "By and large the calculations and appraisals are satisfactory. On the other hand, we found mistakes in the course of summing up and carrying the balance forward."[43] Or in other cases, bookkeepers began to keep records current on livestock but continued to do a poor job of keeping track of fixed assets and rough fodder.[44]

County officials regularly visited rural communities to provide training in bookkeeping. Competitions among accountants, comparable to labor competitions, were promoted across the country as a means of motivating workers to adopt new skills. In addition to offering classes at village level, the party/state also established schools to teach bookkeeping. Cooperative farm presidents and bookkeepers were sent off for six- to eight-week training seminars; villagers identified as having the potential skills to be a good bookkeeper were also dispatched to training sessions. Unfortunately, the scarcity of qualified accountants made them valuable assets to be poached by other government agencies, as the following complaint makes clear.

> Although undoubtedly agriculture must provide a significant number of cadres to industry and administration, this is not compatible with the cadre policy pursued to date by the OSZH [*Országos Szövetkezeti Hitelintézet*, National Cooperative Credit Bank] . . . when they had been attempting to lure away bookkeepers—bookkeepers who had been trained to work in cooperative farms—with a variety of promises. The [not politically enlightened] workers at our cooperative farms easily forget how much the cooperative needs them. They weren't sent to bookkeeping courses so that afterwards they could end up behind a desk at the OSZH. Although this practice of enticing cadres away is waning, it nonetheless resulted in the fact that of the approximately 400 cadres we trained as bookkeepers to date, only about 190 work at cooperative farms.[45]

In 1955 Zala County was mounting monthly "bookkeeping days"; 80 percent of the county's accountants attended, appreciating the fact that the meetings had been moved from the district offices to cooperative farms "because they learn much more when they see problems solved in practice, e.g., having books agree, etc."[46] Ironically, many times cooperative farms faced problems with their books because the one person knowledgeable enough to solve the problem was away at school.

It is important to emphasize that the problem here was not one of numerical illiteracy, nor a lack of experience with money. Performing complex

FIGURE 8.1. "Education for peace! Let's study arithmetic, geometry, spelling, grammar at the courses on fundamentals. Those who complete the course successfully will receive a fourth grade certificate. Apply at the local town council."

calculations with numbers was in fact a daily affair for agricultural workers. Laborers were in the habit of estimating acreage sown or the volume of a haystack or figuring yields. Poorer peasants had to sell their produce to get by, as did manorial workers, who made their only money by selling livestock at the market. Day labor had long been paid with cash. So the problem the party/state faced was one in which the numerical fluency of one set of tasks did not (and could not) move easily to another context (Lave 1988). In short, the problem with bookkeeping was not its numerical or calculative character; it was the requirement that accounts be recorded on paper within specific formal guidelines set by standardizing bodies, in this case the socialist state bureaucracy. Having the skill to perform calculations alone was insufficient as a condition of proper accounting methods; learning how to be a bookkeeper in practice—not in some abstract world of accountants' dreams—would alter the skills villagers were required to have to work.

Villagers' reluctance to keep written records was complicated by two further problems: unqualified trainers and the absence of proper forms. Bureaucrats employed at the county or district level were not necessarily trained in particular skills, such as accounting, and yet they were expected to take these tasks on as part of their job. While it wasn't hard to determine that including entries in the work unit book on the 30th and 31st of February was an obvious falsehood (*Hajdú-Bihar Néplap*, 8 May 1951, 1), other cases were more complicated. In a report from Hajdú-Bihar County, it was noted that "With three exceptions, the district auditors are incapable of overseeing farm bookkeeping and preparation of the financial plan. Therefore the plans are bad and the management of cooperative farm finances is also bad."[47] This was not only true at the local level. "Many times the skilled auditors at the Ministry of Agriculture are insufficiently informed. As one can determine from auditing the books, the directives are incomplete and rarely is concrete help provided."[48] The situation was even worse when the right materials were not available; documents record the case of a county bookkeeper being criticized for neglecting to provide cooperatives with the proper bookkeeping texts.[49] General confusion also meant that simple problems like not distributing the new forms in time for bookkeepers to work up the yearly report were common. To make matters even worse, planning forms were redesigned, perhaps for good reasons, but that didn't help novice accountants.

> The planning forms delivered were good, simpler and more understandable than last year. However, it is our conclusion that whereas some of our planning workers are able to follow the most developed Soviet plan work, they are

frightened by summing up or multiplication problems of some figures and commit the error of leaving these forms empty. When we explained it to them they understood and completed the work easily. This was particularly so when we suggested they carefully read through and use the study aids.[50]

In a separate incident, the forms sent by banks did not correspond to those the cooperative farms used. "The various bookkeeping notifications sent by banks play a large role in this, since they differ from the regular printed material and unsophisticated bookkeepers can't figure them out."[51] These difficulties, combined with the problems caused by the ongoing confusion over lines of authority within various sectors of the national and local bureaucracy, led to chaos. "Cooperative farms [in Hajdú-Bihar] don't believe the district council any more, since, as the auditor from Polgár complained, [the people at the district] have practically become mailmen. They have no scope of authority. He has been a bookkeeper for two years now, but he has never been so disheartened about his work, because the management of finances at cooperative farms has never been so anarchic."[52]

Confusion about lines of authority within the party/state bureaucracy was paired with discomfort over changing patterns of authority at the farm. New forms of knowledge, like bookkeeping, challenged views of skill in farming, skills that were the source of both authority and prestige. Finally wresting authority as managers of the farm away from aristocrats and wealthy peasants, formerly impoverished members would be reluctant to be tutored in agricultural work by lowly office workers. This was an affront to their social aspirations to become legitimate farmers in the community. In this context, county officials had to explain to cooperative farm managers that they shouldn't be offended if they were given tasks by bookkeepers.[53] The class dimensions of this struggle were further complicated by gender politics, a point I touched on earlier. Bookkeepers were predominantly women in this period, whereas farm managers were men. No doubt this division of labor rested on the simple assumption that women were to be assigned less prestigious jobs within the enterprise. It is also significant that women managed the purse strings in village homes, so their supervision of budgets at home meant their role as accountants at cooperative farms made sense. Yet their authority at home did not travel easily into the halls of management.

Over time, the problem was no longer incompetence or insufficient training. Now discrepancies surfaced as a result of ideological disagreements. Patterns of systematic misrepresentation appeared, primarily concerning disagreements with central authorities over farm earnings. The party/state insisted that the state be paid first from a farm's earnings, leaving members whatever was left. This policy, called the "remainder principle" (*maradék elv*), was strongly criticized by farm members, as they questioned the legitimacy of the state's claim to their surplus, just as Russians had in early days of Soviet *kolhoz* (Lewin 1985). This bred resentment and spawned a variety of strategies to circumvent the state's control over their budget. Two means of subverting the remainder principle were common. The first was a simple refusal to pay anything to the state, distributing all surplus among the membership.[54] The second strategy was to sell produce or livestock on the side and distribute the monies among the membership without recording the transaction in official ledgers. Of course, cases of personally motivated embezzlement or malfeasance were also common. For example, in one egregious case of mismanagement in Zala County, not only were the books in poor order; the cooperative president spent money recklessly, emptying the farm's coffers.[55] Attempts by one person or a small group within a cooperative farm to drain the funds were comparable to the ways work unit records were finagled to suit kin or friends within the farm. On the other hand, when the budget was protected by local scheming, the entire membership had often agreed to the strategy. In the final analysis, the issue of well-kept records was about more than state oversight of finances; it was also about fairness. "The annual report demonstrates deficiencies that arise from not following the bylaws, planning out of proportion, and lags in bookkeeping. Auditing plan fulfillment, the fair division of income, and protection of the cooperative's communal property are not guaranteed where accounting is inadequate" (*Hajdú-Bihar Néplap*, 17 Dec. 1954, 5). Having the party talk about fairness may have rung hollow, but disputes between farm members and their former colleagues revolved exactly around this issue.

Class Tensions in the Countryside

The changes proposed for agriculture in the new government program were extensive. Sustained investment in the agricultural sector was planned to offset years of neglect. The simple numeric increase of cooperative farms, without regard for the conditions for farming, was halted, as was the constant increase in lands worked by state farms. The coercive tactics used to drive people into cooperative farms were stopped, to restore freedom of choice. As a result, people were now allowed to leave cooperatives; if the majority of members were in favor of disbanding the farm entirely, this would also be possible, according to the new regulations. Significant debts cooperative farms and private producers had accumulated would be excused: "the various forint indemnities, debts from the older machine station and other debts in kind, and last but not least, certain tax arrears" (Orbán 1972, 134). The requisition system was recalibrated, reduced by 10 to 24 percent, and made permanent, so farmers would no longer be subject to frequent and arbitrary changes in their obligations to the state (ibid., 134). Private farmers were no longer required to enter contracts for goods at the low prices set by the state. "Production contracts must be made more attractive by providing greater numbers of industrial goods" (Balogh 1986, 505). Private farmers were to be encouraged by developing more favorable policies supporting their activities, such as making credit available to finance new construction or renting land currently fallow (Orbán 1972, 134). The services of veterinarians were now free for everyone, and the costs of the machine station services were lowered (Balogh 1986, 505). And finally, and significantly, kulaks were no longer to be harassed; the infamous kulak list in every village and town would be destroyed. Party operatives, on the other hand, were chastised by central authorities. They were now being held responsible for not paying more attention to private farmers, even though they had been following party policy that had been sanctioned only a few months before. A bitter pill to swallow, this about-face in policy forced local officials to repudiate their previous actions. Peasants once jailed for refusing to farm according to new socialist regulations were now lauded as skilled experts.

> Neither the leadership nor the party organizations sufficiently nurture middle peasants. They don't listen to their opinions or advice and they don't take their needs into consideration. . . . At a cooperative farm this spring, for example, the middle peasants were against machines trampling on soft soil because it would spoil the crops and the soil. They knew so from decades of experience. [Officials] listened to none of this, and later it was proven that the middle peasants were right.[56]

Lowering requisition levels and giving cooperative farm members the option of returning to private production was a watershed development in post-Stalinist agrarian policy in Hungary. A precipitous drop in cooperative farm membership ensued, a shift that had serious consequences for class relations in the countryside. In the course of six months, the numbers of families in cooperative farms dropped from 225,000 to 136,000 (Donáth 1977, 142); more than 83 percent of farm members who left had been peasants with small or middle-sized farms (ibid., 155).[57] According to figures from the Central Statistical Office, between June and December of 1953, the numbers of cooperative farms dropped by 13 percent, the number of cooperative farm families by 33 percent, the percentage of field land decreased by 32 percent, and the overall acreage of type III cooperative farms by 30 percent.[58] After Nagy's annoucement on policy changes, the party/state did continue to tout the value of cooperative production but also made strong efforts to court the favor of the "private sector" in their capacity as private farmers. "The most important task is to get the working peasantry to realize that it's now their turn. Make the best of the possibility that has been given by the help of the party and the government. Increase their crops in the fields and intensify the yields from animal husbandry."[59] Private farmers heard the call. Since the new program made it possible for private farmers to rent land, 800,000 kh (456,000 ha) of land that had lain fallow was brought back into cultivation.[60]

Reports prepared for party and government meetings on the state of agriculture in midsummer were somber reading. A particularly dismal picture was painted of state farms and machine stations.[61] There were not enough workers' hostels for employees and even fewer accommodations for families living at the farm. Women were often not given separate sleeping accommodations or lavatories.[62] "In most places the kitchens and cafeterias are unhealthy. They have dirt floors, it's pretty much impossible to keep them clean, and they are in the immediate vicinity of stables and other farm buildings. Workers can't protect themselves from the harmful bugs and flies that endanger their health."[63] There were not enough linens or spoons, and people's clothing was stolen from the few cupboards available. Sick workers were forced to walk great distances to find a doctor, for lack of medical services. Workers complained about always receiving their pay late and were only given money, nothing in kind. This seriously hampered their ability to provide for their families. "In many places we ignored the principle of material interest when developing the wage systems for agricultural workers. We moved too quickly from a system of payment in kind to a system of monetary wages. We disregarded the fact that before the liberation [by Soviet forces in 1945] agricultural workers had received a significant portion of their wages in kind for decades."[64] It was not surprising, therefore, that villagers did not seek employment at state farms. Looking back on this time in later years, Hegedüs, who had been minister of state farms, described the situation darkly. "The most valuable manpower . . . gave wide berth to the state farms, where the increasingly larger scale employment of prisoners and the manpower of the population displaced from their homes [i.e., class enemies sent down to the countryside] reduced the labor shortage" (Estók et al. 2003, 297).

Cooperative farms, presumably the darlings of the regime, were fairing poorly as well. In part this was also due to labor problems. The reluctant migration of private farmers into cooperative production often meant little more than surrendering land to collective production. Elderly family members joined the cooperative; younger folks sought work in industry, which guaranteed a more stable income. In other words, private farmers had downsized, and diversified the occupations of their kin. The energies expended by newer members were limited almost exclusively to working house plots allocated by the cooperative (*háztáji*), leaving the larger field work to those who had willingly joined the farm in earlier times. The bylaws meant little to latecomers. In the absence of cooperative farm barns, it was common for members to keep livestock in their own barns. This inclined them, as officials saw it, to continue to treat the animals as private property, using them to earn money privately by hauling goods. Some members increased the amount of acreage they were tending to privately by renting land laying fallow in neighboring communities, occasionally doing so in their own name or a relative's. They then proceeded to use the cooperative's draft animals housed in their barns to work the land they farmed on their own.[65] All this was indicative of an inability to identify with the cooperative farm as a collective body. "He didn't think of the cooperative farm as his own . . . he felt as if he were a servant at the cooperative. They give him orders and it is precisely for this reason he didn't try to finish the tasks that had been allocated to him."[66]

The poor performance of cooperative farms forced state agencies to offer farms credit to hold them over from season to season. But the minister of agriculture could not simply direct funds to this purpose without going through the proper channels, which in this case meant appealing to Gerő, who was head of the People's Economic Council. Explaining that cooperatives across the country were facing insufficient funds to carry families into the next agricultural season, which was even the case at so-called "model farms" (*mintaszövetkezet*), Erdei asked for permission to use funds intended to pay for produce on contract with the state a month earlier than usual. "At most cooperative farms the members have not received any cash since last fall, and in most cooperatives there weren't enough shares in kind distributed at the annual meeting to ensure the members' provisions until the next year."[67] Erdei was given a cold reception. In those instances in which cooperatives did gain access to state credit, there were strings attached. In April of 1953, a cooperative farm in Győr-Moson-Sopron County submitted a request for credit. Members were in dire straits, without money to buy bread, a situation caused by having run out of flour earlier. The county replied that it was only permitted to provide credit to those farms that had fulfilled all their obligations to the state; the only solution was to advance the farm money that would be given for the produce the farm had contracted with state authorities to deliver later that year.[68] Of course, having problems ensuring income for its members was precisely the reason many farms did not pay the state what it considered its due. And in light of the higher incomes of their neighbors in the village, doing without was even harder.

> It was difficult because there was little cash to disperse right at the time when individually farming working peasants got money and went shopping. Moreover, in many cooperative farms, as a consequence of the weakness of the leadership and the demagoguery of those who took little part in the work, important measures could not be complied with, such as the distribution of grains and feed according to work units or the restriction on the distribution of animal products.[69]

The indifference of newer members to the cooperative farm did not endear them to members who had joined the farm in its early years. The founders had sincerely hoped to make good of collective farming, even though their fortunes had fallen short of expectations. Of course, the decision to abandon the group was not motivated solely by animus to collective production. After all, private farmers had always had a higher income than cooperative farmers. The clash of attitudes betrayed fissures that had long existed between poorer and wealthier members of agrarian communities, tensions that erupted to the surface with the rapid exodus of newer farm members following Nagy's policy changes. Ironically, Nagy's loosening of state strictures opened the way for long-standing animosities between the two groups to grow, sowing the seeds of the kind of class warfare the party/state had tried to foster in earlier years. Now, however, with the economy in such a shambles, the party/state couldn't simply side with cooperative farmers against private producers, as they now formed the backbone of agricultural production. This was a delicate balance. "Within the group rely steadfastly on the former poor peasantry, but at the same time we must fight so that that the older members treat the middle peasants as equal members in the cooperative farm."[70]

There had already been tensions brewing between poorer and wealthier farm members. Before the land reform wealthier peasants had often criticized their poorer brethen for being lazy or profligate, seeing their own comfort as evidence of hardwork and integrity. The poor, on the other hand, bristled at the idea that they were lazy or that their poverty was a result of irresponsible behavior. They may not have been working diligently on land they had inherited or had bought themselves; they worked hard on land owned by others. These discrepant views of work and moral investment manifested themselves in very different attitudes toward the value of land or tools brought to the cooperative farm, that is, what constituted capital in production. As Donáth explained, understanding land to be a form of capital would have meant "actual recognition of individual ownership, that a portion of income would be divided up in the cooperative farm not based on work alone but on the amount of land brought in. Former agrarian proletarians not only had a grievance with this land annuity. In many cooperatives, where [proletarians] were a majority in the leadership, they didn't even pay it" (1977, 143). In a similar vein, the attitude toward the use of household plots also divided wealthier and poorer peasants. Pushing the boundaries of type III farm bylaws, former landowners had done their best in earlier years to expand household plot production by making use of jointly owned equipment, draft power, and produce on their family allotment. In other words, they were treating farm assets as if the cooperative farming contract only extended to the joint use of tools. Poorer farm members had far fewer livestock, and virtually no farm buildings, making it much more difficult for them to take full advantage of household production on their own (ibid.). When the wave of "leave-takers" inundated cooperative farms, the farm members who had not contributed property took their revenge by not returning the land wealthier peasants had brought to the farm, or giving them clearly inferior plots instead. Nor would they turn over equipment and livestock, especially horses (Orbán 1972, 133). The most consistent complaint submitted to local authorities concerned the refusal of cooperative farm members to reimburse those who had left the farm for the work they had completed during the spring and early summer. Newly private farmers considered it their due for labor performed, while farm members still working collectively interpreted their withdrawal as a rejection of the cooperative enterprise entirely. "An example of the mistrust toward the middle peasant is the opinion of X. Y., an old construction worker who said that one can't educate the middle peasant. Only the grave will persuade him (*majd csak a sir agitálja meg*)."[71]

With the shift in government policy giving greater voice and support to private farmers, loyal cooperative farm members grew concerned that their honest efforts at building a socialist farm were being repudiated by the party/state. Their concerns were not without foundation. "We consider it especially important to notice that since the announcement of the new government program, the measures to raise the living standard of working peasants have increased. In comparison real income [has increased] the most for wealthier middle peasants and the least for poor peasants."[72] In November, the concerns about alienating poorer villagers were a topic of conversation at the national meeting of the Communist Party's Agricultural and Cooperative Department devoted to the topic of the (so-called) worker-peasant alliance. "Our experiences show that the large mass of agrarian proletarians and poor peasants did not and do not understand our party's peasant policy. I'm thinking here primarily about the positions taken on middle peasants. (For example, treating the middle peasant as a kulak, the friction between middle and poor peasants in the cooperatives, the disparagement of the middle peasant, etc.)."[73] Cooperative farm members were not entirely abandoned by the party/state, judging by rallying cries published in the newspaper. "[L]ast fall enemy forces tried to tear us apart. The honor of socialist agriculture is at stake, comrades!" (*Zala Néplap*, 30 Mar. 1954, 3). Sympathetic words, no doubt, but there was a catch. "Therefore we are all honor bound, but especially you ... members of the cooperative farm, to fight and struggle, protect [socialist agriculture] and expand it. Triumph will prevail in everyday productive labor" (ibid.).

Administering Conflict

The flood of complaints filed at county offices in the months following the proclamation of the New Course spoke to more than just class tensions in the village. They also underscored the problems caused by the blatant disregard of paperwork that had plagued the introduction of planning and work units at cooperative farms. The grand vision of modernizing the economy through planning rested on a complex infrastructure of accountability and transparency: standardized accounting techniques, carefully prepared business plans, and regular practices of oversight and review. As we saw earlier, none of the structures intended to perform these goals were in place in the early 1950s. Party committees, such as they existed in villages, were few and far between. The administrative capacities of district and county offices were extremely limited, and the talents and patience of staff stretched far too thin to guarantee results. Without a local arm of the party in place, it was impossible to enforce oversight; there was no heavy hand forcing compliance. In 1949 the Ministry of Agriculture mandated that county agencies conduct audits of all farm records, but this was impossible to ensure.[74] In 1953, the ability to fulfill the ministry's expectations was still falling short of expectations, as an article from the county newspaper in Hajdú-Bihar attests. "In the interest of effective work, fleeting oversight must cease. A 5–10-minute visit is useless. They won't get to know the local problems and cannot provide practical assistance" (*Hajdú-Bihar Néplap*, 9 Sept. 1953). So there was a convergence between the party/state's push to ensure accounting procedures be followed and be transparent—obviously to ensure the government capture its due—and the insistence on the part of former cooperative farm members that they had been deprived of their rightful income. In other words, complaints swirled around records everyone had been to that date resisting; written documents were now the means to resolve disputes.

The most common allegation was related to the redistribution of assets.[75] The situation was extremely confused. Charges flew in all directions. County and district authorities were instructed to get up to speed with the regulations, to ensure that they implemented them properly.[76] Indeed, in February 1954, the Cooperative Department of the Ministry of Agriculture prepared a circular describing common complaints being submitted and differentiating between those that were legitimate and those that were not. It was hoped that the clarification would speed up resolution of the disagreements. The ministry expected every case to be personally attended to by the village council staff, listening to the complainant and cooperative farm management's sides of the story. The result of the dispute had to be communicated to the parties in writing.[77] It was nigh impossible for the central authorities to adjudicate these claims, however, since final resolution was left to the parties involved.

Petitions to disband were being falsified and records at farms were in horrible shape, lending credence to claims that the amounts available to reimburse former members had been misrepresented. It was common that people leaving cooperative farms demanded a portion of the farm's assets, even though the remaining members swore that those leaving had not contributed anything to the farm when they joined, taking the position that everything owned by the farm had been acquired collectively.[78] Since members who were leaving the cooperative were required to bear their portion of the debts the farm had accumulated, serious disagreements arose over just how much the member wanting to leave still owed the cooperative. Former members demanded to be paid their portion of the farm's income when they themselves had not yet cleared their debt obligation. On the other hand, cooperative farms were found guilty if they held onto any valuables brought to the cooperative farm when members first joined. One discriminatory strategy used against those leaving was to charge them a ridiculously high market rate for livestock and equipment they claimed to have contributed, or else the cooperative chose not to subtract the debt still owed the member from the initial acquisition of the livestock or tools when the member first joined.[79] Another permutation of this problem was for former members to demand they be given livestock and equipment that were indispensable for the cooperative farm, which was interpreted as a move to dismantle the wealth (*vagyon*) of the farm itself and so was considered a hostile act.[80]

The issue was more complicated than simple greed or revenge. Since the New Phase began in the middle of the summer, when a substantial amount of work to bring in the harvests remained to be done, figuring out how much someone who had left the farm deserved was extremely difficult. (Moreover, people left at different times in the course of the summer and fall.) Type III farm bylaws stipulated that a certain number of work units had to be earned in the course of the year for the farm member to be reimbursed for his/her labor. The measure was clearly intended to force farm members to work collectively if they wished to earn any income from the farm. By this logic, many of the cooperative farm members who had left had not reached the level justifying compensation, so one could easily deny them recompense, a justification that authorities thought did not hold water.[81]

> Some working peasants rightly complain that *the cooperative farm doesn't distribute the shares—in kind or in cash—that are due from the work units completed even if* the peasant in question doesn't owe the cooperative anything, or the debt is much smaller than the value of his share that remains at the cooperative. In many places the working peasants who have left the farm are willing to pay their legitimate debts that are still outstanding, but in these situations the cooperative farm is still reluctant to distribute the produce in kind being held by the farm, wishing to settle their debt for work units in cash.[82]

Recognizing the problems brewing as early as July 1953, the Central Committee stepped in. "By the end of July, the Ministry of Agriculture should figure out the method of distributing income (advance, disbursement, etc.) that best guarantees the members' maximum participation in the work in the fall. In particular, one must determine the date at which sufficient work units should have been acquired to constitute the basis of income distribution when year-end incomes are settled."[83] Another way in which the value of goods was contested concerns a case in which those leaving the farm would be given the produce being grown on lands they had recovered. "They didn't want to start harvesting the potatoes, but I told them, anyone who doesn't dig up potatoes won't get any."[84] Some former farm members insisted that they be given their original plots, even those that had now been incorporated into collective fields of the state farm or cooperative. These complaints were not honored, as they were seen to be a direct challenge to the institution of the cooperative farm itself.[85]

In the new situation, bookkeeping and records of farm activities became extremely important when complaints were being adjudicated. The issue of oversight loomed ever larger on the horizon. From the State Auditing Office to the Hungarian National Bank to the lowest level enterprise, much greater attention had to be paid to accounting procedures and the rules overseeing cooperative finances. In a report Hegedüs compiled for the Central Committee in July 1953, the issue of finances was top of the list of problems in agriculture. It was imperative to consolidate the cooperative farms still in operation.

> By November 30 every cooperative farm must be examined financially. How much of its debt needs to be paid off this year must be determined, as well as what the repayable sum will be from year to year after restructuring the debt in future years. . . . Next year's investment plan for distributing income must be worked out in every cooperative farm by the general meeting, and the source of the necessary means for investment must be clarified.[86]

Just how this was to be achieved when farm members were barely recording expenses consistently was left unsaid. Nor was the problem of having this substantial task imposed on cooperative farms in the busiest period in the agricultural season addressed. In more general terms, the investment in human capital for the state was enormous. As an agronomist explained, "While preparing the annual report it was established that there had been work unit inflation at many cooperatives. To prevent this, a method has been suggested, which is already in use in many districts, that an independent inspector be put in place for recording work units. This is the single means by which work unit inflation will not occur in the future."[87] It was not only cooperative personnel who were being subjected to serious oversight. The head of the Agricultural Department of the Executive Committee of the Hajdú-Bihar Communist Party announced an immediate policy change. All employees in the office would now be required to keep a work book, recording every day what kind of work was done and where.[88]

Working in Brigades

Recording and rewarding work units was at the center of disputes in 1953–54. An equally contentious battle shadowed the work unit question from the very first days of collectivization: the proper organization of work at the farm. The founding principles of type III cooperative farms stipulated that members be distributed into brigades, which could also be subdivided into work groups. A truly modernized farm required a stable and efficient labor force, a point that was explicit in policy discussions but remained implicit outside these circles. For example, in a committee meeting of the Agricultural and Cooperative Department of the Communist Party on December 2, 1948, the difference between working in groups and working collectively was hammered home in the deliberations over the bylaws of type III cooperatives. "In regard to group tasks the rules don't say that the group must work in a cooperative labor organization, but in a common cooperative firm and collective labor organization."[89] Moreover, the brigade system, which was differently structured in animal husbandry than in grains, was meant to be fixed, with stable membership that long outlasted the agricultural season or even the year.[90]

Two elements crucial to modernized firms were skill and mechanization. The significance of a formal labor structure was voiced by the Ministry of Agriculture's representative at a meeting of Győr county officials involved in cooperative farm affairs. "It is extremely important that in contrast to individual or manorial farming, when everyone was forced to work at the task assigned him, now he can work in an area suited to his abilities, and by this means production increases."[91] The other consideration was to mold the manual labor force to the needs of machines, rather than the other way around. "In the course of the lecture the delegate from the Ministry of Agriculture pointed out that in the cooperative farm the organization of the brigade is the most important, since machines cannot be employed properly without brigades."[92]

The purpose of creating a brigade with a fixed membership structure escaped new farm members. After all, resistance among manorial estate managers before the war to industrial management had been premised on the difficulty of standardizing agricultural jobs in light of the great diversity of tasks and the rapidly shifting character of work performed at the farm. More important was the absolute disinterest among farm members in adopting these organizational rules. An ethic of egalitarianism (*egyenlősdi*) at the cooperative farm—everyone working together as equal partners in production—did not only endanger the individuation of the wage structure at the heart of new socialist wage policies; it also undermined the principle of creating permanent work groups within the larger structure of farm management.

> If we were to divide the firm plan down to individuals, then that would mean that they wouldn't acquire a substantial perspective. There's nothing to force them to overproduce. He knows that he cultivated this completely by himself. What will happen if in the fall he will come forward saying I cultivated it the whole year through, why shouldn't the yield be all mine? So here's the critical thing, it doesn't provide perspective, doesn't motivate cooperative farm members. . . . If they were to divide up the area at the cooperative farm it would endanger work unit calculations, the "egyenlősdi" principle would develop, and the members wouldn't try to get more and more work units.[93]

So alongside the constant struggle to get cooperative farm members to work on collectively owned land, the party/state also fought the tendency to dismiss working collectively. Without embedding individual effort within a group structure, such as the brigade, then the grander purpose of transforming farm members into politically self-conscious socialist citizens would be defeated.

Initially, arguments for brigade structure emphasized the efficiencies to be gained from the group's mobility.[94] With time, however, basic problems of disciplining labor eclipsed efficiency in policy circles, leading the party/state to tie brigades to a specific section of the farm's land.⁹⁵ This led to a lot of confusion, because the difference between working a set piece of land as an individual and as a group was not self-evident.⁹⁶ "It was a mistake for cooperative farms to place the emphasis on creating work teams. As a result, they divided up the work areas improperly, among individuals for a year, so that the same person always went to the individually divided areas. But if another person was behind, then he didn't go over to the other's area to help, but instead he left the work undone and went home."⁹⁷ So working together but having land distributed individually still did not make the grade.

> In comparison to last year, there's been improvement in the area of labor organization. The fields are parcelled out to individuals, but there are still deficiencies. At the cooperatives in Biharnagybajom and Nádudvar [Hajdú-Bihar County], they parceled out land to individuals. They think that they can work well, if they hoe together, it doesn't matter that the land is divided up, they work together. They don't understand the significance of dividing up into parcels, so this must . . . be brought out in the press . . .⁹⁸

On the other hand, brigade organization was not a guarantee of a collective spirit, especially when paired with the socialist training tool of competition. In one instance, brigade members refused to help another brigade with gathering wheat, because they were keen to win the competition by finishing theirs first.⁹⁹

Faced with farm members' complete indifference to brigade organization, the party/state recognized the need to confront the facts on the ground. Sharecropping contracts were considered legitimate for the first time. In 1951, the Communist Party's Agricultural and Cooperative Department was adamantly against any form of sharecropping at type III farms. "It slackens work discipline, destroys the collective spirit, and increases individualism. It is a small-scale form in large-scale farming. It is precisely for this reason that sharecropping must be prevented. We must move in the direction of progressively eliminating individual land division."¹⁰⁰ By 1954, the need to ensure land be cultivated overshadowed matters of principle. In July, the Ministry of Agriculture sent out a list of questions inquiring into the current state of labor and firm organization. Included among these questions was one asking whether sharecropping was being done at the farm, how frequently, and primarily for which types of work and for what reasons.¹⁰¹ A county official in Zala estimated that 95–98 percent of row crops had been divided up among individuals. Dividing up plots of land was very attractive and strongly supported by the farm membership. "The advantage is that the same member completes the tasks of tending the crops from beginning to end and treats them as if they were his own. Individual ability, diligence, and skill are conspicuous. Moreover it encourages the broadening of competition, increased production, and guarantees premiums better."[102] And he continued, "At present, dividing up plots is a guarantee of increased yields. From the economic point of view, it is advantageous. From the political point of view it is detrimental because it resembles farming on small parcels so closely."[103]

It was becoming increasingly clear that sharecropping was widespread and popular among farm members, and, more importantly, it brought results. As Ministry of Agriculture officials contemplated a change in policy, newspaper articles kept to the party line, criticizing the practice of surrendering cooperative acreage to sharecroppers. "[Offering alternative solutions] at Dózsa and numerous cooperative farms in the county is against the bylaws. It is inappropriate, since the members' incomes decreases this way. Yet Dózsa did have a way out of the difficulties. They should have had every wife and family member over the age of sixteen join the farm sooner" (*Hajdú-Bihar Néplap*, 31 May 1955, 3). In August of 1956, an article in the same newspaper proclaimed that cooperative farms county wide had improved the organization of labor and increased yields, eliminating any need for sharecropping (ibid., 3 Aug. 1956). In a quick about-face a month later, the first secretary of the Berettyóújfalu District [in Hajdú-Bihar County] openly advocated distributing row crops among individuals for families to cultivate, pointing to the failure of the premium system to ensure these crops be tended to appropriately (ibid., 7 Sept. 1956, 2).

Concessions to the advantages of sharecropping were also made for state farms, despite their fully socialist pedigree as state owned enterprises. A series of recommendations prepared by the Research Institute in Animal Husbandry in April 1954 strongly recommended the use of sharecropping contracts paid in kind rather than in money as a means of dealing with the scarcity of labor at state farms. "It is difficult to explain that the principles of output wages appropriate for sharecropping row crops conflicts with socialist wages."[104] The report also urged a serious rethinking of the brigade organization at state farms. "The current form leads to the dissipation of power, which is against the principles of Stalinist strategy. It makes the concentration and redeployment of forces difficult. This state of affairs must be changed in the interest of production."[105] The report continued by criticizing the entire wage structure at state farms as too complex.

> Seventy percent of agricultural wages should be established in a base wage, and the 30 percent premium should depend on one or two factors. Even a math teacher can't understand the current system, much less a man who milks

cows.... Rather than improving agricultural production, the current wage system guarantees jobs for officials and payroll clerks in the labor relations departments.[106]

As of May, new regulations were issued by the Ministry of Agriculture permitting sharecropping contracts under certain conditions. Row crops (corn, potatoes, onions) could be designated for sharecropping contracts but not grains, and only if the plan's goals for the year were in danger. Nor would sharecropping contracts be allowed as a way of reducing the amount of work done with machines.[107] By the end of July, the Labor Relations Department of the Ministry of Agriculture was advocating several modifications to state farm regulations for the coming year. The manual labor expended on grain crops—done either by individuals or groups—was to be remunerated with piece rates based on the norms and piece rate levels from 1954. Sharecropping of row crops would be permitted if labor was scarce, and any grains that could not be cut by machine, such as the grasses alongside roads or in a ditch, or grains that the farm had not planned to harvest by machine could also be contracted out to sharecroppers by the state farm.[108]

All of these measures were an indication of the crisis being faced in agricultural production. A report issued in April 1954 described the state of agricultural production six months after the push to improve transparency and efficiency in administration.[109] Credit was taken for an improvement in the worker-peasant alliance, though evidence for a change was the better mood among working peasants, enhanced by a greater sense of security in private farming. Many features of the report are familiar: the need to streamline the bureaucracy and bolster the profile of model farms to serve as inducements to higher levels of cooperative membership nationwide. Propaganda efforts were intensified: increase the number of experts offering talks in villages, add more programs on the radio showcasing agricultural experts and scientists, initiate a regular news reel about agriculture, and appoint an editor at the *Szabad Nép* solely responsible for covering agrarian issues. Major shifts in personnel were envisioned. Engineers and skilled workers were to be drawn in the thousands from industry to agriculture. Agronomists were to be moved from machine stations to cooperative farms and a separate network of agronomists whose primary obligation was to assist private farmers was to be established, based at the village council.[110] Of central importance was changing the way plans were designed. "How might we improve planning work in agriculture, so that plans not be barriers but aids for improving agriculture?"[111]

Unsurprisingly, the major reason cited for the slowness of implementation was the poor performance of the Ministry of Agriculture and local administrative offices.

> The Ministry of Agriculture did not strengthen the county and district agricultural organs, and for this reason the lower organs lack initiative. Even though the ministry has assigned the implementation of tasks that had been taken care of at the national level to the authority of lower organs, they still don't take care of these affairs. As a result the lower organs inundate the Ministry of Agriculture with a myriad of issues that they haven't taken care of, which increases the bureaucracy and prevents the heads of the ministry from dealing with the organization of implementing regulations and substantial issues during most of their work day.[112]

Other agencies were also called to task.

> The Ministry of Industry and the entire state apparatus switches over extremely sluggishly to the tasks of improving agricultural production. In addition, it is a serious problem that the National Planning Office, the Ministry of Agriculture, the Ministry of Domestic and Foreign Commerce, as well as other state organs, did not estimate the needs of agriculture appropriately and so the shortage of certain goods of extreme importance for agriculture hasn't ceased in the course of implementing current plans.[113]

Introducing substantial shifts in planning, administration, and production was a tall order. And as before, the party/state's demands on government agencies to actively manage everyday affairs clashed with goals of streamlining the bureaucracy and reducing paperwork.

RETHINKING WORK UNITS

In October 1954 Erdei was reappointed minister of agriculture, having spent a year as justice minister in Nagy's reorganized cabinet. Sobering meetings among party leaders about the state of the economy in summer and early fall prompted serious reform measures to be initiated. The Institute of Economics was brought back to life (its former incarnation being the Institute for Research on the Hungarian Economy) under the leadership of István Friss, with Ferenc Donáth, who had just been released from prison at the end of July, appointed assistant director (Rainer 1999, 71, 86). The Agricultural Economics Department, closed in March of 1952, was also rejuvenated.[114] The first order of business was revising the work unit. Erdei specifically requested that old hands like Kuzmiak and Pálinkás be recruited to work on the project, alongside qualified personnel from the ministry and academy.[115] Erdei also sought out the advice of people directly engaged in collective agriculture: cooperative farm presidents, agronomists, and machine station directors.

Themes raised in the university critique written in 1952 were front and center: how to encourage higher productivity and improve the quality of production. "Much greater care must be devoted to harmonizing the individual and collective interests of the members."[116] Called on to interpret the implications for agriculture issuing from the resolution issued by the Central Committee of the Soviet Communist Party on the 17th of September, Hegedűs gave a lecture to the highest authorities of the Agitation and Propaganda Department of the Communist Party on October 12, 1953. He spoke directly to the problem of payment schemes and pulled no punches.

> Until they put into practice the premium for overfulfilling production plans according to brigades, and within brigades the work teams at cooperative farms, we can't aim for the permanent elimination of "*egyenlősdi*" from the system of income distribution.... It should be suggested that at several strong cooperative farms, where planning, accounting, and record keeping are on a firm basis, they should get rid of work units as an experiment and shift to a system in which the wages of brigades and work teams depend on output.[117]

Officially scrapping the work unit system was simply a recognition of its widespread failure. A group of farm presidents and agronomists described the situation honestly to Erdei in a meeting he called in Szolnok to appraise the situation on the ground.

> In any case they confessed that there is not one cooperative farm, without exception, where measuring work according to work unit regulations and continuous recording is done according to the rules. It is especially worth noting that throughout the county a method has developed that before completing any work the brigade leader and the farm member engage in bargaining.... For example, to unload a cart or dig up a certain area of sugar beets the member stipulates X work units [regardless of acreage], and on this basis they decide upon an agreed number of work units.[118]

In discussing alternatives, remuneration systems based either on yields or on time spent were considered. Using output as a measure was desirable, except insofar as it caused problems with figuring out midyear advances.[119] This was no small matter.

> Counting work units would improve immediately, if they would stipulate by law that 30 percent of the farm's collective crops—all the things members need for private plots (corn, barley, rough fodder, etc.)—absolutely must be distributed among the members, regardless of the needs of the collective farm. The reasoning behind this is that the members don't have confidence in the

work unit, or rather they don't feel as if they share in the common produce in proportion to the amount of work they have completed.[120]

Participants at the meeting in Szolnok did not agree on the status of sharecropping: some were in favor, others not. But there was consensus about having individuals work a specific plot of land. "This is the basis for consolidating work discipline, and if we connect it up with individual premiums, then it's possible to reach enormous results."[121]

The work of revising the distribution of income at cooperative farms began in earnest in January.

> *The income distribution method in force now*—with respect to its essentials—emerged by taking into consideration the method of income distribution at artels from the Soviet Union. Different organs worked out various parts and specifications of the method implemented here at home at different points in time without there being a possibility to ensure appropriate proportionality. As a result, the distribution method is far more complex than necessary and contains disproportionalities.[122]

The Agricultural Organization Institute (formerly the Organization Institute within the Ministry of Agriculture) prepared a report setting out the various alternatives and the work required to evaluate their relative merit.[123] Possible configurations included shares based on yields, on labor time, or on a combination of yields and labor time, taking into consideration whether the system fairly distributed shares according to work performed and whether it would insure a living wage and not impose unduly complicated bookkeeping measures. It was decided that a simplified version of the current system be introduced in 1956, in which the relative values of tasks were calculated better, and missing tasks be incorporated. For later years, a new system combining a base wage and output measure should be introduced. Work would be recorded, which would provide members with a set number of work units based on labor expended, to which would be added an additional number of work units based on yields achieved. Yields were not to be measured against initial projections in the plan, which had been proposed as a possible solution but was judged unfair. "This method of using relatively simple planning, accounting, and recordkeeping makes the membership interested in producing quantities of produce, while at the same time ensures that their livelihood is based to a significant degree on the work completed."[124] Cooperative farms would be given the option of adopting this system alongside the revised work unit system from 1956.

Two other components of members' income remained unresolved: premiums and sharecropping arrangements. Premiums had been worked out after the work unit system was completed and, like the work unit system, carried all sorts of disproportionalities. Marked differences in the ease of earning premiums in one branch of production over another, such as animal husbandry, had to be eliminated to ensure a balance between motivating workers and discouraging them. Sharecropping was to be discouraged, but it was widespread to make up for a scarcity of workers and to guarantee a stable income for members. In the course of revising the system of distributing income, studies had to be conducted to determine when sharecropping could be allowed: in what circumstances and as a solution to what sorts of difficulties. Fully aware that keeping accurate records and dealing with accounting forms had posed significant impediments to proper remuneration, a bifurcated system was proposed to lessen bookkeeping problems. At those farms where the personnel was fully competent to record and figure out how to balance the books, work units as measures of labor input would be retained; work units would be based on the combined value of output norms (e.g., amount of acreage stipulated per day) and the relative significance of a task. In the case of the numerous farms where these expectations could not be met, a simpler "labor day unit" would be adopted.[125] "The work day unit originates in the length of time of the work completed and the classification of the task. The basis of measuring the length of time is ten hours of actual work; the unit of measurement is the length of one hour."[126] Changes to recordkeeping were also proposed, requiring a study of various means of presentation (columns, book form, or filing cards) suited for different branches of production and under what conditions and adjusted to the size and developmental level of the cooperative and the accounting system deployed.[127] In the final analysis, the goal of revising the work unit system would be to standardize and simplify recordkeeping to ensure that income be distributed fairly and in a comprehensible manner. It was also important to make accommodations within the system for the needs of both large farms and small.[128] All of this was to be completed by the end of November.

Much needed to be done: a thorough review of the current system, extensive studies of the labor demands of each branch of production for every day in the agricultural season, a reexamination of target norms (minimum, maximum) for all tasks, calculations of premiums, rules on the use of sharecropping, and the development of a new accounting system. The work was to be divided among research institutes, ministerial offices, universities, colleges, and technical schools, primarily directed by the Agricultural Organizational Institute. The list included the Academy of Sciences, the Agricultural University, the College of Horticulture and Viticulture, the College of Mechanical Engineering in Agriculture, all agricultural colleges and technical schools, the training center for cooperative presidents, the Cooperative Farm Council, the Cooperative Department of the Ministry of Agriculture, the Inspectorate of the State Agricultural Machine Factory, the Research Institute in Animal Husbandry, the Pesticide Research Institute, and the Institute of Economics. Research trips were also planned to the Soviet Union, Poland, Czechoslovakia, East Germany, Romania, and Bulgaria.[129] The work schedule indicated who was responsible for various tasks, a list so detailed that even the process of proofreading the final manuscript was included.

In addition to the sustained efforts revising the entire work unit system, the Ministry of Agriculture also entertained occasional proposals to experiment with an alternative payment scheme. In one instance in Hajdú-Bihar County, a different system of rewarding the management of machine stations was initiated, in which premiums for the director, head agronomist, head accountant, and the business economist, among others, would be figured on the basis of the cooperative farm's entire business, not solely on the yields of lands worked by the machine station. If the value of the work unit in the district where the experiment was to take place exceeded a certain amount—20, 22, 25 forints, etc.—then management would receive a premium. The head of the Wage and Labor Relations Department was willing to allow the experiment to continue, even though studies had shown that income per acre was not correlated to the number of work units accrued working the land.[130]

Conclusion

Rákosi wrested control from Nagy in spring of 1955 with Moscow's support and proceeded to reverse many of the economic policies that had been introduced, most notably the emphasis placed on agriculture in the economy. Revising the work unit continued unabated, and other reform plans being developed by economists freed from the narrowly interpreted Marxist-Leninist approach were carried on as well. Nonetheless, economic and political conditions worsened once more and the public mood grew darker and more impatient.[131] Despite Rákosi's attempts to clamp down on dissent, a growing number of party members and intellectuals involved in policy making started to voice their disagreements publicly. A series of meetings were held in May and June 1956, addressing a variety of issues, from philosophy and history to journalism and economic affairs. An entire afternoon in June was devoted to agrarian questions, in which Ferenc Donáth played a central role. In a few short months, the 1956 revolution erupted, prompting what Béla Király has called "the first war between socialist states" (1984). As the regime installed by the Soviets consolidated control over the next six months, serious attention was paid to the dismal state of agriculture. In July 1957 the Communist Party announced a new agrarian policy, giving villagers greater control over their land and permitting nascent commodity markets to develop. But these measures were short lived, quashed under Soviet pressure in late 1958 to fully collectivize agricultural production.

The three years between Stalin's death and the outbreak of the 1956 revolution are marked by two countravening processes: the increasing administrative stability of the party/state and increasing conflict over the ways the party/state defined value and collective production, struggles waged as much in bureaucratic offices as in villages. The processes involved in stabilizing the administrative structure of the party/state—what I have referred to in part as the mundane acts of modernizing the governing apparatus—were evident in three domains: curtailing violence, incorporating "bourgeois" expertise into policy making, and reducing gaps between policy goals and bureaucratic procedures. Recourse to hasty violence as a means of dealing with political problems on the part of authorities decreased, though the heavy fist (and ever-present threat of violence) of Communist Party rule was not relinquished. A number of prominent "bourgeois" experts—which often meant simply those trained before 1949—who had been shelved for several years were brought on board to guide policy making, assisted by a flock of freshly minted graduates from the Economics University. Rather than craft new policy schema, however, their energies were devoted to patching up policies that had not brought the results expected. Fixing the work unit—by introducing metrics to gauge quality and premiums to spur productivity—could not alter the basic contradiction inherent to the payment system. Two different principles had been combined: rewarding effort with piece rate wages, on the one hand, and sharing the farm's surplus among the members at year's end on the other. In short the work unit was essentially a piece rate system, but return on effort depended on the final gains the entire labor force was able to produce that year. It mattered little if one put in a hard day's (week's, month's) work if the farm had failed to produce a surplus to distribute among its members after the harvest. All the while, government agencies overseeing fiscal affairs became increasingly intrusive in monitoring bookkeeping and accounting practices, less tolerant of mistakes and illegalities in procedure.

Difficulties faced by the party/state in the initial years of cooperative farming resulted from the widespread unwillingness of landowners to farm collectively, except when subjected to unrelenting duress, and the pervasive misunderstanding among cooperative farm members about work units and why they were necessary. By 1953, however, villagers began to contest more fundamental principles of cooperative farm organization, that is, what accounts for the accumulated value of the farm and what it means to work collectively. Cooperative farm bylaws stipulated that members who had contributed land when they joined were to be paid rent each year for the use of their formerly private property. As cooperative farms lost members, or were dismantled entirely, the notion that land constituted a distinct source of value contributed to the farm was challenged. Formerly proletarianized workers—manorial workers and day laborers—only recognized labor as a source of value in farming, refusing to return land and livestock to people leaving the cooperative behind. They felt no obligation to recognize other stores of value, since they had been denied—as a happenstance of class position—the chance to invest their labor into acquiring property while working at, indeed enhancing the value, of manorial estates or the properties of wealthier peasants. Their view was that a history of unjust distribution of goods within society would no longer dictate who had a proprietary right to the farm's assets. Former property owners were incensed at the audacity shown by cooperative farm members who directly and unashamedly contravened the legal statues of cooperative farming. Law suits followed. Legal disputes were also caused by disagreements over how many work units were required to be recognized as a full participating member of the farm. The timing of Nagy's announcement about voluntary membership in cooperative farms—he took the reins of government in the middle of the summer—complicated the process, since much of the work of tending to crops and harvesting had not yet taken place. The landslide of law suits submitted to county and national authorities relating to property restitution or work unit calculation contributed to the stabilization of administrative procedure, thereby strengthening the legitimacy of the party/state, since officials were expected to follow the law when adjudicating these suits rather than resolving them arbitrarily. These disputes made all too clear that the animosity between villagers bred by decades of class prejudice and abuse—ill feelings the Communist Party had thought it could mobilize for its own purposes—would surface when basic notions about the integrity of labor and the legitimacy of property were at stake.

Confusion about what it meant to farm collectively was rampant in the early 1950s among bureaucrats as well as farm members. One might think of this as a classic case of villagers resisting Communist Party rule. I propose a more complicated reading. The criteria for collective production—working in a brigade but being held individually responsible for work completed—bewildered farm members. If plots of land were to be apportioned to individual workers, then what difference did make if one worked in a brigade or not? The goals envisioned by ideologues of breeding a more enlightened working class through group effort were baffling. What was the difference from before, when people worked in large groups as migrant workers or manorial servants? And who cares about collective worker's consciousness anyway? But as was the case with work units, initial bafflement was replaced over time by a more concerted effort to turn the system to farm members' benefit. The increasingly common recourse to sharecropping arrangements at cooperative farms guaranteed that work would be completed, so it was a pragmatic solution to members' continued boycotting work at the farm. I would suggest, however, that insisting on sharecropping and other means of tying effort to reward also served to break the Gordian knot at the heart of work units, ensuring that the family's efforts to improve yields and the quality of produce would be directly rewarded. Income, or at least a good portion of income, would no longer be bound up with dividends alone, freeing members from suffering the precarious fortunes of the farm as a whole.

A final point: the work unit has often been talked about as a move away from monetary wages, as an alternative configuration to reward effort designed for socialist cooperatives. Marxist-Leninist ideologues envisioned the day when money would no longer be a necessity, when accounting and planning would be conducted by using measures of value in kind and labor time would be the measuring stick for work (Kulcsár 1947, 440). But those days, possible only with the arrival of communism, were long in coming. Until then, socialist states would depend on money for three of its fundamental functions: as a medium of exchange, as a unit of account, and as a store of value. Money would continue to play a role in keeping track of labor inputs and rewarding work, planning the economy, assessing the level of productivity, assembling budgets, and standing watch over communal assets, as well as ferreting out enemy forces bent on derailing economic trends (ibid., 440–42). In other words, Hungarian socialist policy makers needed money, but not markets. By designing an effective wage system, they would be able to discipline workers to follow the core principle for wages in socialism: "Each according to his work"—nothing more, nothing less. Over time, members of cooperative farms would reconceptualize their definition of labor value and make it their own. Now they would deploy a new calculus for figuring the value of their work, disarticulating tasks into distinct activities of a specified temporal duration and relative value based on levels of difficulty and skill. One might still consider the work unit system a form of non-monetized wages. Doing so, however, would tear the historical ligaments to its initial design and application and cripple our ability to recognize that techniques designed to rationalize labor and monetize its value built into the work unit could have such transformative effects.

Conclusion

How does one go about estimating the value of an event? I posed this question at the outset to frame my inquiry into the commodification of labor so as to foreground the arduous work of work science and convey the fleeting quality of labor. I proceeded to explore just what was entailed in constructing the modern technology we call a wage through the lens of Hungarian history. I have referred to this project as a prequel to my earlier book on collectivization, because it provided the back story I yearned to know about the design and initial implementation of the work unit. Zeroing in on the construction of a wage system—commensurating tasks, measuring effort, judging skill, ranking jobs—brought into sharper focus the historical and cultural specificities of modernist social engineering. The humble work unit's conception, birth, and first baby steps chronicled here also made clear, primarily by its absence, the crucial role of infrastructure—institutions of learning, standards of practice, government services, competent professionals, and modernized workplaces—in making these changes possible.

I have long been dissatisfied with twentieth-century histories of rationalization and scientific management. I had no quarrel with the notion that society has been dramatically altered by the many social projects we group under the term rationalization; Weber's insights had drawn our attention to these processes early on (Weber, 1978). I was puzzled, rather, about why we did not know more about the specifics, the techniques whereby these changes were accomplished. After all, "how" is at the center of rationalization, standardization, and engineering. These processes require enormous amounts of work—coming upon an idea, realizing its value, representing it as necessary, building a system, negotiating its parameters, implementing it (trying to force it down people's throats), revising it, chucking it, and starting over. The actual making of a rationalized workplace or scientifically designed wage system is completely invisible if one simply reviews initial policy formulations, as if they were the beginning of a process, or describes a working policy as if it were a final product. Doing so neglects the substantial efforts invested in assembling and coordinating practices that constitute a new social formation but which have yet to coalesce. As a consequence, the necessarily emergent dynamics of formalizing practices are completely overlooked, rendering our comprehension of the processes of commodifying labor shallow at best.

The work unit did not last; I provide a brief postscript on its eventual demise. I then address specific historiographic contributions I have made here, that is, alternative readings of Stalinist political dynamics, the transition, and state building in Hungary. I end with a few remarks on the broader theoretical issues I have considered.

Postscript

In 1961, 96 percent of agricultural production had fallen under the control of the party/state (Orbán 1972, 218). More than four thousand cooperative farms joined state farms in the landscape (Donáth 1977, 172). Cooperative farms were found in every village, often more than one to accommodate the divisions between class and neighborhood in the community. In the early 1960s, most farmers still harvested wheat with a scythe, spurring the party/state to invest in machinery to modernize production nationwide. Nonetheless, cooperative farm members continued to shun work on collectively shared properties, doing the least amount of work possible to qualify for a yearly dividend. Meanwhile, they devoted all their energies to the small family plots they were allotted as members of the cooperative. It became common for soldiers and school children to be mobilized at the height of the agricultural season to bring in the crops. Conceding to people's discomfort with farming collectively, the party/state openly sanctioned sharecropping contracts with cooperative farms to guarantee that the land not be left fallow.

For most of the decade, the work unit continued to be the means of distributing farm income among members, money delivered ceremoniously in an envelope at the annual meeting as before. By the 1970s, however, the work unit had been abandoned entirely. Cooperative farm members continued to receive a portion of their share in grains over the course of the year but the rest of their income—distributed as before at the annual meeting—was paid according to a new wage scheme. For most, base pay was calculated by the hour, although a few select groups at the farm—such as the tractor brigade—were paid by piece rates, substantially increasing their salaries over others of those

CONCLUSION 267

consigned to the measly hourly rate. Remunerating workers on the basis of piece rates finally restored the crucial component of the idealized standard wage agrarian work scientists had envisioned before the war, payment according to output. Since this connection was limited to a small number of favored groups at the farm, the new monetary wage system did not substantially increase labor productivity. As I recounted in the introduction, the shift to hourly wages was not to everyone's liking. No longer a burdensome device foisted on farm members by the party/state, work units had become naturalized; the elaborate system of numbers and decimal points was understood to be a sensible way of keeping track of everyone's contribution at the farm. With hourly wages, all the differences between the way people worked day in and day out had been washed away. People had come to appreciate the fine-tuned differentiations among farm members that work units had made possible. Commodifying time, skill, and effort had become common sense, even if the conceptual link with money and markets was completely absent.

Writing a Different History

Adopting a different periodization in order to study the transition to Stalinist socialism in Hungary has made it possible for me to offer alternative interpretations of the dynamics of regime change at the national and local level, as well as question the assumptions about the role of the Soviet Union in molding institutions in the new People's Republic of Hungary. The substance of these interventions address three topics: (1) the role of the Soviet Union in Hungary's early collectivization project; (2) reasons for difficulty in implementing the new wage system; and (3) the character of the new party/state.

THE ROLE OF THE SOVIET UNION

I began my archival research assuming that there were strong continuities between pre-World War II work science and socialist wage forms, a view that from the outset inclined me to expect that my understanding of role of the Soviet Union in early Hungarian socialist policy on cooperative farms would change. An exciting prospect, especially if the research confirmed my suspicion that interwar German agricultural economics would loom far larger in the history of collectivization policy in Hungary than has been previously acknowledged. Crucial features of the work unit—tying the wage to shares of the farm's surplus and the scale of valuation adopted—were directly taken from Soviet practice. But the speed with which the complex new matrix of value was assembled—a mere six weeks—attests to the fact that techniques for delineating and ranking tasks so neatly were close to hand. Just when and how various elements of Soviet wage policy were adopted or modified in Hungary shifted over the course of the eight years examined here. Sometimes the pressure to conform to Soviet practice was primarily political, for example, in 1951 when Rákosi tightened the screws on the bureaucracy so he could boast of being Stalin's best disciple; at other times, consulting Soviet resources for alternative solutions was purely pragmatic, as was the case when the Soviet system of innovations incorporating measures of quality as well as quantity into the work unit system was considered.

Recent scholarship on collectivization (Iordachi and Bauerkämper 2014; Iordachi and Dobrincu 2009; Kligman and Verdery 2011) offers extensive evidence of differences in the ways collectivization was implemented in the region, providing a far more nuanced discussion of collectivization than was previously available. "[W]hile the power relations between the Soviet Union and Eastern European countries were highly unequal, one should not rule out the existence of practices of unequal negotiation, adaptation to local conditions, or innovation in Marxist-Leninist practices in the satellite countries" (Iordachi and Bauerkämper 2014, 22). Writing histories of socialism in this light opens avenues for comparative analysis of culturally nuanced political histories of imperialism, colonialism, and postcolonialism elsewhere around the globe.[1] Incorporating the role of interwar work science and agricultural economics into the history of collectivization also allows us to compare a phenomenon usually limited to the history of socialist states with other schemes for modernizing agriculture at the time, such as colonial plantations and capitalist latifundia.

IMPLEMENTATION

Examining the wealth of information housed in the archives permits us to set aside the standard view of the Stalinist period as one in which the stark opposites of compliance and resistance were the modus vivendi of everyday life. The emotional and political terrain was far more complicated than this view would allow. The role of simple confusion as a factor in the fits and starts of policy implementation are vastly underestimated, as we have seen. When villagers resisted party dictates—whether it be business plans for one's farm or the elaborate calculations involved in work units—they did so because they were perplexed. The new wage system did not make sense, so why should they follow its principles? And when the reasons were made clear—or simply asserted—the question then became whether and how to follow these dictates. The learning curve was steep, compounded in this period by the

CONCLUSION   269

frequent changes in regulation, the absence of proper forms, and the lack of qualified personnel in the town hall or county seat. It is important to point out that reluctance to change, as well as simple indifference, also characterized the response of owners and managers of manorial estates in the 1920s and 1930s. They too balked at the demands agrarian work scientists made on their time and resources. The innovations business economists hoped to see were substantial: disaggregating the activities of each branch of production to assess its profitability, introducing complex forms of bookkeeping to track these differences, hiring staff to oversee new techniques promoted to rationalize labor, and perhaps even paying workers in cash. The discrepancy between the visions of agrarian modernizers and the everyday experiences of manorial personnel was immense.

Describing the political dynamics of the Stalinist era as one of oppression and resistance erases salient class and religious differences. James Mark has questioned the longstanding view that the only two options available for professionals and intellectuals during the Stalinist period were complicity with or rejection of Communist Party politics (2005a, 2005b). As James argues convincingly, the view that keeping one's distance from politics was the only moral choice had a long history in conservative Catholic circles but was a minority perspective. The far more common approach was to work out one's personal accommodations in the political realm, decisions based on ideological conviction, profession, and employment opportunities. Significant class differences also characterized villagers' attitudes toward collectivization. Emphasizing peasant resistance to collectivization fails to capture the reasons why some citizens were actually enthusiastic about the changes taking place. By focusing on the majority's rejection of cooperative farming, however, we lose sight of those in villages who were keen to join cooperatives and their reasons why. Impoverished day laborers and manorial workers welcomed the opportunity to join a group farm in 1948. Despite having received property in the land reform, they were otherwise quite poor: no tools, few animals, and no outbuildings or barns to process and store goods. More important to poor villagers may have been the opportunity to run a farm to prove to their critics that they could succeed, if given the chance. Landowners frequently justified their rights to property by claiming to have nourished the land with their sweat and tears. New landowners were adamant that they too had nourished land with their sweat and tears but had been denied the right to its harvests. This disagreement about the source of value, and the rights to dignity it conferred, was at the heart of battles over cooperative property in 1953 and 1954.

Poorer villagers hoped to benefit from the party's support of collective production, but that did not mean they followed the dictates of party policy.

Working as a group meant different things to different people. Exhortations to farm collectively were driven by the Communist Party's insistence that workers rethink their relations of production, that they recognize the superiority of collective effort. No one cared about these noble aspirations; most cooperative farm members shunned working with the group at all, so the fine-grained discussions of brigade membership and individuation of effort taking place among policy makers were completely irrelevant. On the other hand, entering into sharecropping arrangements, also forms of group labor, was welcomed, precisely because the disconnect between effort and reward built into the work unit could be avoided, and direct compensation—often in kind—was far more valuable than iffy promises of reward at year's end. In other words, the struggle was not whether one worked with one, two, or ten others; it was over how one benefitted from the shared activities overall.

The struggles over property and labor organization in cooperative farms in the early 1950s reveal core assumptions villagers held about the source of value and the character of work and the degree to which these differed within communities. In this sense, they mirror the debates among agricultural economists and work scientists who debated the value of labor in the 1930s. Forcing people to work differently and introducing new metrics to calculate the value of their work in the 1950s opened up to public debate issues that would otherwise have been inaccessible to social analysis. Now villagers were drawn into open contests with each other about fundamental principles of social life: who works and doesn't, and how that is defined; who is deserving of reward and who isn't, and why; and who is authorized to decide, and why.

A MODERNIZING STATE

Valuable studies of the bureaucratic chaos and inefficiencies of the Stalinist party/state have long been available. In the words of one commentator, "It was an economy that was overplanned to a state of planlessness and overorganized to the point of collapse. Although natural plan indicators arranged the last producible nail and deliverable egg into its militarily rigid, plan-command system, it was, in the final analysis, disorganized, clumsily deformed, and anarchic" (Zavada 1984, 139). What has escaped notice, however, are the substantial efforts devoted to the task of modernizing the bureaucracy that were also taking place. I have chronicled numerous examples of anarchy, arbitrariness, and incompetence, but I would argue that this description of socialist state building is inadequate. Reviewing numerous state documents discussing mundane issues of administrative procedure, new organizational structures, and the implementation of policy strategies provides ample evidence of efforts to construct a reliable bureaucratic system to serve the party/state.[2] I was also struck by the consistently negative and critical tone of party/state documents. Though accomplishments were acknowledged, much attention was paid to tasks left unfinished or poorly executed. This incessant internal criticism seemed at odds with the public face of the party/state, where accolade followed accolade in the press. Minutes of meetings held at county offices chronicled lengthy discussions on how to implement policies that were poorly conceptualized, confusing, and sometimes contradictory. The tenor of these reports may have been an artifact of internal bureaucratic communication, common in any complex organization. I believe that more was at stake than quotidian bureaucratic haggling. This genre of intense critique and high expectations fits the public image of Marxist-Leninist party politics, which was premised on the potential infallibility of the Communist Party. Inadequacies had to be constantly ferreted out and addressed, precisely because of the unrealistic ideals of success embedded in party ideology. In this case, therefore, it was important to attend as much to what had been neglected along the way as to what had been achieved.[3]

Modernizing the Hungarian bureaucracy required investments of time and money in the new infrastructure to meet those goals. The scale of these investments was huge. If we consider the educational sphere alone, we can see the enormity of the project: universities were reorganized to train a new generation of economists and agronomists with skills appropriate to central planning and collective agriculture, the Academy of Sciences and several research institutes were reconstituted for similar purposes, technical schools were established to teach accounting, evening classes explaining the governing tactics of a Marxist-Leninist Communist Party were developed, and villagers were called to winter seminars to teach them how to use the work unit. In the 1920s and 1930s, replacing civil servants trained in law with professionals—experts in economics or other relevant fields—had been a prominent goal of agrarian modernizers. For them, rationalizing the state did not only mean making the work of bureaucracies more efficient; it also meant staffing bureaucracies with knowledgeable experts to guide the formation of policy. The barriers to broadening expertise in the 1950s were different, however, as the disdain for pursuing a career in economics so common in the 1920s had vanished. Now the problem lay in the glaring lack of expertise in Marxist-Leninist political economy, among either university faculty or research scholars. Nonexistent or inadequate translations of Soviet texts meant training a new generation of economists was delayed; it also caused problems for policy makers wishing to keep current on Soviet innovations in particular domains, such as agronomy or forestry. Visiting Soviet advisors mitigated some of this gap, but they could only do so much in the face of intransigent Hungarian intellectuals who refused to discard their previous views.

The particular dynamics of Hungarian party/state politics resemble in many ways those of other contemporary governments: fissures between ministerial agencies, intra-agency conflicts, complex problems caused by delegating authority to lower-level agencies, and frequent disagreements within the governing elite of the party. On the other hand, centralized economic planning constituted a novel challenge; wartime planning, or even increased state control over industry in the postwar era, paled in comparison to the degree of state intervention in the most minute economic concerns by the Stalinist party/state. Planning was no longer simply the judicious allocation of resources to meet reasonable forecasts; it would now oversee all productive activities in every single branch of the economy. In some interpretations, every person, not just every factory, would have an explicit set of planning goals to fulfill (Pittaway, personal communication). This strained the administrative capacities of the state beyond measure, difficulties that could not be easily rectified by the strong arm of the state's coercive apparatus, since it too was only as effective as its reach. These problems were exacerbated in rural communities. Pleas and exhortations rained down from Budapest, calling upon the party faithful to shoulder responsibilities in mobilizing community members. But in the absence of any active party organization, these calls fell on deaf ears. Villagers were swept up by the party/state's aggressive policing—fined and imprisoned for minor infractions—but punitive measures could never substitute for an active party cell. In short, the technical infrastructure required for centralized economic planning was woefully inadequate, its weaknesses mirrored in and exacerbated by a poorly organized party apparatus. Violent retributions were often the approach chosen by ineffectual bureaucrats.

The Science of Commodification

My analysis of commodification and infrastructure has benefitted immensely from the debate over the performativity of economics and the wider discussion of accounting, finance, and markets that has blossomed in science studies over the last decade (see the introduction). That said, I actually began this project assuming that economists would play a significant role in designing cooperative farm wages. In fact, it was precisely the active role of agricultural economists and work scientists, and then socialist policy makers, in crafting this tool that intrigued me. In contrast to examples in the performativity literature where the initial innovation was not intended to change economic

CONCLUSION 273

practice, as in the history of the Black-Scholes equation (MacKenzie 2006), in this instance pioneers in scientific management boldly espoused a new economic order. Their enthusiasm was constantly on display, expressed in the numerous publications they issued explaining not only how, but why, they would propose rationalizing production to alter behavior so substantially. My analytic task then became exploring how these scientific models would actually take shape in the form of policies and practices, paying keen attention to all the steps along the way, whether they be technical specifications or bureaucratic maneuvers.

I was fascinated by the history of scientific management because I could never figure out how these apparently elegant models promising higher productivity were to be adopted for use in factories and firms. The reigning assumption, of course, was that the models expressed the causal mechanism linking effort to reward, so their adoption in actual wage plans would be straightforward. But as I argued at the outset, scientific models aren't intended to be implemented in business plans or policy statements. They are tools to think with, puzzle about, play with. So the wage schemes were merely sketches to design creative management reforms. In other words, the formidable engineering of people and places scientific management demanded was always built from the ground up. Erected on preexisting foundations that lent strength to its frame, the new edifice was intended to house those living in the vicinity, whose past behaviors were well known. Class and ethnic stereotypes, along with other parochial features, were taken into account in the planning stages. Therefore, work science formulae and metrics of labor value only meant something in context; they were indexical, expressed in local terms for specific purposes. Numbers were no longer disembodied symbols on a page but revealed to be plump figures serving local interests. This meant, however, that elegant metrics of labor value could not stand on their own; they required the support of institutional bodies that shared their interests and would develop their capacities to reach this goal. In short, commodifying labor by engineering precise wage systems would only be possible in the end if an infrastructure supporting the science of work and the modernization of production had been carefully constructed and was fully operating.

My approach has been to examine the commodification of labor as an historical accomplishment, a process whose features vary in time and place. I have argued that the role of markets in commodifying labor has been exaggerated, overshadowing other possible means of defining and assessing the value of labor. In the history recounted here, social scientists—work scientists, agricultural economists—took an active role in defining and evaluating labor value, developing techniques that were incorporated into wage systems at cooperative farms. They also played an important role in campaigning for and setting the groundwork for the institutional infrastructure needed to apply the techniques they had championed. Though not intended initially to sidestep market forces, these techniques—embedded in the matrix of labor value called the work unit—did in fact contribute to a transformation of people's attitudes about work, time, and money, in a centrally planned economy where market forces were actively surpressed. Paying attention to the technicalities of commensuration and valuation in an analysis of wage forms—that is, studying the formalizing practices, material investments, and infrastructural requirements of commodification—strengthens our ability to explain the contingency of historical processes. Technicalities cannot be consigned to an afterthought. Attending to these complexities provides us with tools to explain why the adoption of apparently similar formalizing practices can result in substantially different historical trajectories.

Archives[1]

Politikatörténete Intézet Levéltára	PIL	Archive of the Political History Institute
Politikatörténete Intézet Levéltára MKP	PIL 274 f.	PIL Hungarian Communist Party
MKP Állampolitikai Osztály (1945–1948)	12 cs.	Department of State Policy
MKP Falusi (Agrárpolitikai) Bizottság, Osztály	13 cs.	Village (Agrarian Policy) Committee, Department
MKP Szakszervezeti Osztály (1945–1948)	20 cs.	Trade Union Department
MKP Szövetkezetpolitikai Osztály	14 cs.	Cooperative Policy Department
Szociáldemokrata Párt	PIL 283 f.	Social Democratic Party
Vidéki Titkárság	16 cs.	Rural Secretariat
Szakszervezeti Osztály	18 cs.	Trade Union Department
Szövetkezeti Osztály	24 cs.	Cooperative Department
Gazdaságpolitikai Osztály	32 cs.	Economic Policy Department
Mezőgazdasági és Agrárpolitikai Osztály	33 cs.	Agricultural and Agrarian Policy Department

1. Abbreviations used in archival citations include:
   - f.   fond
   - cs.   csoport (group)
   - d.   doboz (box)
   - fcs.   fondcsoport (fond group)
   - i.sz. iratszám (number of the document)
   - o.   osztály (department)
   - ő.e. őssegység (unit)
   - sz.   szám (number)

Magyar Dolgozók Pártja (MDP)	MOL 276 f.	Hungarian Workers' Party
Politikai Bizottság	53 cs.	Policy Committee
Gerő Ernő titkári iratai	66 cs.	Secretary Ernő Gerő's Papers
Hegedüs András titkári iratai	74 cs.	Secretary András Hegedüs's Papers
Rákosi Mátyás titkári iratai	66 cs.	Secretary Mátyás Rákosi's Papers
Központi Vezetőség Államgazdasági Osztály 1949 február—1952 májusig	116 cs.	Central Directorate of the State Economy Department
Központi Vezetőség Államgazdasági Osztály 1948–1949	115 cs.	
Mezőgazdasági és Szövetkezeti Osztály	93 cs.	Department of Agriculture and Cooperatives
MDP ülésanyagai	85 cs.	Meeting Materials of the Department of Agriculture and Cooperatives
Titkárság jegyzéke	54 cs.	Notes of the Secretariat
Földművelödési Minisztérium (FM), pre-1945	MOL K184	Ministry of Agriculture
Földművelödési Minisztérium (FM), post-1945	MOL XIX-K-1	Ministry of Agriculture
Munkaügyi Fősztály	XIX-K-1-j	Department for Labor Issues
Agrárpolitikai Főosztály (1946–1948)	XIX-K-1-z	Department of Agrarian Policy
TÜK iratok, Visszaminősített TÜK Minisztertanács Előterjesztései	XIX-K-1-c	Secret Papers, Downgraded Secret Papers Reports of the Council of Ministers
Titkárság	XIX-k-1-b	Secretariat
Erdei Ferenc min iratai	XIX-K-1-ah	Ministerial Papers of Ferenc Erdei
Hegedüs András min. iratai	XIX-K-1-ai	Ministerial Papers of András Hegedüs
Termelőszövetkezeti Osztály	XIX-K-1-bb	Cooperative Department
Országos Tervhivatal	MOL XIX-A-16-a	National Planning Office
Moszkvai Nagykövetség Bizalmas Iratai Magyar-Szovjet Mezőgazdasági Kapcsolatok	MOL XIX-J-42-a	Confidential Materials from the Embassy in Moscow: Hungarian-Soviet Agricultural Relations
Nagy Imre és Társai Vizsgálati Iratok	MOL XX-5-h	Investigation Materials of Imre Nagy and companions
Budapest Fővárosi Levéltára	BFL	Archive of the Capital City of Budapest

A Jogszolgáltatás Területi Szervei	XXV	The District Organs of the Administration of Justice
Országos Széchenyi Könyvtár Kézirattár	OSzK	Manuscript Collection of the National Széchényi Library
OSzK Plakáttár		Poster Collection
Magyar Tudományos Akadémia Levéltára	MTA	Archives of the Hungarian Academy of Sciences
II. osztály: Közgazdaságtudomány	II.o.	Department II: Economics
IV. osztály: Agrártudományok osztály	IV.o.	Department IV: Agrarian Sciences
Győr-Moson-Sopron Megye Győri Levéltára	Gy-M-S. M. GYL	The Archive of Győr-Moson-Sopron County, Győr office
MDP Győr-Sopron Megyei Bizottság Mezőgazdasági és Szövetkezetpolitikai Osztály Fegyelmi munkaterület Agitációs és Propaganda Osztály	30 f. 2 fcs.	Hungarian Workers' Party Committee for Győr-Sopron County Agricultural and Cooperative Policy Department Disciplinary Branch Agitation and Propaganda Department
Győr-Sopron Megyei Tanács Végrehajtóbizottság Mezőgazdasági Osztálya	XXIII. f., 9. fcs.	Executive Committee of the Győr-Sopron County Council Agricultural Department
Hajdú-Bihar Megyei Levéltár	HBML	The Archive of Hajdú-Bihar County
MDP Megyei Bizottság	XXXII., 41 f., 2 fcs. 42, 43, 46 f., 2 fcs.	The County Committee of the Hungarian Workers' Party
Hajdú-Bihar Megyei Tanács Végrehajtóbizottság Mezőgazdasági Osztály	XXIII.9.b.	Executive Committee of the Hajdú-Bihar County Council Agricultural Department
Zala Megyei Levéltár	ZML	Archive of Zala County
MDP Megyei Bizottsága Agitációs és Propaganda Osztály Mezőgadasági és Szövetkezetpolitikai Osztály	57 f., 2 fcs.	County Committee of the Hungarian Workers' Party Agitation and Propaganda Department Agricultural and Cooperative Policy Department
Zalamegyei Tanácsa Végrehajtóbizottságának Mezőgazdasági Osztály	XXIII. f., 8 fcs.	Executive Committee of the Zala County Council Agricultural Department
Tszek iratai	XXX., f.	Cooperative Farm Papers

Notes

Introduction

1. Two literatures are missing from my analysis of the value of labor and commensuration, by virtue of my focus on the ways Hungarian economists and policy makers defined value during this historical time period. I am not discussing the concept of labor value, in Marx's work or Marxist scholarship. For provocative discussions of this topic, see Elson 1979 and Postone 1993. Neither will I be discussing the rich and voluminous scholarship in anthropology on value and valuation, though it has clearly influenced my thinking. See, for example, Guyer 2004; Hart 1981; Munn 1986.

2. A square fathom (*négyszögöl*) measures 38.32 square feet or 3.57 square meters.

3. There is a voluminous literature on the social and economic dynamics of the transition away from socialism in the 1990s in Eastern Europe and the former Soviet Union. I have discussed the transition and specifically the possible consequences of decollectivization and the privatization of public goods elsewhere (Berdahl, Bunzl, and Lampland 2000; Lampland 2002). A few notable book-length monographs on this period include Creed (1998), Dunn (2004), Eyal, Szelényi, and Townsley (1998), Fodor (2003), Gille (2007), Haney (2002), Róna-Tas (1997), Stark and Bruszt (1998), and Verdery (2003)

4. The Stalinist era in the history of Hungarian socialism refers to the period between 1948 and 1956.

5. Callon's intervention and his interlocutors have certainly influenced the degree to which the social sciences have become a legitimate site of investigation for science studies scholars, who previously were focused almost exclusively on the natural sciences.

6. The term *üzem* in Hungarian means firm, akin to the German word *Betrieb*. Therefore the term *üzemtan* means the study of the firm but specifically in relation to the internal workings of the firm. The closest English equivalent is industrial engineering, but the term *mezőgazdasági üzemtan* would be translated as agricultural industrial engineering. That circumlocution confuses rather than clarifies. Agricultural economics is probably a better catchall, but it doesn't convey the emphasis on engineering the management and organization *within* one firm or enterprise. In the interests of style, I use several different terms throughout the book to refer to this discipline: firm economics, studies of the firm, business economics, and agricultural economics.

7. Gillespie's fascinating book about the Hawthorne experiments (1991) is an exception, though his focus is on experimental technique and management rather than wages.

8. Leon Pratt Alford was a prominent industrial engineer in the first half of the twentieth century. He was active in the American Society of Mechanical Engineers and the Institute of Management. He eventually became a professor of administrative engineering and chaired the Department of Industrial Engineering at Columbia University. I am indebted to Sanford Jacoby of the UCLA Anderson School of Management for helping me find these materials.

9. There is also the danger of provisional numerical values being inadvertently understood to be fixed indicators. This occurs in scientific modeling, when a particular element of a model is transformed over time, and unwittingly, into a fixed metric (Naomi Oreskes, personal communication). This also occurs in the development of technical systems in business accounting. Discussing the development of flexible machining systems (FMS) in an aircraft company, Thomas (1994, 61) notes, "On the one hand, corporate review required [return on investment] figures in support of the proposal; this requirement, in a sense, encouraged FMS proponents to play games with the numbers. On the other hand, divisional management and R&D took the corporate view seriously enough to make bold claims despite their fragile numbers."

10. I am not making the point Scott (1999) describes in broad strokes about abstract plans and local practices here. I am arguing that this disjuncture characterizes all situations in which provisional models are portrayed as necessary interventions. This is as true for policies that are successfully implemented as for those that fail. The analytic task then becomes explaining how the chasm between formal schemes and social practices is bridged over. This is as important when policies become effective as when they fail. In brief, a symmetrical account is warranted, not the sort of Whiggish melancholy Scott's book presents.

11. Asking what formalizing representations do, rather than asking what they are, situates my inquiry in relation to longstanding debates over referentiality and practices of meaning (Lucy 1993; Peirce 1972; Rotman 2000).

12. Çalişkan (2010) and Muneisa (2007) have made similar arguments about indexicality in relation to markets and prices.

13. Daniel Breslau makes the same point about of the role of economists in designing state policies in his book *In Search of the Unequivocal: The Political Economy of Measurement in US Labor Market Policy*. Discussing the diffusion model often deployed in studies of innovation, he says, "There is an increasing awareness in research on technology and applied science that transfer of inventions or ideas involves a transfer of the conditions under which the innovation is produced to the setting of its application" (1998, 22).

14. In his monumental study of Magnitogorsk, Stephen Kotkin also noted the great advantage of studying a centralized state-controlled economy. He found it much more difficult to acquire the archival materials owned by a private foreign firm that helped to build the city in the 1930s than Soviet documents (1995, 370).

15. Borhi's recent book and Roman's earlier volume challenge longstanding sympathetic accounts of the role of theUnited States in the early years of the Cold War in Eastern Europe (2004; 1996).

16. Heinzen notes the same phenomenon in the early years of the Commissariat of Agriculture in the Soviet Union. "Sometimes the words 'Ministry of Agriculture' were simply crossed out and 'People's Commissariat' was scrawled above them, though busy officials usually had no time for such prettifying" (2004, 31).

17. The Soviet Commissariat of Agriculture established after the Bolshevik takeover in 1917 also held out the US Department of Agriculture as a model of scientific research and professionalism to be emulated (Heinzen 2004, 42).

18. I distinguish between party agencies and government offices in the Stalinist state, even

NOTES TO CHAPTER 1

though the degree of independence exercised by government organs was circumscribed. I do so because a careful reading of documents has shown how these wings of the party/state could find themselves at odds, either for simple problems arising from divergent mandates or as a result of complex problems bound up with personnel, mode of policy implementation, and questions of scale within the state.

19. In other research projects since then, I have found myself searching for comparable kinds of materials, realizing only in hindsight that I had gotten used to reading government publications were that little more than how-to manuals.

20. The frequent recording of local rumors, such as expectations of an American/English invasion, are fascinating sources to gauge both what may have been circulating in local communities and, just as importantly, what the secret police considered important to convey to their superiors.

21. The State Audit Center (Állami Ellenörző Központ) was an extremely important body overseeing budgets and business practices more generally. Its counterpart in the Soviet Union wielded enormous power (Shearer 1996).

Chapter One

1. The customary unit measuring land in Hungary was the cadastral *hold* (kh, or 0.57 hectares), which approximates a US acre (0.4 hectares). Since virtually all sources cite acreage in these terms, I will do so as well, followed by the equivalent in hectares.

2. This description is based on a composite drawn up by István Szalay for comparative purposes in his study of the amount of labor required to run an estate and its cost (1931, 11–16).

3. Somogyi's book of the same name is the classic discussion of this agrarian platform (1942).

4. The debate over the size of landholdings also took on a strong confessional strain. Among certain vocal Protestant leaders, the inability of Hungarian Protestants to make a living off the land could be traced directly back to the Counter-Reformation and the Hapsburg restoration after the rout of the Turks in 1686 (Hanebrink 2006, 130–31). Wealthy Catholic aristocrats who had allied themselves with the Hapsburg crown were rewarded with huge estates to the detriment of Protestant landowers, they argued, while Catholic Germans were actively recruited by Maria Teresa to settle in the ravaged country. By the twentieth century, the consequence of losing so much land to Catholics, they claimed, resulted in Hungarian Protestants being left with miniscule and unviable farms.

5. András Heller, later founder of the Agricultural Work Science Institute, wrote an entire book on the life of manorial servants, based on his years as a constable in Fejér County (1937). Since he had lived alongside them, and followed their situation as a career, he felt confident his perspective was valid, even though as a high-level county official, few would describe him as having lived among peasants. More to the point, he conducted research on ninety-nine middle-sized and large farms in his district: gathering statistics; visiting farms; interviewing village authorities, doctors, veterinarians and priests; examining earlier legal disputes over labor contracts; and collecting the personal views of many manorial servants (1936, 455).

6. An article appeared in *Köztelek* in the spring of 1940 discussing the difficulties manorial estates were facing in contracting workers for the coming year. One strategy to win over the workers was to improve housing, increase yearly income, and provide help for families. In addition the pressures of the government's planned economy also led to greater attention to social services, suggesting that these were perhaps less generally offered than Scherer's article would suggest (*Köztelek*, 24 Mar. 1940). Hence we find a different explanation for the provision of so-

cial services. Rather than the extension of generosity and good will, the need to comply with the government's plan resulted in the provisioning of better services.

7. The new property owners mentioned here were beneficiaries of the limited land reform in 1924, many of whom were war heroes.

8. Heller claimed that manorial servants ate more poultry and pork and drank more milk than villagers. If this were true, it could be explained by their distance from town and markets where goods could be sold (1936, 460).

9. Kovács stipulated that the highest level of human exertion was fifteen kilogrammeter, or being able to lift fifteen kilos one meter high in one second, or one kilo fifteen meters high in one second (1935, 219).

10. Protestant youth also spoke out against the injustice the neglect of land reform represented, but as Balogh points out, their position on the land reform did not turn them against their own religious elite (1998, 46).

11. Faber, who was the secretary of the Farm Steward Circle in Lower Dunántúl, made similar arguments on behalf of the profession of farm managers as "preservers of the nation" (*nemzetfenntartók*). Faber cited numbers showing that, in comparison to a variety of other professional groups such as merchants and railroad officials, farm managers had larger families, a sign of their dedication to the nationalist agenda. They were only surpassed by mine officials, independent craftsmen, and manorial servants (Faber 1932, 15–16). Faber would go on to become the managing director of the National Association of Hungarian Estate Managers (A Magyar Gazdatisztek Országos Egyesülete) and be a vocal advocate on their behalf.

12. Even though peasants could also benefit from the labor of family members, the countervailing pressures of inheritance weighed against that choice.

13. Scherer mentions two examples of church properties in *egyke* areas and the obvious difference in family size evident between the estate and surrounding family farms.

14. The term *puszta* in Hungarian is usually assumed to refer to a wide swath of the Great Plain in eastern Hungary. In the world of manorial estates, however, *puszta* refers to the small settlements scattered across manorial estates, isolated from the nearby villages.

15. As a shorthand, the historiographic literature tends to speak of three "Jewish Laws" in the late 1930s and 1940s: the initial law passed in 1938 stripping Jews of equal rights as citizens, the second Jewish Law of 1939 forcing Jews into labor service, and the third in 1941 barring marriage between Jews and non-Jews. In 1942, Law XV deprived Jews of the right to own property, land first and foremost; Erényi refers to it as the Fourth Jewish Law (1997, 10–12). This shorthand only describes the tip of the iceberg: laws and regulations numbered in the hundreds (for a thorough review, see Vértes 1997).

Chapter Two

1. "From 1910 to 1913 the Taft Commission on Economy and Efficiency studied the office practices of large firms in order to recommend more efficient office methods in government" (Yates 1989, 48).

2. With the corporate scandals of the first years of the twenty-first century still fresh, these debates seem far less arcane than they might once have been.

3. These are Austria, Belgium, Czechoslovakia, Finland, France, Germany, Great Britain, Italy, the Netherlands, Poland, Russia, Spain, Sweden, and Switzerland (74–90).

4. See David Shearer's provocative study *Industry, State, and Society in Stalin's Russia, 1926–1934* (1996), in which he argues that important aspects of the state-centered economic structure

NOTES TO CHAPTER 2   283

implemented in 1930 in the Soviet Union were modeled upon the German industrial cartel system (see in particular 117–21).

5. For further examples, see ILO 1926, for a discussion of a state farm in Australia, and Mehos and Moon 2011, analyzing plantation agriculture in Indonesia in the 1950s.

6. *Köztelek* was the weekly newspaper published by OMGE for wealthy landowners. Much of the newspaper was devoted to introducing new products, cutting-edge research, advice columns, and strident editorials defending manorial production.

7. Siegescu's extensive discussions of the US Department of Agriculture focused in particular on the significance of research to the mission of the department. The value of research, Siegescu argued, was demonstrated by the fact that every official at the Department of Agriculture conducted scientific work. "An official who doesn't produce scientific results cannot stay in his position for long" (Siegescu, *Köztelek*, 30 Jan. 1930). Russian experts working in the Department of Agriculture before the revolution, and who continued on in the People's Commissariat of Agriculture established by the Bolsheviks, also looked to the USDA as a model of a more progressive institution, in which scientific research was supported and specialists respected (Heinzen 2004, 41–43).

8. In pre–World War I Russia, Chaianov also used the metaphor of the agronomist as physician who would cure the peasantry, which he referred to as an "ailing body" (Kotsonis 1999, 102). In *Making Peasants Backward*, Kotsonis discusses Russian economists' and agronomic theorists' interests in scientific management and Taylorism at the beginning of the twentieth century (see chapter 4 in particular).

9. While an innovation as a research institute closely associated with the Ministry of Agriculture, this new institute followed closely in the steps of earlier scholars hoping to introduce a systematic approach to the estimation of the value of goods, labor, and property at farms and manorial estates. The most well known of these specialists is Árpád Hensch, whose study of agricultural business studies was published in 1901.

10. I have translated the Hungarian term *gazdatiszt* both as estate manager and farm steward, as they apply equally. The Hungarian-English dictionary (Országh 1974, s.v. "gazdatiszt") defines it as a manager who has finished technical school as an agricultural engineer.

11. Close to a quarter of estate managers were Jews (Faber 1932, 14), a surprising number if one considers the historical barriers for Jews to own land. Faber also comments on this unusual statistic. "The relatively large number of Jewish managers is striking, since . . . 22.3 percent is nearly four times their national proportion" (ibid.). On the other hand, although Jews had only recently acquired the rights to buy land, they were very well represented among renters of manorial estates.

12. Titles of presentations include: "The role of credentialed farmers in building a network of producers' cooperatives," "Machine and grainery cooperatives," and "Selling animals and animal products through cooperatives" (Walleshausen 1993, 143).

13. Johann Heinrich von Thünen was a well-known German economist and landowner who lived in the first half of the nineteenth century. His name is mentioned in Hungarian writings on agricultural economics in the interwar period in relation to his deductive theory of marginal productivity (see chapter 3).

14. Loft begins her analysis of cost accounting in the United Kingdom with a quote from a conference on cost accounting in June 1919 that defines accounting in very similar terms. "Costing is the 'X-ray' of commerce. Properly employed, it will penetrate the externals, reveal obstructions and irregularities, focus on danger-points, and provide a permanent evidence of unquestionable fact" (1986, 137).

Chapter Three

1. "During my apprenticeship [at an estate] when my supervisors trusted me to oversee farming work, I myself performed time studies with a watch in hand" (1925, 24). His project resulted in a publication in the 1913 issue of *Köztelek* entitled "Calculations concerning carrying tasks" (ibid.).

2. See Biernacki (1995) for a fascinating discussion of labor power in German history; see also Wise and Smith 1989a, 1989b, 1990.

3. Since sugar beets were a major cash crop for large estates, rationalizing cultivation was an urgent concern. Three separate articles in *Agricultural Work Science* (*Mezőgazdasági Munkatudomány*) addressed various aspects of the problem (two articles by Heller 1944; Szakáll 1944). In the United States, the National Research Project of the WPA (Works Progress Administration) sponsored comparable studies, examining the role of technology and its effects on labor needs for both sugar beets and corn (Macy et. al., 1937, 1938).

4. In Reichbenbach's estimation, the costs involved in "acquiring" a group of migrant workers made it impractical for smaller farms to employ them (1930, 210).

5. Wheat harvesters could, in principle, be defined as sharecroppers, since they received a portion of the crop they harvested. Since they did not oversee the entire process of planting and growing the crop, however, they were unable to influence yields and so could not moderate their results. The quality of wheat did depend on prompt harvesting, but any difference in quality was not reflected in their share, since it was stipulated at the outset of the wheat harvest season, for example one-eleventh or one-tenth of the harvest.

6. Szalay created a composite category, the work day unit or day labor day (*munkanapegység*, *napszámnap*), for his study of manual labor needs of medium and large-size farms (1931, 36, 15).

7. Disagreement about what is the best metric or analytic category is a recurring issue in science (Wise 1995).

8. Keszler makes it clear that Taylor was not the first to measure labor output but was the best known in this field (1941, 879). Along the same lines, Balogh emphasized that scientific management (*tudományos üzemszervezés*) was a generic term, Taylorism not capturing all its meanings (1930, 277).

9. Manorial estates differed substantially in the degree of capital investment. Since labor was cheap and abundant, there was little pressure to mechanize.

10. Badics never explained directly why gold crowns and wheat are suited as indices. It is worthy of note, however, that his monetary instrument to illustrate price was not contemporary Hungarian currency but the value of gold crowns as quoted in Zürich (1929, 9n; see chapter 4 for a discussion of the recurring problem Hungarians faced when national currency was unreliable). One might offer a reasonable guess that wheat became the stand-in for all other goods since this was the crop he chose to represent the buying power of goods migrant workers earn (15–16). It is clearer why wheat constitutes a valuable index for judging farm income, since as he explains, "the sale of wheat constitutes a large part of the farmer's entire earnings, while poppyseed for example is tiny" (9).

11. Kodar preferred a typology that included race psychology (*Fajpszichológia*), also known as *Rassenpsychologie* (1944, 18).

12. The term "become americanized" in Hungary meant to become a reluctant or recalcitrant worker (Révai Nagy Lexikona 1911, s.v. "amerikázás"). This is based on the active resistance American workers made to the introduction of Taylorism and scientific management, a history well enough known in Hungary at the time to coin a term.

13. Alford's description of the ways a job specification must be compiled makes explicit reference to comparable characteristics: "Type of worker most desirable for task: American or foreign, white or black; heavy and strong or light and agile; young, middle-aged, or old. This part of specification should also state temperament, intelligence, and education which best fit workman for place" (1938, 1356).

14. Skepticism regarding whether rationalized labor systems would actually lead to worker satisfaction is illustrated in an exchange of letters in late 1928 between Béla Kovrig and Gyula Szekfű concerning editorial changes for an article being published in *Magyar Szemle*. Kovrig objected to editors replacing the phrase "rationalization of work" with "sensible work" (*értelmes munka*). "The term rationalization is a generic term for certain procedures that the adjective 'sensible' replaces in no way whatsoever. Some work can be sensible and still irrational." Szekfű justified the decision by explaining that the editorial board was of the opinion that implementing rationalization in economic life would cause misery for workers. Apparently rationalization had a bad connotation (OSZK Irattár, F7/1119).

15. Some believed that workers could be untaught as well, as an editorial in early 1919 opined. "A government has never worked with greater results and effort than to teach the people to live without work" (*Köztelek*, 16 Jan. 1919).

16. Some tasks performed by manorial servants at large-farm estates could be redefined based on task, such as tending livestock.

17. Similar sentiments were voiced in an editorial in *Köztelek*: "The value of an agricultural worker's work varies depending upon what it is used for. On the other hand, by the nature of farming one cannot always devote work to where it best bears fruit. In the course of the year for the farm servant it disperses and balances out. One day's work is worth two to three times another" (*Köztelek*, 11 Feb. 1940).

18. Attempts to improve the expectations and cultural level of workers were a crucial element in national policy for rural communities, lessons that would include instilling a desire for bettering their lives. "It is definitely in the interest of agriculture that the culture of the agrarian population be higher, not only because agricultural work requires a certain degree of intelligence but also because the illiterate, unschooled worker is easier to influence politically, more quickly misled in extreme directions, than a more learned worker who has obtained proper upbringing and guidance in a good school" (Reichenbach 1925, 15).

19. Fighting against gendered prejudices in agrarian work was important in Germany after World War I when the nation suffered a great loss of manpower.

20. The dangers of money and its association with immorality are further underscored by the manner in which day laborers are often treated in these texts, as unreliable and lazy. Essentially they constitute the lumpen proletariat in the minds of business scientists.

Chapter Four

1. Aristocrats relied upon their inferiors to deal with their monetary affairs. A common tradition was to maintain what was called a "house Jew," that is, an agent employed to see to monetary affairs. "Jews were engaged in running pubs, in commerce and in many places with raising money; one could say that almost every noble man and lord had a confidential relationship with a Jew with whom he did business" (*Magyar Gazdák Szemléje*, March 1896, 15).

2. When World War I ended and the Hapsburg Empire crumbled, Hungary was thrown into disarray. In the fall of 1918, a republic was established, but in mid-March 1919 the government was toppled, replaced by a socialist republic modeled on the Soviet Union. Quickly crushed by

international forces, the People's Republic was succeeded in August 1919 by a conservative government that ruled until the end of World War II.

3. The evidence for this section of the chapter comes from a number of agrarian newspapers published between 1849 and 1896. These include *Agricultural Papers (Gazdasági Lapok)*, 1849, 1854; *Hungarian Farmer (Magyar Gazda)*, 1859; *Village Farmer (Falusi Gazda)*, 1865; *National Economic Review (Nemzetgazdasági Szemle)*, 1877, 1880; *Agricultural Review (Mezőgazdasági Szemle)*, 1885; *Hungarian Farmers' Review (Magyar Gazdák Szemléje)*, 1896; and *Agricultural Gazette (Földmívelési Ertesítő)*, 1898. I also include references to pamphlets and books written in the same time period.

4. Only two-fifths of former serfs actually received land in 1848 by virtue of the abolition of their usury obligations to the aristocracy. The rest of the agrarian labor force who were landless were deprived of their rights to property by a series of clever machinations by aristocrats redefining centuries-long contractual obligations in the 1820s and 1830s (Für 1965).

5. In this formulation, homeless (*hontalan*) meant both without a house and home and without a nation or homeland.

6. During the recent Wall Street crash, the term paper money was in use, obviously conveying the same kind of skepticism about the value of the assets being bought and sold (http://paper-money.blogspot.com/; accessed on March 6, 2010).

7. The author is referring to the practice of ennobling wealthy Jews, a practice that had become common in Hungary in the late nineteenth century but nowhere else in the region (McCagg 1972).

8. The 2008 currency crisis in Zimbabwe did not exceed Hungary's. According to the figures provided by Hanke and Kwok (2009), the monthly inflation rate in Zimbabwe was highest in mid-November 2008, 79,600,000,000 percent, in comparison to Hungary's inflation rate in July 1946, which hit 4.19×10^{16} percent.

9. Siklos offers a valuable analysis of trade between Germany and Hungary during the war, weighing the evidence against the usual hyperbole about the irreparable damage the Germans inflicted on Hungarian resources (1991, 43–52).

10. Though the general depiction of black marketers in the press and public opinion was of the immoral, greedy speculator, lots of average folks also participated. "The theft of goods such as fuel, or even raw materials from local factories by impoverished workers was endemic.... For many artisans, participation in the black market was essential to survival, to secure materials. Sometimes it involved the sub-contracting of work to skilled workers employed in large factories, who stole raw materials and used factory machinery to earn supplementary incomes, or obtain goods in kind" (Pittaway 2012, 72).

11. Bomberger and Makinen estimate that "[t]he combination of reparations and [Soviet army] occupation costs accounted for 25–50 percent of monthly expenditures by the Hungarian government during the hyperinflation" (1983, 804).

12. A variety of fiscal instruments was also contemplated, including the deposit pengő (*betét-Pengő*), farmer pengő (*gazdaPengő*) and the credit pengő (*hitelPengő*) (Botos 2006, 186).

13. In the fall of 1939, the Hungarian Economic Research Institute developed an index based on an imaginary food basket designed to provide an adult man approximately 75 percent of his weekly requirements. By this means it was able to keep careful track of price fluctuations in the cities and countryside all throughout the war, but it also provides a valuable insight into what is considered a weekly diet. As we have seen in chapter 1, the basic caloric needs of workers in various strata had already been calculated before the war.

14. PIL 274 f., 12 cs./19 ő.e., p. 154.
15. Miners received their premiums in coal as well, which gave them valuable goods to exchange on the black market (Pittaway 2012, 61).
16. PIL 274 f., 12 cs./19 ő.e., p. 153.
17. Ibid., p. 161. Family size dictated the size of a worker's calorie wage. "Due to the calorie allowance for family members, the wages of an unskilled worker with many family members often significantly surpass that of skilled workers" (ibid., p. 153).
18. Ibid., p. 159.
19. Ibid.
20. In some instances, workers were able to increase their income by doing piece work (*akkordmunka*), working overtime, and earning premiums. As elsewhere in the report, these discussions were clearly limited to industrial workers (ibid., p. 158).
21. PIL 283 f., 32 cs./6 ő.e., pp. 33–33a.
22. Ibid., p. 34.
23. "A trillion banknote consists of as many 10 Pengő notes as it would take to stack them end to end closely alongside one another to circle the earth 25 million times at the equator" (Büky 1946, 8).
24. PIL 283 f., 24 cs./31 ő.e. 1948, pp. 107–8.
25. The name of the new currency was selected from Hungarian history. It was named for the gold florin in use during Károly Róbert's reign in the fourteenth century (Botos 2006, 197).
26. PIL 274 f., 20 cs./66 ő.e., p. 112, 1946.dec.6.

Chapter Five

1. My analysis of the Ministry of Agriculture Trial is based on Anna Hamar's thorough analysis of the circumstances surrounding the trial and the personnel caught up in its wake (1997), as well as primary documents from the trial itself (BFL XXV.1a.3417/1948). After the fall of the Communist Party, a review of the case was mounted in February 1990, with the result that virtually all of the defendants were exonerated. The charges brought against them were deemed by the court not to be relevant, because the actions had not been illegal at the time, or they were judged simply as not having had any basis in fact.
2. The tension between the need for expertise and the restrictions party commitments imposed on political and economic practices plagued the new socialist state in the Soviet Union and in China, especially during the Cultural Revolution. See Haber 2013 for a discussion of this problem in the Soviet Union and Chai 1981 and MacFarquhar and Schoenhals 2006 for China.
3. In rural communities, these attitudes were clearly expressed in the dismissive attitudes people had about what were called "men in pants" (*nadrágos ember*), or outsiders to village life.
4. See Péteri's history of the changes wrought in the Academy of Sciences during this period, which included a major shift in the disciplines represented (1991).
5. Soviet forces continued fighting for four months, the Germans only being fully vanquished from Hungarian soil on April 4, 1945.
6. Two generals, Vörös and Faragho, "had passionately served the pro-German cause. Vörös had been chief of staff of the army. Faragho was a particularly odd choice. He had been the commander of the gendarmerie, an organization that had been exceedingly brutal in the extermination of the Jews and in the struggle against political oposition, in particular, Communists" (Kenéz 2006, 26).

7. I rely substantially upon György Gyarmati's excellent scholarship on politics and administrative reform during the 1940s and 1950s for my analysis (1981, 1989, 1996).

8. http:/www.1000ev.hu/index.php?a=3&param=8374, accessed on September 19, 2009.

9. The authorities didn't recognize unsigned letters as legitimate documents in screening (Palasik 2000, 72). This policy had not been in effect during the war, with horrendous consequences.

10. James Heinzen points out the contradictory attitudes the Bolshevik Party held about the state. "It is one of the great paradoxes of the Bolshevik revolution that the new regime was fully committed to using the power of the state to oversee the massive reconstruction of the social and economic spheres, yet it remained at the same time deeply suspicious of the bureaucracy that lay at its disposal" (2004, 45).

11. Ibid., 64.

12. MOL 276 f., 85 cs./ 2 ö.e., p. 7.

13. Ibid.

14. The notion of a "professional status hierarchy" was related to the corporatist vision promoted in conservative circles during the interwar period. It harkened to a feudal hierarchy of fixed status, but one based on professional competence rather than aristocratic privilege.

15. While village level committees were only legally empowered to make recommendations to the county and national level, they did far more. "The Land Claims Committees not only inventoried the lands to be expropriated, and not only did they take a position on the confiscation or redemption of properties, they actually expropriated them and distributed them immediately. They did not merely judge who was entitled to land, but they also determined who would receive how much land. They did not wait for their plan on apportionment to be sanctioned. Thus they stepped far beyond the sphere of authority established in the law!" (Donáth 1977, 52).

16. BFL XV.1a.3417/1948, 1 d., p. 31.

17. Ibid., p. 32.

18. Hamar points out that historians now agree that the primary motivation of the land reform, at least on the part of the Communist Party, was political (Hamar 1997, 44). My favorite description of the Communist Party's motivations comes from Ferenc Donáth's magisterial analysis of the land reform published in 1977, an account based in part on his own participation in Communist Party decision making at the time. "The poor peasant conception of the land reform . . . was not directly linked to the Communist Party's political strategy, which aimed at the abolition of capitalist relations of production, thus the abolition of the private property of the means of production. Surely it was not logical; in fact, it appeared theoretically mistaken and opportunistic (see the position of R. Luxemburg and others on the 1917 Russian Revolution) simultaneously to fight for the abolition of social classes and to increase immeasurably the number of smallholders. But then, in politics the geometric rule that a straight line is the shortest route between two points does not always prevail" (Donáth 1977, 41).

19. In Maya Haber's recent dissertation on post-World War II Soviet social sciences (2013), I was surprised to find a number of parallels between the reorganization of scientific research in Hungary after the war and activities occurring at exactly the same time in the Soviet Union. For example, Soviet institutions of higher learning and research after World War II were also staffed by the old guard, "bourgeois" researchers, people I would have assumed would have been replaced by then by young specialists trained in Marxist-Leninist economics. But since social science research was effectively shut down in the 1930s, the only people competent to do social research were those trained in the late tsarist period who had survived the purges. Soviet econo-

mists were only learning the basics of Marxist-Leninist scholarship in the 1950s, at the same time as young economists being trained in Hungary for the same task.

20. As we saw in chapter 2, what was considered valuable book knowledge and what was considered grounded experience of pragmatic value varied within the relatively small community of landowners and manorial personnel. This fueled recurring battles over the character of education at high schools, technical colleges, and universities.

21. Bringing professional education under the aegis of the Ministry of Agriculture was a long term goal of 1930s intellectuals and OMGE (Buday, *Köztelek*, 29 May 1929, 831–32).

22. MOL 276 f., 93 cs./104 ő.e.

23. Ibid., p. 34.

24. Ibid.

25. MOL 276 f., 54 cs./33. ő.e, p. 18.

26. As the one-time minister of agriculture explained in 1957, "'A fairly large portion of the teachers at the Agrarian University presented themselves as opposed to the objectives of the Communists, but this can be said about a substantive portion of the students as well. . . . Therefore the party resolved to conduct a thorough census of teachers and students alike at rural departments, and in the interests of making centralized party control more effective, summoned the rural departments to Budapest'" (Walleshausen 1993, 187).

27. MOL 276 f., 54 cs./10 ő.e.

28. MOL 274 f., 13 cs./15 ő.e.

29. MOL XIX-K-1-ah, 1 d., MB/53; 1949.nov.17.

30. As late as 1955 a report was issued criticizing the quality of economics training for upper-level cadres and calling for reform in higher education. "Implementing a proletarian dictatorship, the ensuing changes in the Hungarian people's economy and the building of socialism has created a great demand for economics cadres with strong professional preparation in the basics of Marxism-Leninism. This need has appeared in every branch of the people's economy. There has been a particularly strong need for economist cadres in higher government agencies—the National Planning Office, the Central Statistical Office, economic departments—who in the beginning, in the absence of a new economic intelligentsia, were forced to work almost exclusively with old experts" (MTA Lt. II. o., 183. d., 2. dosszié, p. 1E).

31. Mihály Kalocsay, interview by author, tape recording, August 8, 1997.

32. MOL 276 f., 85 cs./11 ö.e.

33. Ibid.

34. Ibid.

35. MOL XIX-K-1-ah 6 d., N-42.

36. The absence of any sort of statistical information in the early 1950s became a major stumbling block for training the new cadre of economists. Soviet economists made the same complaint (Haber 2013).

37. Ibid.

38. MOL 276 f., 93 cs./348 ő.e., pp. 89–90. It was difficult to become acquainted with Soviet policy and scholarship, because only in 1951 did the Ministry of Agriculture establish a department dedicated to translating Soviet literature.

39. Ibid., p. 92.

40. Ibid.

41. MOL 274 f., 13 cs./3 ö.e., p. 10.

42. MOL 276 f., 93 cs./81 ő.e., p. 259; 1950.jan.13.

43. Ibid.

44. This point is not only germane to the history of the Second World War. Rathenau's famous Office of War Raw Materials in World War I was the crucible for significant strategies for cartellization in 1920s Germany. War Communism, and Trotsky's militarization of the economy, also come to mind as central instruments for institutionalizing economic planning in the early Soviet Union.

45. See Borhi's excellent, thorough analysis of the Soviets' encroachment on the Hungarian economy (2004).

46. Insofar as the Soviets chose to define Austrian properties as German owned—by virtue of the 1938 Anschluss—many Austrian-owned companies in Hungary were also appropriated.

Chapter Six

1. This is not András Hegedüs, the prominent Communist Party member and later minister of state farms.

2. Movement studies, or units of movement, on the other hand, were not recorded for the purposes of determining wages, only as a part of work analysis itself (Mártonfi 1949, 65).

3. Ferenc Kalocsay, interview by author, tape recording, November 10, 1997. A former employee of this institute told me they renamed it, half jokingly, "The Work Science and Irrationalization Institute."

4. PIL 274 f., 12 cs. /160 ö.e., pp. 1–4, 9–12.

5. Ibid, p. 2.

6. Kalocsay, interview, November 10, 1997.

7. PIL 274 f., 12 cs. /16 ő.e., p. 12.

8. Ibid, p. 1.

9. Goverment edict made demands of the institute it was unprepared to take on. The happenstance character of early staffing decisions is well illustrated by the following anecdote Kalocsay told me. Kalocsay often came into the office with riding boots on, having gone horseback riding before work. Noticing his attire one morning, the head of the institute approached him to ask if he would tackle the agricultural sector now in their bailiwick. Kalocsay admitted he had grown up in the countryside and had dealt with horses, but his professional training was at a technical school for business organization. He knew nothing about agrarian economics. This small technicality did not dissuade his boss from designating him the new staff member in charge of the agricultural sector. Kalocsay's fate was sealed. After attending the Agrarian University at the behest of the institute, he went on to a stellar career in agrarian economics, publishing extensively in the field and rising to the top of the profession (e.g., Kalocsay 1961, 1968, 1970). Kalocsay, interview, August 8, 1997.

10. Kalocsay, August 8, 1997

11. O.L. XIX-K-1-j 1. d. iratszám 8140/27/4/1950, 1950.I.23.

12. Kalocsay, August 8, 1997

13. György Latkovics, interview by author, tape recording, June 18, 1997.

14. János Rainer's magisterial biography of Imre Nagy is a cornerstone of the scholarship on the 1950s, valued for its rich analytic account and careful documentation. I have relied heavily on Rainer's book throughout.

15. MOL 276 f., 53 cs. /35 ő.e.

16. Ibid., p. 8.

17. Ibid, pp. 7–8.

18. Ibid.

19. Ibid., p. 23.

20. The expression "narodnik kulak party" describes a party dedicated to populist peasant politics, a stance disdained by the Soviet Communist Party as backward and counterrevolutionary.

21. The expression "Bukharinist tendencies" is Soviet shorthand for those who oppose a fast-paced process of collectivization; it falls into the category of Rightist aberrations. It is based on the rift between Stalin and Bukharin on how quickly to implement collectivization in 1929–1930.

22. In 1953 cooperative farms were collectively working 14.5 percent of the land, of which 20.3 percent was arable land (Orbán 1972, 112). The numbers were a bit higher—18.3 percent and 26 percent—if if we include family plots alloted to them through the cooperative. As of 1956 the numbers were virtually unchanged: 14.9 and 19.8, respectively. At the same time, private farms constituted 69.4 percent of those employed in agriculture, with the remaining 30 percent split almost evenly between cooperative farms and state farms (ibid., 131).

23. The details of taxation and requisitioning in the 1950s are fairly well known, in part due to Erdmann's extensive study of these figures and processes (Erdmann 1987, 1994; see also Zavada 1986).

24. For example, families with less than 1 kh paid no land tax. Those with 1–5 kh arable acreage paid 4 kg of wheat per gold crown, the amount of requisitions they were obligated to submit to the state. By fulfilling this obligation, their land tax was waved. From there land tax levels increased, not just in how much wheat was figured per gold crown value, but so that above 6 kh, the size of the entire property—not just arable fields—was included in the calculations (Erdmann 1987, 379).

25. Extending the policy of state requisitions from agriculture imposed during the war into the socialist period was common throughout Eastern Europe.

26. All farms were required to produce a daily per capita bread ration (wheat)—for themselves and for those not engaged in subsistence farming, such as workers and city dwellers. In 1948, this amounted to 25 dkg (Erdmann 1987, 381).

27. There were possibilities of reducing the level of one's wheat requisitions by substituting other crops.

28. Families owning more than 8 kh were required to submit milk as part of their requisitions, even if they did not own a cow. In such cases, people were forced to buy the milk to fulfill their obligations (Estók et al. 2003, 292). In Nógrád County, people told me that they had to travel south to the Great Plain to buy enough wheat to fulfill their obligations.

29. MOL 276 f., 85 cs./7 ő.e., p. 2.

30. MOL 276 f., 85 cs./17 ő.e., p. 5.

31. The disdain wealthier villagers felt toward these "early adopters" of collective production was captured succinctly in the abbreviation used to refer to the third variety of farm: t-sz-cs. The shortened term "cooperative farm group" (*termelő szövetkezeti csoport*, or t-sz-cs) that was widely used for type III farms was reinterpreted to mean "a manorial servant continuing to serve" (*tovább szolgáló cseléd*), that is, early members of cooperatives still depended on others for their livelihood rather than standing on their own two feet as landowners.

32. MOL 276 f., 93 cs./33 ő.e., p. 143.

33. The fact that these statutes were being institutionalized at exactly the same time that Nagy was being censured for promoting a comparable vision of collectivization—a slow and gradual process—underscores the extent to which the debate over Nagy's proposals was not motivated solely, or even primarily, by worries over agricultural policy.

34. The flowering of alternative associations was no doubt promoted by the changes in agricultural policy initiated by Imre Nagy in the summer of 1953. At the same time thousands of cooperative farm members also returned to private farming. After Nagy was replaced by Rákosi again in 1955, and the level of coercive taxation was raised once more, the trend toward diversity was quashed. Erdei judiciously avoids any mention of Imre Nagy here, for obvious reasons.

35. Approximately 300,000 of those employed in agriculture—every eighth person—in 1949 had no land. Land was distributed to families, not individuals, in 1945, so that adult children in a household were excluded from consideration. In some instances, people had not received land because they were in prisoner-of-war camps, primarily Soviet, when land was allotted. Roma were completely shut out of the land reform, even though they constituted 25 percent of agrarian wage workers, according to Stewart's estimate (1993, 199n2).

36. MOL 276 f., 85 cs./10 ő.e.

37. MOL 276 f., 85 cs./13 ő.e., p. 3.

38. Ibid., p. 14.

39. Latkovics, interview by author, June 18, 1997.

40. Kalocsay, August 8, 1997.

41. Kalocsay, November 10, 1997.

42. Mihály Kuzmiak, interview by author, tape recording, September 10, 1997.

43. Ibid.

44. Reichenbach cautioned against generalizing the numbers he listed in earlier versions of the manorial work books (*gazdakönyv*) for the same reason. The 1944 issue of the work book encouraged estate owners to keep their own statistics in order to evaluate the numbers the book provided. This was the dream of work scientists: the promise of bookkeeping and work science was always on the horizon, if estate owners would just do their part.

45. See Lampland for a discussion of various categories of migrant labor and their temporal duration (1995, 67).

46. Latkovics, June 18, 1997.

47. I suspect that these nationalist sentiments were emphasized in the post-1989 period. This is not to say that there weren't strong prejudices against Russians, just that in the postsocialist era Hungarians may have given greater credence to these views because of their successes in late socialism.

48. Kalocsay, August 8, 1997.

Chapter Seven

1. Gail Kligman and Katherine Verdery make an important argument in their recent book on Romanian collectivization, that of the central role of the secret police in building up the party/state and achieving collectivization. "The significance of the Securitate . . . both for transforming Romania's political institutions and for implementing collectivization cannot be overstated. Its repressive apparatus was the principal weapon of political change during the early period, when the Party-state had not yet been fully institutionalized and did not control economic life" (2011, 57). While the Ministry of the Interior in Hungary played an important role in the oppressive politics of the early 1950s, it did not play the same role in local politics or collectivization as the Securitate in Romania.

2. HBML XXIII. 9./b., 43O d./388 szám, 1952.április 21.

3. An irony of bureaucratic rationalization was the practice of issuing circulars on a regular basis explaining how official papers must be dealt with and how to respond to requests. Encour-

NOTES TO CHAPTER 7   293

aging bureaucrats to use the phone instead of adding to existing correspondence was useful but not followed in this case.

4. MOL XIX-K-1-ah 3. doboz, Mb/144, 1952.május 12.
5. Ibid., 1952.május 30.
6. Ibid., 1952.jul.10. Even a minister could fall short of fulfilling the plan.
7. HBML XXIII. 9./b., 429 d., 1950 okt.
8. HBML XXIII. 9./b., 429 d./iratszám 2693, 1951.jan.7 .
9. HBML XXXII., 41. f., 2 fcs./106 ő.e., 1950.május 8–15.
10. This notion bears a strong resemblance to calls for a "bread day" among Soviet peasants in the early days of collectivization (Davies 1989, 262, 266; Viola 1999, 216). A report issued by the Bulgarian Ministerial Council and the Hungarian Central Committee of the Communist Party in June 1951 documented very similar problems with cooperative farms. No attempt was made to distinguish the quantity and quality of work completed by individual farm members. Everyone was paid the same amount, regardless of the character of their contribution (PIL 276. f., 93 cs./299 ö.e., pp. 197–98).
11. GY-M-S. M. GYL. 30 f., 2 fcs./10 ö.e., 1950.nov.8.
12. It is interesting to note that a farm in Győr-Moson-Sopron County chose not to assign each member their own individual plot to work—presumably the advantage of joining a less "socialist" cooperative—but preferred to pay each other according to the hours worked based on their own system of norms (GY-M-S. M. GYL. 30 f., 2 fcs./2. ö.e., iratszám 186, 1948.okt.29).
13. GY-M-S M. GYL. 30 f., 2 fcs./2 ö.e., iratszáma 186, 1948.okt.29.
14. HBML XXXII. 41 f., 2 fcs./104 ö.e., 1950.I.5.
15. ZML 57 f., 2 fcs./7 ö.e., 1950.feb.5.
16. ZML 57 f., 2 fcs./44 ő.e., 1949.jan.5.
17. HBML XXXII. 42 f., 2 fcs./21 ő.e., 1950.dec.24.
18. HBML XXXII. 46 f., 2 fcs./41 ö.e., 1950.okt.4.
19. ZML 57 f., 2 fcs./7 ő.e., 1952.jan.13.
20. GY-M-S. M. GYL. 30 f., 2 fcs./17 ő.e., iratszáma 489, 1949.nov.16.
21. MOL XIX-K-1-ah, 2 d. Mb/102, 1951.
22. MOL 276 f., 93 fcs./342 ő.e., p. 12, 1951.aug.22.
23. MOL 276 f., 54 cs./56 ő.e., p. 36, 1949.aug.3.
24. Letter from István Kovács to Mátyás Rákosi, PIL 276 f., 65 cs./248 ő.e., p. 13, 1949.dec.2.
25. MOL 276 f., 54 cs./79 ő.e., p. 17, 1949.dec.30.
26. ZML 57 f., 2 fcs./6 ő.e., p. 9, 1952.dec.14.
27. ZML 57 f., 2 fcs./61 ő.e., 1950.feb.28.
28. ZML 57 f., 2 fcs./6 ő.e., p. 9, 1952.dec.14.
29. HBML XXIII. 9/b., 430 d., iratszáma 366, 1952. május 15.
30. MOL 276 f., 93 cs./373 ő.e., p. 294a, 1951.március.
31. MOL 276, 93 cs./210 ő.e., p. 37, 1950.III.31.
32. MOL 76 f., 54/56 ő.e., p. 35, 1949.aug.3. According to a letter sent to Erdei in September 1949, a call to mount a labor competition for collecting oak tree seeds was the first actual mass movement of the People's Republic ("*népi demokráciánk első tömegmozgalom*"). It was the first step in a fifteen-year plan for repopulating forests, modeled on a comparable program in the Soviet Union (M.O.L. XIX-k-1-ah 6. d. N-257).
33. The apparent contradiction between being held accountable individually for one's work but at the same time being forced to work in brigades organized by the farm leadership will be more extensively explored in relation to battles over labor organization discussed in chapter 8.

34. No matter how familiar to industrial workers, akkord wages and other forms of piecerate systems were not popular. Workers fought hard against the new norm rates introduced to speed up production in the 1950s (see Pittaway 2012).

35. GY-M-S. M. GYL. 30 f., 2 fcs./2 ö.e., iratszáma 387, 1949.ápr.15.

36. The fix was in from the start. The general principles for competitions specifically stipulated that the large majority of awards and prizes would be handed out to the collective sector (PIL 276 f., 93 cs./90 ő.e., p. 132, 1949.feb.25). Two years later the Ministry of Agriculture was criticized for not supporting competitions sufficiently and for not mobilizing private farmers in particular (MOL 276 f., 93 cs./348 ö.e., p. 264, 1951.feb.2.).

37. MOL 276 f., 93 cs./90 ö.e., p. 132, 1949.feb.25.

38. HBML XXIII. 9/b, 430 d., iratszáma 2705, 1951.ápr.17.

39. MOL 276 f., 93 cs./210 ő.e., p. 141, 1950.jul.1.

40. MOL 276 f., 93 cs./264 ő.e., p. 135, 1951.feb.28.

41. MOL 276 f., 93 cs./348 ő.e., p. 264, 1951.feb.2.

42. MOL XIX-K-1-ah, 2 d., Mb/22, 1951.feb.20.

43. MOL 276 f, 93 cs./348 ő.e., pp. 263–264, 1951.feb.2.

44. MOL 276 f., 93 cs./104 ő.e., pp. 252–53.

45. ZML 57 f., 2 fcs./7 ő.e., p. 14, 1950.feb.5.

46. Large numbers of aristocrats, former army officers, business owners, and other unreliables were exiled from the city in the early 1950s. Entire families were banished to the countryside, sent to specific settlement camps, prison farms, state farms, or villages to live in horrible conditions. One fellow recalled how, as a child, his ear froze to the ground since his accommodations in the midst of winter were a simple tent. He then joked that if he were ever to be deported again, he would know now what to pack for the trip: blankets, warm clothing, and a shovel. People deported in the early 1950s constituted a separate class from those sent to prison camps or gulags for crimes.

47. GY-M-S. M. GYL. 30 f., 2 fcs./29 ő.e., iratszáma 39, 1951.dec.30.

48. MOL 276 f., 85 cs./31 ő.e., p. 25b, 1949.jul.29.

49. ZML 57. f., 2 fcs./4 ő.e., 1950.május.9.

50. Iconic class enemies were represented as collective abstractions—the kulak, the black enemy (the priesthood), the black marketeer—but when charges were brought in specific cases, then the whole force of the propaganda machine was aimed at one individual, describing in detail the pernicious ways he had eroded the moral fabric. (I use the male pronoun deliberately, as the number of men charged with offenses was disproportionately greater than that of women.)

51. "Living in the closed off area of the Hortobágy there were 7,244 persons, among whom 2,057 were kulaks and declassé elements" (Hantó 1993, 114).

52. HBML XXIII. 3/a, 73 d., iratszáma 13–31/1952, 1952.jun.9, 1952.nov.12. It seems absurd that the depth of furrows or the spaces between plants could constitute a crime against the state. But this was the case. As Pál Romány, who was Minister of Agriculture from 1975 to 1980, recalled, "Mostly here [in Hungary] they declared deep tillage to be 'socialist.' The so-called shallow tillage—which G. Adolf Manninger, the former manager of the manorial estate in Fürged, had adapted from American Dry Farming to local conditions—was branded as a *capitalist procedure*. (They didn't consider preserving the soil's moisture and the smaller costs of farming a socialist advantage)" (Romány 1996, 490).

53. MOL XIX-K-1-ah, 9 d., N/1515, 1951.november.

54. In a classic case of different agendas clashing within the government, a letter (1951.okt.18)

from the Finance Ministry to the Agricultural Department of the MDP Central Committee describes a plan to insist the cooperative farm members pay back taxes—in cash and in kind—despite being relieved of these obligations when they joined the cooperative farm. The only solution to satisfy the Finance Ministry was to kick the new members out of the cooperative in order for them to pay their taxes, a clear violation of the Agricultural Department's mandate of increasing the size of cooperative farms (MOL 276 f., 93 cs./ 264 ő.e., p. 419.)

55. MOL XIX-K-1-ah, 9 d., N/1515, 1951.november.
56. MOL 276 f., 66 cs./68 ő.e., p. 41.
57. The term gold crown is the nineteenth-century designation of the quality of land—referring to its monetary value—instituted in the cadastral survey of 1853. How plots of land were evaluated was heavily influenced by the politics following the defeat of the Hungarians' War of Independence in 1848–49. As a result the gold crown value could vary widely depending on one's political allegiance or relative wealth. To the present day, these numbers are used as metrics for the quality of land, even though more than 150 years have passed since the land was first evaluated (see Lampland 1998).
58. MOL XIX-K-1-ah 9 d., N/1515. Szabó complained in his letter to Erdei that, having acquired a good sense of the mood in the countryside, he was having a hard time writing his newest novel. He was not alone. A number of populist writers were shaken by the conditions in villages in the early 1950s and struggled to find a way of describing the reality outside the strictures of Communist Party propaganda.
59. HBML XXXIII, 41 f., 2 fcs./106 ő.e., 1950.ápr.16.
60. MOL 276 f., 93 cs./348 ő.e., pp. 153–2?, 1951.feb.7.
61. MOL 276 f., 74 cs./10 ő.e., pp. 220–221, late 1952.
62. HBML XXIII. 3/a, 73 d./iratszáma 13–31/1952, 1952.szept.10.
63. ZML 57 f., 2 fcs./53 ő.e., 1950.jul.29.
64. HBML XXXII. 46 f., 2 fcs./41 ő.e., 1950.okt.4.
65. HBML XXXII. 46 f., 2 fcs./93 ő.e., 1950.júl.12.
66. ZML 57 f. 2 fcs./20 ő.e., 1949.aug.1.
67. HBML XXXII, 46f., 2 fcs./41 ő.e., 1950.szept.9-
68. HBML XXIII. 3./a., 1 d., 1951.júl.17.
69. ZML 57 f., 2. fcs./54 ő.e., 1951.júl.13.
70. GY-M-S M. GYL 30 f., 2 fcs./1. ő.e., 1950.júl.12.
71. ZML 57 f., 2 fcs./6 ő.e., 1952.feb.28.
72. GY-M-S M. GYL 30 f., 2 fcs./29 ő.e., 2925 i.sz., 1951.feb.22.
73. ZML MDP.mg 57 f., 2 fcs./7 ő.e., 1952.jan.13.
74. HBML XXIII. 9/b., 429 d., 1950.okt.10.
75. HBML XXIII. 9/b., 430 d./420 i.sz., 1952.máj.16.
76. Ibid.
77. ZML57 f., 2 fcs./54 ő.e., 1952.nov.15.
78. HBML XXXII 41 f., 2 fcs./93 ő.e., 1950.jul.21.
79. GY-M-S M. GYL 30 f., 2 fcs./5 ő.e., 422 i.sz., 1949.ápr.4. For many years when I heard the term "administrative measures," I thought it referred to benign, bureaucratic techniques. In fact, it refers to the excesses of the regime: use of force, intimidation, terror, and so on during the Stalinist period and later in Hungary. It is important not to confuse the analysis of bureaucratic rationalization I am discussing here with the specific meaning of "administrative measures" more commonly used in the documents to mean an intrusive and punitive state action.

80. GYMS 30 f., 2 fcs. /1 ő.e., 2804 i.sz., 1950.ápr.6.
81. HBML XXXII. 41 f., 2 fcs. /49 ő.e., 21. d., 1949.dec.15.
82. HBML XXXII. 41 f., 2 fcs. /93 ő.e., 1950.ápr.11.
83. GY-M-S M. GYL 30 f., 2 fcs. /5 ő.e., 421 i.sz., 1949.március.28.
84. HBML XXXII. 46 f., 2 fcs. /41 ő.e., 1950.szept.9.
85. ZML 57 f., 2 fcs. /53 ő.e., 1950.júl.29.
86. A weak and ineffective party apparatus clashes with the usual image of the Hungarian Stalinist state as all powerful and oppressive, that is, truly authoritarian. It also would seem to be at odds with the pervasive cruelty of class warfare being waged incessantly. This is not, however, a contradiction, since the absence of a well-functioning and competent party apparatus made it more difficult to control policy implementation locally than would otherwise have been the case, leading to extensive abuse.
87. ZML 57 f., 2 fcs. /7 ő.e., 1952.jan.13.
88. HBML XXXII. 46 f., 2. fcs. /42 ő.e., 1950.május.16.
89. HBML XXIII. 3/a., 73 doboz, 18-1(46), 1952.jul., 1952.márc.10.
90. MOL XIX-K-1-ah, 7 d., N/682.
91. The term lumpen was a shorthand for the dregs of society: criminal elements, malingerers, and, often, Roma or gypsies.
92. GY-M-S. M. GYL. 30 f., 2 fcs.,/11 ő.e., 1950. julius.17.
93. GY-M-S. M. GYL 30 f., 2 fcs. /18 ő.e., 1951.XII.23.
94. The ambitious were not deterred from using this dangerous weapon, circumventing their superiors by writing directly to Rákosi to expose a dangerous cabal in their midst. Erdei took a different tack. He was regularly asked to vouch for people who had been active in the National Peasant Party, confirming or denying their leftist credentials. (Erdei had been publicly active in the NPP before 1948, but he is now classified by commentators as a cryptocommunist, a communist party member who hid his affiliation by aligning himself with another political party.) Erdei's response was always the same, explaining that the before 1948, the person in question had exhibited narodnik (populist) leanings—a no-no among Marxist-Leninists—but had now come around to the Communist point of view.
95. ZML XXIII. f., 8 fcs. /27 ő.e., 1952.apr.16.
96. Soviet economists voiced exactly the same complaints. In 1956, economists appealed to Krushchev to deal with the "'completely abnormal situation' induced by the inaccessibility of the 'state reports for scientific generalization and economic analysis. . . . Our economic science cannot successfully develop without numbers and facts'" (Haber 2013, chap. 2, p. 19).
97. Ibid.
98. The same sort of tensions characterized the situation of the first Commissariat of Agriculture in the Soviet Union vis-à-vis other government agencies. "Since the peasantry had seized (and the state subsequently had nationalized) nearly all state, private, and royal family land in 1917 and 1918, the Commissariat of Agriculture served the function, in essence, of a commissariat of the peasantry, a ministry representing a teeming 'petty bourgeoisie' in the world's only country where private property had been permanently outlawed. The Commissariat's leadership took their assignment very seriously, openly stating that they believed that they represented, and would labor on behalf of, the economic interests of the peasantry" (Heinzen 2004, 5).
99. MOL XIX-K-1-ah, 3 doboz, Mb/197.
100. Ibid, N/198.

NOTES TO CHAPTER 8   297

Chapter Eight

1. "The criticism was well founded since Rákosi et. al. substantially augmented the number of ministeries. The situation is well illustrated with the following list: there was a separate ministry for general machine industry and medium machine industry, just as there was a ministry for construction and for the construction materials industry. The management of education also developed in interesting ways within the framework of various ministeries for public education, primary education and adult education" (Estók et al. 2003, 296).

2. MOL 276 f., 85 cs./31 ő.e., pp. 4–5, 1949.aug.1.

3. MOL 276 f., 53 cs./90 ő.e.

4. An ideological battle over idealism versus materialism in biology was waged in the Soviet Union during the 1920s and 1930s, a Stalinist campaign we usually associate with Lysenko. "The idealists—following in the tradition of the Germans Gregor Mendel and August Weismann and the American T. H. Morgan—isolated the key to evolution in chromosomes. The materialists—a label Lysenko assigned to himself—believed that environment played the central, if not sole, role in evolution. The materialists also rejected outright the attempt to identify hereditary material in cells" (Pollock 2006, 42).

5. MOL 276 f., 85 cs./55 ő.e., p. 7, 1950.jul.10.

6. MOL XIX-K-1-ah 4 d., N/204, 1952.

7. The Soviet party/state also leveled these criticisms against social scientists at home in the period following World War II as the social sciences were being rejuvenated (Haber 2013, 9).

8. MOL XIX-K-1-ah, 10 d., N/4, 1952.jan.3.

9. ZML 57 f., 2 fcs./6 ő.e., pp. 2–3, 1952.dec.14.

10. Ibid., p. 5. This sentiment is remarkably Lamarckian.

11. Ibid.

12. MOL 276 f., 85 cs./54 ő.e., p. 6.

13. HBML XXIII. 9/b., 149 cs./14-FM-34/953, 1953.jun.1.

14. A Ministry for State Farms was established for two short periods in the 1950s to complement the Ministry of Agriculture: January 1952 to July 1953 and again from October 1954 to December 1956 (Bölöny 1978, 183).

15. MOL 276 f., 93 cs./szki 420 ő.e., p. 286, 1953.ápr.11.

16. Ibid.

17. Ibid, p. 287.

18. MOL XIX-K-1-ah, 5 d., 1953.feb.17.

19. Reichenbach had also been a member of the presidential council of the Hungarian Economics Research Institute.

20. Like Reichenbach, Heller also changed his name after the war. Among other experts who were featured in earlier chapters, we know that Kovrig emigrated to the United States and Világhy, once dean at Mosonmagyaróvár, was demoted to a teaching position in agricultural economics at the new agricultural university in 1949. Czettler died in custody, having fallen victim to charges of conspiracy against the state in the notorious priests' trial in 1952. Magyary, Kenéz, and Badics all died within a year of the end of World War II.

21. MOL 276 f., 66 cs./328 ő.e., p. 118, 1953.nov.11.

22. Knowing the rudiments of Marxism-Leninism did not guarantee that one's interpretation of concepts would be shared with all members of the Hungarian Communist Party. One of the former heads of the institute, Péter Erdős, committed a serious ideological faux pas and was expelled from the party. "The legend in Budapest claims that at one of his lectures in economics

he illustrated the Marxist economic concept of improductive labor with the example of Mátyás Rákosi, about which—if true—the 'wise leader' surely wasn't pleased" (Péteri 1998, 201). See Péteri's fine analysis of the restructuring of research institutes and the Academy of Sciences in this period (1993, 1997).

23. Magyar Tudományos Akadémia Levéltar, II. osztály (Közgazdaságtudomány), 183 doboz, 6 dosszié, p. 4.

24. Other institutions also gained a reputation as refuges for problematic individuals. "The specialized farms were the gathering place for enemy elements who were politically unreliable and had been dismissed from the Ministry of Agriculture and the domain of state farming. They guarantee high salaried jobs for various 'phony' experts. For example, [at one specialized farm] the director had been the village notary . . . The agronomist leading the brigade had had a farm of 2,000 kh [1,140 ha]" (MOL 276 f., 93 cs./488 ö.e., p. 2, 1954.nov.19).

25. MOL 276 f., 93 cs./289 ö.e., p. 412, 1951.jul.5.

26. MOL 276 f., 93 cs./289 ö.e., pp. 412–13, 1951.jul.5.

27. MOL 276 f., 93 cs./289 ö.e., pp. 223–26, 1951.szept.26.

28. "It is common that the young people in gardening earn 2 1/2, 3 work units for tasks requiring no special expertise, while members whose work in other branches of the firm is more difficult and requires greater expertise are given 1–1.2 work units" (ibid., p. 224).

29. Ibid, p. 223.

30. GY-M-S. M. GYL 57 f., 2 cs./17 ö.e., 1949. nov.11–14.

31. MOL 276 f., 93 cs./289 ö.e., p. 223, 1951.szept.26.

32. The textbook was published as the product of a working group, without the authors specified. Based on the fact that Imre Nagy lectured regularly at the university, and exercised significant influence over the students in economics relating to agriculture, I am fairly confident in assuming that Nagy played a role in drafting this text.

33. T-376-12.339-52-28, p. 42, 1952.

34. Ibid.; emphasis in the original.

35. Ibid., p. 43; emphasis in the original.

36. Ibid.

37. ZML 57 f., 2 cs./1 ö.e., p. 62.

38. GY-M-S. M. GYL 30 f., 2 fcs./31, 1952.jan.27.

39. "It is only possible to dismiss or fire a bookkeeper if he did not perform his duties the way it is written in the by-laws (embezzlement, etc.), and even then only with the consent of the district Agricultural Department" (HBML XXIII. 9/b., 429 d., iratszáma 1338, p. 2–3, 1951.dec.8).

40. GY-M-S. M. GYL XXIII. f., 9 fcs./31-28-5/1953, p. 1, 1953.okt.15.

41. HBML XXIII. f., 9b fcs./73 d., 14-49/952.VI., p. 1, 1952.ápr.11.

42. HBML XXXII. f., 41 fcs./106 ö.e., 1950.jun.12.

43. ZML 57 f., 2 fcs./64 ö.e., 1951.május 21.

44. GY-M-S. M. GYL XXIII. f., 9 fcs./31-28-5/1953, p. 1, 1953.okt.15

45. HBML XXIII. f., 9b fcs./546 sz., p. 3, 1952.nov.26,.

46. ZML 57 f., 2 fcs./65 ö.e., P-624-10-c/1955, p. 3, 1955.nov.1.

47. HBML XXIII., 9/b, 430 d., 389 sz., 1952.ápr.18.

48. Ibid.

49. HBML XXIII., 9/b., 13-i/20/953, p. 2, 1953.dec.21.

50. ZML XXIII. f., 8. fcs./27 ö.e., T-20-1-16-a-III., 1952.ápr.16.

51. GY-M-S. M. GYL XXIII. f., 9 fcs./31-28-5/1953, p. 1, 1953.okt.15.

52. HBML tanács.vb 388 sz., p. 2, 1951.ápr.21. In Romania, government officials and cooperative farmers dealt with the same sorts of problems in the 1950s: the fits and starts around standardizing forms (Kligman and Verdery 2011, 166–67) and the avalanche of paperwork overwhelming bureaucrats (ibid., 175).

53. GY-M-S. M. GYL 30 f., 2 fcs./31 irat., 1952.jan.27.

54. Withholding money from the state was a widespread phenomenon, as illustrated by the frequency with which putting the state first was a criterion for excellence in bookkeeping contests. "The cooperative farm completely fulfilled its obligations—in kind and monetary obligations—to the state" (GY-M-S. M. GYL XXIII. f., 9 fcs./31-67-1/1953, 1953.dec.14).

55. ZML XXIII. f., 8. fcs./58 ő.e., száma E-624-1-13-c-VII/1954., jun.17, jul.12.

56. MOL 276 f., 93 cs./441 ö.e., p. 27–28, 1953.dec.14.

57. Orbán describes the members who exited in 1953 as primarily those who had joined cooperative farms late, under duress, after August of the previous year (Orbán 1972, 133).

58. MOL 276 f., 65/69 ö.e., p. 26, 1954.febr.6.

59. MOL 276 f., 65 cs./69 ö.e., p. 9, 1954.jan.25.

60. Ibid, p. 8.

61. MOL 276 f., 93 cs./420 ö.e., pp. 173–94.

62. This is reminiscent of the housing arrangements for migrant workers before World War II and caused the same concerns about compromising morality. "This harms the farm's moral authority and results in a bad mood among honorable workers" (ibid., p. 186).

63. Ibid, p. 174.

64. Ibid, p. 178.

65. GY-M-S. M. GYL 30 f., 2 fcs./19 ő.e., 1953.XI.17.

66. Ibid.

67. MOL XIX-K-1-ah 10 d., N/405, 1952.ápr.1.

68. GY-M- S. M. GYL XXIII. f, 9 fcs./23-17-3, 1953.ápr.19 and 29.

69. MOL 276 f., 93 cs./289 ő.e., p. 324, 1951.szept.26.

70. GY-M-S. M. GYL 30 f., 2 fcs./19 ö.e., 1953.aug.26.

71. MOL 276 f., 93 cs./441/c ő.e., p. 27, 1953.dec.14.

72. MOL 276 f., 53 cs./180 ő.e., 1954.jun.9.

73. MOL 276 f., 93 cs./441/c ő.e., p. 33; 1953.nov.10.

74. ZML 57 f., 2 cs./1 ő.e., p. 62

75. Kligman and Verdery discuss similar cases of villagers petitioning the party/state to rectify unfair treatment, in response to the "Party's invitation to reject as unjust their assignment to [kulak] status and to request reclassification. In this way, the Party created a form of dialogue that signaled the 'democratic' give and take of the regime, paternalistically concerned about its citizens' grievances" (2011, 352–53). Enrolling the populace *emotionally* in the party/state's projects was a pervasive feature of Stalinist governance, whether it be in the form of competitions at work, public meetings, or identifying class enemies (see chapter 7). While the documentary record suggests that these attempts were regularly rebuffed, it is important to recognize when villagers might have seen these overtures as a means of calling the regime to account, to stand by the Communist Party's promises. Such appears to have been the case with the legal suits surrounding the dissolution of cooperative farms in 1953–1954.

76. ZML 57 f., 2 fcs./6 ő.e., 37/8-999, 1953.aug.6.

77. ZML XXIII. f., 8 fcs./57 ő.e., IX.Ált.626-14/1954, 1954.feb.9.

78. ZML 57.f., 2 fcs./7 ő.e., 1953.IX.27.

79. ZML XXIII. 8/57 ő.e., IX.Ált.626-14/1954, 1954.feb.9.

80. Ibid.

81. In some instances, the cooperative farm members came to a resolution about how to solve the labor issue. At Rákóczi cooperative in Ságod, everyone got their portion in kind, whether staying in or leaving. With regard to work done after the annual meeting, the former members did not have their work chronicled in work units, but they were paid a daily wage (15 forint). "This happened at the request of the members who had left, because it was the position of those who had left that a work unit is worthless, so they'll only work from now on for cash" (ZML XXIII. 8/57 ő.e., S-624–6-c-VII/1954, 1954.ápr.5.). This action brought censure from county authorities.

82. ZML XXIII. 8./57 ő.e., IX.Ált.626-14/1954, 1954.feb.9; emphasis in the original.

83. MOL 276 f., 53 cs./127 ő.e., 1953.VII.22.

84. ZML 57 f., 2 fcs./7 ő.e., 1953.sept.27.

85. ZML XXIII. 8/57 ő.e., IX.Ált.626-14/1954, 1954.feb.9.

86. MOL 276 f., 53/127 ő.e. , p. 11–12, 1953.jul.20.

87. HBML XXIII. 9/b. d., 149 cs./13-i/20/953, 1953.dec.21.

88. HBML XXIII. 9/b. d., 149 cs./13-i/9/953, 1953.maj.29.

89. MOL 276 f., 85 cs./5 ő.e., p. 3.

90. ZML 57 f., 2 cs./7 ő.e., 1952.jan.13.

91. GY-M-S. M. GYL 30 f., 2. fcs./2 ő.e., 387é 1949.ápr.15., p. 1.

92. HBML XXXII. f., 41 fcs./106 ő.e., 1950.május.22–27.

93. HBML XXXII. f., 41 fcs./90 ő.e., p. 8, 1950.május.27.

94. GY-M-S. M. GYL 30 f., 2 fcs./2 ő.e., 387, 1949.ápr.15.

95. GY-M-S. M. GYL 30 f., 2 fcs./29 ő.e., 39, 1951.dec.30.

96. I must admit that it took me a long time to figure out this distinction as I plowed through county documents.

97. HBML XXXII f., 41 fcs./106 ő.e., 1950.majus.22–27.

98. HBML XXXII f., 41 fcs./87 ő.e., p. 3, 1952.május.8.

99. HBML XXIII. 9/b., 150 cs., 625-H-2/81/953, 1953.aug.3.

100. MOL 276 f., 93 cs./507 ő.e., 1951, p. 313.

101. ZML XXIII f., 8 cs./57 ő.e., T-624-26-c-VII/1954, 1954.jul.24.

102. Ibid.

103. Ibid.

104. MOL 276 f., 93 cs./510 ő.e., p. 89, 1954.ápr.22.

105. Ibid., p. 90, 1954.ápr.22.

106. Ibid., p. 89.

107. MOL XIX-K-1-j 38 d.

108. MOL XIX-K-1-j 36 d., VI-322–24/1.

109. MOL 276 f., 93 cs./510 ő.e., pp. 75–87, 1954.ápr.23.

110. The numbers to be reached within the year are staggering: 2,500 agronomists were to be moved from machine stations to cooperatives, 1,800 agronomists would be sent to village councils, and another 1,500 young agronomists should train alongside agronomists still working at machine stations and state farms (ibid., 84–85).

111. Ibid., p. 84, 1954.ápr.23.

112. Ibid., p. 79, 1954.ápr.23.

113. Ibid., 80.

114. Erdei's research interests had long inclined to firm-level analyses, a strong contrast to

Hegedüs's predeliction for taking a more macro approach to economic policies in agriculture. Operations management in agricultural economics were put on the backburner in the dark period of "sovietizing" agricultural policy in 1951 and 1952 (Latkovics, personal communication).

115. MOL XIX-K-1-ah 5 d., 1954.nov.20.
116. MOL 276 f., 65 cs./69 ö.e., p. 11, 1954.jan.25.
117. MOL 276 f., 93 cs./620 ö.e., p. 615.
118. XIX-K-1-ah 5 d., Mb/26/1, p. 3, 1954.nov.20.
119. Counting the number of days worked (*munkanap*) as a proxy was proposed to tide members over until the final results of the harvests were known.
120. MOL XIX-K-1-ah 5 doboz., Mb/26/1, p. 4.
121. Ibid.
122. Ibid., p. 162; emphasis in the original
123. MOL 276 f., 93 cs./477 ö.e., pp. 161–77, 1955.jan.24.
124. Ibid., pp. 163–64.
125. This is the term Szalay had used for his analysis of labor costs (see chapter 3).
126. MOL 276 f., 93 cs./477 ö.e., p. 167, 1955.jan.24.
127. The need to simplify wage systems was not an issue just at cooperative farms. A proposal was put forward by the Ministry of Agriculture's Planning Department in July 1956 suggesting all state farm workers shift onto a time-based wage system, combining wage funds. The Wage and Labor Affairs Department rejected this proposal on the grounds that it would lead to a drop in output and productivity. The department had already suggested introducing time wages or time wages plus premium for appropriate areas of production (MOL XIX-K-1-j 48 doboz, VI-322-Ált.-9, 1956.jul.23).
128. Ibid., 9.
129. Donáth went to the Soviet Union, a bizarre turn of events after having been in prison for nearly two years. But a swift move between a position of prestige and one of disrepute was common, as was the reverse.
130. MOL XIX-K-1-j 47 d., VI-322-P-25, 1955.feb.2, jun.4.
131. I have always wondered whether the 1955 withdrawal of Russian troops stationed in Austria since World War II could have given Hungarians hope that their situation as an occupied country might change as well. If so, this could have fueled political ambitions leading to the revolution in 1956. Once the Soviets invaded to crush the revolution, any lingering notion that Hungarians could alter their status in Cold War politics died.

Conclusion

1. A lively debate concerning the value of postcolonial approaches to studies of the postsocialist transition has been very productive (Atanasoski 2013; Cervinkova 2012; Chari and Verdery 2009).

2. I recognize that the sample of documents I have reviewed may have skewed my perceptions of the Stalinist party/state, leading me to emphasize the normality of everyday bureaucratic affairs as its usual depiction. Interviews conducted were intended to offset this potential problem. Nonetheless, I am fully aware that other domains besides economic policy in agriculture may have been managed differently.

3. Andreas Glaeser has argued in his fascinating book *Political Epistemics* (2011) that the East German regime lacked the institutional capacity to reform itself from within. "The institutional arrangements making up the party state systematically undercut both the deepening of locally

produced knowledge and its systematic integration into an overarching analysis of socialism within a larger social world" (xvi). This problem did not characterize the Hungarian party/state in anywhere near the same proportion. While it was true that questioning the most basic principles of Marxism-Leninism, and hence the political legitimacy of the leaders in power, was quickly quashed, a number of significant structural criticisms voiced within the party/state and by dissidents made their way into internal debates within the party/state, eventually implemented in reforms years hence. Thus haggling over procedures and policies that I saw in document after document—though justified by the ostensible infallibility of the Communist party trumpeted in public—could and did accrue to change policies over time.

Bibliography

Aereboe, Friedrich. 1930. *Wirtschaft und Kultur in den Vereinigten Staaten von Nordamerika* [*Economy and Culture in the United States of North America*]. Berlin: Verlagsbuchhandlung Paul Parey.

———. 1928. *Agrarpolitik: Ein Lehrbuch* [*Agricultural Policy: A Text Book*]. Berlin: Verlagsbuchhandlung Paul Parey.

———. 1920. *Allgemeine Landwirtschaftsliche Betriebslehre* [*A General Study of Agricultural Business Operations*]. Berlin: Verlagsbuchhandlung Paul Parey.

Agárdy, Jenő. 1943. "A Munkajog Fejlődése" ["The Development of Labor Law"]. *Új Európa* 3 (1): 149–57.

Alford, L. P., ed. 1938. *Cost and Production Handbook*. Eighth ed. New York: The Ronald Press Company.

Ambrus, Béla. 1979. *A Magyarországi Tanácsköztársaság Pénzrendszere* [*The Monetary System of the Hungarian Soviet Republic*]. Budapest: Akadémiai Kiadó.

Arnold, A. J., and S. McCartney. 2003. "'It may be earlier than you think': Evidence, Myths and Informed Debate in Accounting History." *Critical Perspectives on Accounting* 14: 227–53.

Atanasoski, Neda. 2013. *Humanitarian Violence: The US Deployment of Diversity*. Minneapolis: University of Minnesota Press.

"Az Egyházi Vagyon és a Rerum Novarum" ["Church Property and the Rerum Novarum"]. 1941. In *Rerum Novarum: XIII. Leo Pápa Szociális és Társadalomújító Szózatának Hatása Szent István Magyar Birodalmában 1891–1941* [*Rerum Novarum: The Influence of Pope Leo XIII's Sermon on Society and Rejuvenating Society in the Hungarian Empire, 1891–1941*], ed. Rerum Novarum Emlékbizottság, 54–56. Budapest: Rerum Novarum Emlékbizottság.

Babb, Sarah. 2001. *Managing Mexico: Economists from Nationalism to Neoliberalism*. Princeton: Princeton University Press.

Baczoni, Gábor. 2000. "Négy Törvénysertő Pér Utóélete" ["The Afterlife of Four Illegal Trials"]. In *Államvédelem a Rákosi-korszakban* [*State Security in the Rákosi Era*], ed. György Gyarmati, 239–361. Budapest: Történeti Hivatal.

Badics, József. 1929. "Az áralakulás hatása a mezőgazdaság jövedelmezőségére az 1913–1925 években" ["The Influence of Changes in Prices on Profitability in Agriculture During the Years

1913–1925"]. PhD diss., Kir. Magyar Egyetemi Közgazdaságtudományi Kar Mezőgazdasági Szakosztályának Mezőgazdasági Üzemtani Intézete.

Bailes, Kendall. 1981. "The American Connection: Ideology and the Transfer of American Technology to the Soviet Union, 1917–1941." *Comparative Studies in Society and History* 23 (3): 421–48.

Balás, Károly. 1941. "Jólét és Tulajdon" ["Prosperity and Property"]. *Közgazdasági Szemle* 65 (84): 425–34.

———. 1939. "Az Állami Beavatkozás Mértéke a Gazdasági Életbe" ["The Degree of State Intervention in Economic Life"]. *Közgazdasági Szemle* 63 (82): 915–20.

Balogh, Károly. 1930. "A Racionális Önköltségszámítás" ["Rational Cost Accounting"]. *Mezőgazdasági Üzem és Számtartás* 2 (6–7): 142–46.

Balogh, Margit. 1998. *A KALOT és a Katolikus Társadalompolitika, 1935–1946* [*KALOT and Catholic Social Policy, 1935–1946*]. Társadalom- és Művelődéstörténeti Tanulmányok, vol. 23. Budapest: MTA Történettudományi Intézet.

Balogh, Sándor, ed. 1986. *Nehéz Esztendők Krónikája. 1949–1953. Dokumentumok* [*The Chronicle of Difficult Years, 1949–1953. Documents*]. Budapest: Gondolat.

Balogh, Sándor, István Birta, Lajos Izsák, Sándor Jakab, Mihály Korom, and Péter Simon. 1978. *A Magyar Népi Demokrácia Története, 1944–1962* [*The History of the Hungarian People's Democracy, 1944–1962*]. Budapest: Kossuth Könyvkiadó.

Barany, Zoltán. 1995. "Soviet Takeovers: The Role of Soviet Advisors in Mongolia in the 1920s and in Eastern Europe after World War II." *East European Quarterly* 28 (4): 409–33.

Barry, Andrew, and Don Slater. 2002b. "Technology, Politics and the Market: An Interview with Michel Callon." *Economy and Society* 31 (2): 285–306.

———. 2002a. "Introduction: The Technological Economy." *Economy and Society* 31 (2): 175–93.

Bateman, Ian J., and Kenneth G. Willis, eds. 1999. *Valuing Environmental Preferences: Theory and Practice of the Contingent Valuation Method in the US, EU, and Developing Countries.* Oxford: Oxford University Press.

Beissinger, Mark R. 1988. *Scientific Management, Socialist Discipline, and Soviet Power.* Cambridge: Harvard University Press.

Berdahl, Daphne, Matti Bunzl, and Martha Lampland. 2000. *Altering States: Ethnographies of Transition in Eastern Europe and the Soviet Union.* Ann Arbor: University of Michigan Press.

Berend, Ivan T. 1996. *Central and Eastern Europe, 1944–1993: Detour from the Periphery to the Periphery.* Cambridge: Cambridge University Press.

Berend, Iván T., and György Ránki. 1958. *Magyarország Gyáripara a Második Világháború előtt és a Háború Időszakában (1933–1944)* [*The Manufacturing Industry before and during the Second World War*]. Gazdaságtörténeti Értekezések, vol. 2. Budapest: Akadémiai Kiadó.

Bereznai, Aurél. 1943. *Munkaerőtervgazdálkodás a Mezőgazdaságban* [*Planning Labor Power in Agriculture*]. Budapest: Magyar Közigazgatástudományi Intézet.

Biernacki, Richard. 1995. *The Fabrication of Labor: Germany and Britain, 1640–1914.* Studies on the History of Society and Culture. Berkeley: University of California Press.

Bojkó, Béla. 1997. *Államkapitalizmus Magyarországon, 1919–1945* [*State Capitalism in Hungary, 1919–1945*]. Budapest: Püski.

Bokovoy, Melissa. 1998. *Peasants and Communists: Politics and Ideology in the Yugoslav Countryside, 1941–1953.* Pittsburgh: University of Pittsburgh Press.

Bölöny, József. 1978. *Magyarország Kormányai 1848–1975* [*The Governments of Hungary, 1848–1975*]. Budapest: Akadémiai Kiadó.

Bomberger, W. A., and G. E. Makinen. 1983. "The Hungarian Hyperinflation and Stabilization of 1945–1946." *Journal of Political Economy* 91 (5): 801–24.

Borhi, László. 2004. *Hungary in the Cold War, 1945–1956*. Budapest; New York: Central European University Press.

Botos, János. 2006. *A Korona, Pengő és Forint Inflációja (1900–2006)* [*Inflation of the Korona, Pengő, and Forint (1900–2006)*]. Budapest: Szaktudás Kiadó Ház.

Braun, Kathrin, and Cordula Kropp. 2010. "Introduction: Beyond Speaking Truth? Institutional Responses to Uncertainty in Scientific Governance." *Science, Technology & Human Values* 35 (6): 771–82.

Breslau, Daniel. 2003. "Economics Invents the Economy: Mathematics, Statistics, and Models in the Work of Irving Fisher and Wesley Mitchell." *Theory and Society* 32: 379–411.

———. 1998. *In Search of the Unequivocal: The Political Economy of Measurement in US Labor Market Policy*. Westport, CT: Praeger.

Brinkmann, Th. 1925. *Aus dem Betrieb und der Organisation der Amerikanischen Landwirtschaft* [*On the Operation and Organization of American Agriculture*]. Berlin: Verlag von Paul Parey.

Büky, József. 1946. "A Pengőtől a Forintig" ["From the Pengő to the Forint"]. *Budapest* (augusztus): 3–9.

Çalışkan, Koray. 2010. *Market Threads: How Cotton Farmers and Traders Create a Global Commodity*. Princeton: Princeton University Press.

Çalışkan, Koray, and Michel Callon. 2009. "Economization, Part 1: Shifting Attention from the Economy towards Processess of Economization." *Economy and Society* 38 (3): 369–98.

Callon, Michel. 2005. "Why Virtualism Paves the Way to Political Impotence: A Reply to Daniel Miller's Critique of *The Laws of the Markets*." *Economic Sociology: European Electronic Newsletter* 6 (2): 3–20.

———. 1998. *The Laws of the Markets*. Sociological Review Monograph Series. Oxford, UK; Malden, MA: Blackwell Publishers/Sociological Review.

———. 1998. "Introduction: The Embeddedness of Economic Markets in Economics." In *The Laws of the Markets*, ed. Michel Callon, 1–57. Oxford: Blackwell Publishers.

Campbell, Joan. 1989. *Joy in Work, German Work: The National Debate, 1800–1945*. Princeton: Princeton University Press.

Canguilhem, Georges. 1991. *The Normal and the Pathological*. Trans. Carolyn R. Fawcett. New York: Zone Books.

Carruthers, Bruce, and Wendy Espeland. 1991. "Accounting for Rationality: Double-Entry Bookkeeping and the Rhetoric of Economic Rationality." *American Journal of Sociology* 97 (1): 31–69.

Cartwright, Nancy. 2005. "The Vanity of Rigour in Economics: Theoretical Models and Galilean Experiments." In *The "Experiment" in the History of Economics*, eds. P. Fontaine and R. Leonard, 135–53. New York: Routledge.

———. 1983. *How the Laws of Physics Lie*. Oxford: Clarendon Press.

Casper, Monica, and Adele Clarke. 1998 "Making the Pap Smear into the 'Right Tool' for the Job: Cervical Cancer Screening in the USA, circa 1940–95." *Social Studies of Science* 28 (2): 255–90.

Cervinkova, Hana. 2012. "Postcolonialism, Postsocialism, and the Anthropology of East-Central Europe." *Journal of Postcolonial Writing* 48 (2): 155–63.

Chai, Trong R. 1981. "The Chinese Academy of Sciences and the Cultural Revolution: A Test of the 'Red and Expert' Concept." *Journal of Politics* 43 (4): 1215–29.

Champ, Patricia A., Kevin J. Boyle, and Thomas C. Brown, eds. 2003. *A Primer on Nonmarket Valuation.* Dordrecht: Kluwer Academic Publishers.

Chari, Sharad, and Katherine Verdery. 2009. "Thinking between the Posts: Postcolonialism, Postsocialism, and Ethnography after the Cold War." *Comparative Studies in Society and History* 51 (1): 6–34.

Collins, Harry, and Robert Evans. 2009. *Rethinking Expertise.* Chicago: University of Chicago Press.

Connelly, John. 2000. *Captive University: The Sovietization of East German, Czech, and Polish Higher Education, 1945–1956.* Chapel Hill: University of North Carolina Press.

Corrigan, Philip, Harvie Ramsay, and Derek Sayer. 1978. *Socialist Construction and Marxist Theory: Bolshevism and its Critique.* New York: Monthly Review Press.

Cravens, Hamilton. 1978. *The Triumph of Evolution: American Scientists and the Heredity-Environment Controversy, 1900–1941.* Philadelphia: University of Pennsylvania Press.

Creed, Gerald. 1998. *Domesticating Revolution: From Socialist Reform to Ambivalent Transition in a Bulgarian Village.* University Park: Pennsylvania State University Press.

Csizmadia, Andor. 1979. *Bürokrácia és Közigazgatási Reformok Magyarhonban* [*Bureaucracy and Administrative Reform in Hungary*]. Budapest: Gondolat Kiadó.

———. 1976. *A Magyar Közigazgatás Fejlődése a XVIII: Századtól a Tanácsrendszer Létrejöttéig* [*The Development of Hungarian Administration from the Eighteenth Century to the Foundation of the Soviet System of 1919*]. Budapest: Akadémiai Kiadó.

Csizmadia, Sándor. 1896. *A Földmívelő-munkásság Helyzete és Feladata* [*The Situation and Task of the Agrarian Working Class*]. Orosháza: Kellner Albert Könyvnyomdája.

Czettler, Jenő. 1995 [1923]. "Agrárkérdés és Katolicizmus" ["Catholicism and the Agrarian Question"]. In *Mezőgazdaság és Szociális Kérdés* [*Agriculture and the Social Question*], ed. Jenő Czettler, 157–62. Budapest: Századvég Kiadó.

———. 1941. "A Rerum Novarum és a Földkérdés" ["The Rerum Novarum and the Land Question"]. In *Rerum Novarum: XIII. Leo Pápa Szociális és Társadalomújító Szózatának Hatása Szent István Magyar Birodalmában 1891–1941* [*Rerum Novarum: The Influence of Pope Leo XIII's Sermon on Society and Rejuvenating Society in the Hungarian Empire, 1891–1941*], ed. Rerum Novarum Emlékbizottság, 33–34. Budapest: Rerum Novarum Emlékbizottság.

Davies, R. W. 1989. *The Soviet Economy in Turmoil, 1929–1930.* Basingstoke: Palgrave Macmillan.

———. 1980. *The Soviet Collective Farm, 1929–1930.* Vol. 2. Cambridge: Harvard University Press.

Davis, S., K. Menon, and G. Morgan. 1982. "The Images that Have Shaped Accounting Theory." *Accounting, Organizations, and Society* 7 (4): 307–18.

Devinat, Paul. 1927. *Scientific Management in Europe.* Geneva: International Labour Office.

Didier, Emmanuel. 2007. "Do Statistics 'Perform' the Economy?" In *Do Economists Make Markets? On the Performativity of Economics*, ed. Donald MacKenzie, Fabian Muneisa, and Lucia Siu, 276–310. Princeton: Princeton University Press.

DiMaggio, Paul J., and Walter Powell. 1991. "The Iron Cage Revisited: Institutional Isomorphism and Collective Rationality in Organizational Fields." In *The New Institutionalism in Organizational Analysis*, ed. Walter Powell and Paul J. DiMaggio, 63–82. Chicago: University of Chicago Press.

Dimock, Marshall E. 1960. "Management in the USSR.—Comparisons to the United States." *Public Administration Review* 20 (3): 139–47.

Donáth, Ferenc. 1977. *Reform és Forradalom: A Magyar Mezőgazdaság Strukturális Átalakulása*

1945–1975 [*Reform and Revolution: The Structural Transformation of Hungarian Agriculture, 1945–1975*]. Budapest: Akadémiai Kiadó.

Drori, Gili S., John W. Meyer, Francisco O. Ramirez, and Evan Schofer. 2003. *Science in the Modern World Polity: Institutionalization and Globalization.* Stanford: Stanford University Press.

Dunn, Elizabeth C. 2004. *Privatizing Poland: Baby Food, Big Business, and the Remaking of Labor.* Ithaca: Cornell University Press.

Éber, Ernő. 1941. "Mezőgazdasági Cselédbér—Mezőgazdasági Cselédkereset" ["The Wages of Agricultural Servants—The Income of Agricultural Servants"]. *Közgazdasági Szemle* 65 (84): 43–62.

———. 1939. "A Mezőgazdasági Belterjességének Közgazdasági Jelentősége" ["The Economic Significance of Agricultural Intensification"]. *Közgazdasági Szemle* 63 (82): 695–713.

———. 1937. "A Parasztgazdálkodás Javításának Lehetőségei" ["The Possibilities of Improving Peasant Farming"]. *Magyar Szemle* 31 (1): 38–43.

Eckstein, Alexander. 1952. "The Economic Development of Hungary, 1920 to 1950: A Study in the Growth of an Economically Underdeveloped Area." PhD. diss., University of California, Berkeley.

Edwards, Richard. 1979. *Contested Terrain: The Transformation of the Workplace in the Twentieth Century.* New York: Basic Books.

Egyeki, Sándor. 1948. "A Mezőgazdasági Munkabérpolitika Fejlődése" ["The Development of Agricultural Wage Policy"]. In *Szakszervezeti Tanács által Rendezett Bértitkári Tanfolyam Előadássorozata*, 1–20. Budapest: Szabados András.

Elson, Diane, ed. 1979. *Value: The Representation of Labour in Capitalism.* London: CSE Books.

Erdei, Ferenc. 1956. "A Termelősszövetkezeti Mozgalom és a Közös Gazdaságok Felépítése" ["The Movement of Producer Cooperatives and Collective Farms"]. *Társadalmi Szemle* 5:13–33.

———. 1941. *A Magyar Paraszttársadalom* [*Hungarian Peasant Society*]. Budapest: Franklin Társulat.

Erdélyi, Mihály. 1936. *A Pszichotechnika Alapkérdései* [*The Basic Questions of Psychotechnique*]. Budapest: Királyi Magyar Egyetemi Nyomda.

Erdmann, Gyula. 1994. "A Paraszti Érdekképviselet Ügyének Bukása Magyarországon 1945–1947" ["The Failure of Peasant Interest Representation, 1945–1947"]. In *Paraszti Kiszolgáltatottság—Paraszti Érdekvédelem, Önigazgatás* [*Peasant Exploitation—Peasant Interest Representation and Self-Government*], ed. Gyula Erdmann, 121–77. Gyula: Békés Megyei Levéltár.

——— 1987. "A Beszolgáltatási Rendszer Magyarországon 1949–1953" ["The Requisitions System in Hungary, 1949–1953"]. *Agrártörténeti Szemle* 29 (3–4): 379–411.

Esbenshade, Richard S. 2006. "The Populist-Urbanist Debate in Hungary and the Divided Construction of Hungarian National Identity, 1929–1944." PhD diss., University of California, Santa Cruz.

Espeland, Wendy. 1998. *The Struggle for Water.* Chicago: University of Chicago Press.

Espeland, Wendy, and Mitchell Stevens. 1998. "Commensuration as a Social Process." *Annual Review of Sociology* 24: 313–43.

Estók, János, György Fehér, Péter Gunst, and Zsuzsanna Varga. 2003. *Agrárvilág Magyarországon 1848–2002* [*The Agrarian World in Hungary, 1848–2002*]. Budapest: Argumentum Kiadó, Magyar Mezőgazdasági Múzeum.

Eyal, Gil, Iván Szelényi, and Eleanor Townsley. 1998. *Making Capitalism without Capitalists: Class Formation and Elite Struggles in Post-Communist Central Europe.* London: Verso.

Faber, György. 1941. *A Magyar Gazdatisztek Országos Egyesülete Évi Jelentése. 1940. évi Munkasságáról* [*The Yearly Report of the National Association of Hungarian Estate Managers*]. A Magyar Gazdatisztek Országos Egyesületének Kiadványai no. 4. Budapest: Bethlen Gábor Irodalmi és Nyomdai Rt Nyomása.

———. 1932. *A Magyar Gazdatiszti Kar Válsága* [*The Crisis of the Hungarian Profession of Estate Managers*]. Az Alsódunántúli Gazdatiszti Kör Kiadványai no. 2. Kaposvár: Szabó Lipót Könyvnyomda és Könyvkiadóvállalat.

Farkas, Árpád. 1941. *A Mezőgazdasági Kézimunka Módszeres Vizsgálata* [*A Methodical Examination of Agricultural Work Done by Hand*]. Kolozsvár: Minerva Irodalmi és Nyomdai Műintézet R.-T.

Fehér, György. 2003. "A Termelés Változó Feltételei" ["The Shifting Conditions of Production"]. In *Agrárvilág Magyarországon 1848–2002*, ed. János Estók, György Fehér, Péter Gunst, and Zsuzsanna Varga, 91–155. Budapest: Argumentum Kiadó.

Fellner, Kálmán. 1933. "Az Orosz Mezőgazdasági Politika Újabb Törekvései" ["The Newest Ambitions of Russian Agricultural Policy"]. *Mezőgazdasági Közlöny* 6 (6–7): 279–99.

Ferleger, Louis, and William Lazonick. 1994. "Higher Education for an Innovative Economy: Land-Grant Colleges and the Managerial Revolution in America." *Business and Economic History* 23 (1): 116–28.

———. 1993. "The Managerial Revolution and the Developmental State: The Case of US Agriculture." *Business and Economic History* 22 (2): 67–98.

Fischer, Andor. 1936. "Munkatanulmányok és Költségszámítások a Burgonyatermesztés Köréből" ["Labor Studies and Cost Calculations in the Domain of Potato Cultivation"]. PhD diss., Műszaki és Gazdaságtudományi Egyetem Mezőgazdasági Osztályának Üzemtani Intézetéből.

Fitzgerald, Deborah. 2003. *Every Farm a Factory: The Industrial Ideal in American Agriculture.* Yale Agrarian Studies Series. New Haven: Yale University Press.

———. 1996. "Blinded by Technology: American Agriculture in the Soviet Union, 1928–1932." *Agricultural History* 70 (3): 1928–32.

Fleck, Ludwig. 1979. *Genesis and Development of a Scientific Fact.* Chicago: University of Chicago Press.

Fodor, Éva. 2003. *Working Difference: Women's Working Lives in Hungary and Austria, 1945–1995.* Durham: Duke University Press.

Fodor, Ferenc. 1937. "A Falukutató Mozgalom Kritikája" ["The Critique of the Village Research Movement"]. *Magyar Szemle* 30 (május): 23–33.

Földművelésügyi Minisztérium Szakoktatási Főosztálya. 1950. *Munkaegységszámolás és Jövedelemelosztás a Termelőszövetkezetekben. Üzemtan. V. füzet* [*Figuring Work Units and Distributing Income in Cooperative Farms. Business Studies. V Booklet*]. Budapest: Fővárosi Nyomda.

Földmívelésügyi Minisztérium Termelőszövetkezeti Főosztálya. 1949. *Közösen Termelő Szövetkezeti Csoportok Munkaegység Könyve* [*Work Unit Book for Collectively Producing Cooperative Groups*]. Budapest: Szikra Irodalmi és Lapkiadó.

Földművelésügyi Minisztérium Termelőszövetkezeti Ostálya. 1949, 1952. *Munkaegység Könyv* [*Work Unit Book*]. Budapest: Földművelésügyi Minisztérium Termelőszövetkezeti Osztálya.

Foucault, Michel. 1971. "Orders of Discourse." *Social Science Information* 10 (2): 7–30.

Fourcade, Marion. 2011b. "Price and Prejudice: On Economics and the Enchantment (and Disenchantment) of Nature." In *The Worth of Goods: Valuation and Pricing in the Economy*, ed. Jens Beckert and Patrick Aspers, 41–62. New York: Oxford University Press.

———. 2011a. "Cents and Sensibility: Economic Values and the Nature of 'Nature.'" *American Journal of Sociology* 116 (6): 1721–77. ———. 2009. *Economists and Societies: Discipline and Profession in the United States, Britain, and France, 1890s to 1990s*. Princeton: Princeton University Press.

———. 2007. "Theories of Markets and Theories of Society." *American Behavioral Scientist* 50 (8): 1015–34.

Friss, István. 1957. *Népgazdaságunk Vezetésének Néhány Gyakorlati és Elméleti Kérdéséről* [*On Several Practical and Theoretical Questions of the Leadership of Our People's Economy*]. Budapest: Kossuth Könyvkiadó.

Frommer, Benjamin. 2005. *National Cleansing: Retribution against Nazi Collaborators in Postwar Czechoslovakia*. New York: Cambridge University Press.

Für, Lajos. 1965. "Jobbágyföld-Parasztföld" ["Serf Land-Peasant Land"]. In *A Parasztság Magyarországon a Kapitalizmus Korában, 1848–1914* [*The Peasantry in Hungary during the Capitalist Period, 1848–1914*], ed. István Szabó. Budapest: Akadémiai Kiadó.

Gaal, Jenő. 1885. *A Mezőgazdasági Válság Kérdése* [*The Question of the Agricultural Crisis*]. Budapest: Pesti Könyvnyomda-Részvény-Társaság.

Galgóczy, Miklós, and László Berzsenyi-Janosits. 1942. *Gazdaszámok 95 Számtáblázat a Gazdasági Gyakorlat Minden Ágából* [*Farm Numbers, 95: Tables of Agricultural Practice from Every Branch*]. Magyaróvár-Új Vidék: A Szerzők Kiadása.

Gallhofer, Sonja, and James Haslan. 1991. "The Aura of Accounting in the Context of a Crisis: Germany and the First World War." *Accounting, Organizations and Society* 16 (5/6): 487–520.

Gaponyenko, G. 1950. "Munkaszervezés és Jövedelemelosztás a Kolhozban" ["The Organization of Labor and Distribution of Income in the Kolhoz"]. *Magyar-Szovjet Közgazdasági Szemle* 4 (1): 129–33.

Garber, Peter M., and Michael G. Spencer. 1994. *The Dissolution of the Austro-Hungarian Empire: Lessons for Currency Reform*. Essays in International Finance, vol. 191. Princeton: International Finance Section, Department of Economics, Princeton University.

Gergely, Jenő. 1999. *A Katolikus Egyház Története Magyarországon 1919–1945* [*The History of the Catholic Church in Hungary, 1919–1945*]. Budapest: Pannonica Kiadó.

———. 1989. *Katolikus Egyház, Magyar Társadalom 1890–1986: Prohászkától Lékaiig* [*Catholic Church, Hungarian Society, 1890–1986: From Prohászka to Lékai*]. Budapest: Tankönyvkiadó.

Gielen, Anne C., Marcel J. M. Kerkhofs, and Jan C. Van Ours. 2010. "How Performance Related Pay Affects Productivity and Employment." *Journal of Population Economics* 23 (1): 291–301.

Gille, Zsuzsa. 2007. *From the Cult of Waste to the Trash Heap of History: The Politics of Waste in Socialist and Postsocialist Hungary*. Bloomington: Indiana University Press.

Gillespie, Richard. 1991. *Manufacturing Knowledge: A History of the Hawthorne Experiments*. Cambridge: Cambridge University Press.

Glaser, Edward M. 1973. "Knowledge Transfer and Institutional Change." *Professional Psychology* 4 (4): 434–44.

Graeber, David. 1996. "Beads and Money: Notes towards a Theory of Wealth and Power." *American Ethnologist* 23 (1): 4–24.

Guala, Francesco. 2001. "Building Economic Machines: The FCC Auctions." *Studies in the History and Philosophy of Science* 32 (3): 453–77.

Guillén, Mauro F. 1994. *Models of Management: Work, Authority, and Organization in a Comparative Perspective*. Chicago: University of Chicago Press.

Gunst, Péter. 2003. "A Magyar Mezőgazdaság 1919 és 1945 között" ["Hungarian Agriculture between 1919 and 1945"]. In *Agrárvilág Magyarországon 1848–2002*, ed. János Estók, György Fehér, Péter Gunst, and Zsuzsanna Varga, 157–259. Budapest: Argumentum Kiadó.

Guyer, Jane I. 2004. *Marginal Gains: Monetary Transactions in Atlantic Africa*. Chicago: Chicago University Press.

Gyarmati, György, ed. 2000. *Államvédelem a Rákosi-korszakban: Tanulmányok és Dokumentumok a Politikai Rendőrség Második Világháború utáni Tevékenységéről* [*National Security in the Rákosi Period: Studies and Documents about the Activities of the Secret Police after the Second World War*]. Budapest: Történeti Hivatal.

Gyarmati, György. 1996. "A Közigazgatás Újjászervezése az 'Ideiglenesség' Korszakában" ["The Reorganization of Administration during the 'Provisional' Government"]. *Történelmi Szemle* 38 (1): 63–98.

———. 1989. "Három Koncepció és ami Utána Következik: Közigazgatási Reformtörekvések és Kudarcaik Sorozata Magyarországon 1945–1948" ["Three Concepts and What Follows: A Series of Administrative Attempts at Reform and their Failure, 1945–1948"]. *Tér És Társadalom* 4 (1–2): 3–41.

———. 1987. "A Beszolgáltatási Rendszer Magyarországon 1949–1953" ["The Requisition System in Hungary, 1949–1953"]. *Agrártörténeti Szemle* 29 (3–4): 379–411.

———. 1981. "Politikai Szempontok Érvényesülése a Tanácsrendszer Előkészítő Munkálataiban" ["The Predominance of Political Views in the Preparatory Work of the Soviet System in 1919"]. *Történelmi Szemle* 2: 178–88.

Györffy, István. 1942. *Magyar Nép, Magyar Föld* [*Hungarian People, Hungarian Land*]. Budapest: Turul Kiadás.

Haber, Maya. 2013. "Socialist Realist Science: Constructing Knowledge about Rural Life in the Soviet Union, 1943–1958." PhD. diss., University of California, Los Angeles.

Hacking, Ian. 1990. *The Taming of Chance*. Cambridge: Cambridge University Press.

———. 1983. *Representing and Intervening: Introductory Topics in the Philosophy of Natural Science*. Cambridge: Cambridge University Press.

Hajpál, Gyula. 1943. "A Mezőgazdasági Cseléd Természetbeni Járandóságainak Átértékelése" ["The Reevaluation of In-Kind Compensation for Agricultural Servants"]. *Mezőgazdasági Munkatudomány* (márc.-ápr.): 138–48.

Halács, Ágoston. 1928–29. "A Mezőgazdasági Munkaköltségek Csökkentésének Lehetősége" ["The Possibility of Reducing Labor Costs in Agriculture"]. *Mezőgazdasági Közlöny* 1 (1): 42–48.

Hamar, Anna. 1997. "Az FM-Per—Avagy Az Agrárelit Bukása" ["The Ministry of Agriculture Trial, or the Downfall of the Agrarian Elite"]. Unpublished manuscript.

Hamar, Norbert. 1944. "A Testi Munka Hatása a Veseműködésre" ["The Affect of Physical Labor on Kidney Function"]. *Mezőgazdasági Munkatudomány* (2–3): 139–45.

Hanebrink, Paul A. 2006. *In Defense of Christian Hungary: Religion, Nationalism, and Antisemitism, 1890–1944*. Ithaca: Cornell University Press.

Haney, Lynne A. 2002. *Inventing the Needy: Gender and the Politics of Welfare in Hungary*. Berkeley: University of California Press.

Hanke, Steve H., and Alex K. F. Kwok. 2009. "On the Measurement of Zimbabwe's Hyperinflation." *Cato Journal* 29 (2): 353–64.

Hann, Chris. 1980. *Tázlár: A Village in Hungary*. Cambridge: Cambridge University Press.

Hanseth, Ole, Eric Monteiro, and Morten Hatling. 1996. "Developing Information Infrastructure: The Tension between Standardization and Flexibility." *Science, Technology & Human Values* 21 (4): 407–96.

Hantó, Zsuzsa. 1993. "Kulákkérdés a Pártapparátusok és az Államvédelmi Hatóságok Iratainak Tükrében" ["The Kulak Question Reflected in the Documents of the Party Apparatus and National Security Authorities"]. *Gazdaság és Társadalom* 6: 94–121.

Harkai Schiller, Pál. 1937. "Az Intézet Működése" ["The Operations of the Institute"]. *Lélektani Tanulmányok* 1: 17–21.

Harmath, Jenő. 1937. *Új Részestermelési Rendszerek és azok Jövedelmezősége a Mezőgazdaságban* [*New Sharecropping Systems and their Profitability*]. Budapest: "Pátria" Irodalmi Vállalat és Nyomdai Részvény Társaság.

Harrison, J. F. C. 1969. *Quest for the New Moral World: Robert Owen and the Owenites in Britain and America*. New York: Scribner.

Hart, Keith. 1981. "On Commoditization." In *From Craft to Industry: The Ethnography of Protoindustrial Cloth Production*, ed. Esther N. Goody, 38–49. Cambridge: Cambridge University Press.

Hartstein, Péter. 1937. "A Mezőgazdasági Munkásság Kereseti Viszonyainak Megszervezése" ["The Income Relations of the Agricultural Labor Force"]. *Magyar Sors* Offprint.

Hatvani (Hofferik), István. 1935. *A Pszichotechnika Szerepe és Jelentősége a Munka Racionalizálásában* [*The Role and Significance of Psychotechnique in the Rationalization of Work*]. Győr: Baross-nyomda.

Hecht, Gabrielle. 2003. "Globalization Meets Frankenstein? Reflections on Terrorism, Nuclearity, and Global Technopolitical Discourse." *History and Technology* 19 (1):1–8.

Hegedüs, Ede. 1947b. "A Munkabérrendszerről. 2." ["On the Wage System. 2."] *Gazdaság* 2 (36): 40.

———. 1947a. "A Munkabérrendszerről. 1." ["On the Wage System. 1."] *Gazdaság* 1 (20): 24.

Heinzen, James W. 2004. *Inventing a Soviet Countryside: State Power and the Transformation of Rural Russia, 1917–1929*. Pittsburgh: University of Pittsburgh Press.

Heller, András. 1944. "A Cukorrépaszedés Gépesítésének Kérdése" ["The Question of the Mechanization of Harvesting Sugar Beets III"]. *Mezőgazdasági Munkatudomány* (4–5): 264–76.

———. 1941a. "Mussolini Földreformja" ["Mussolini's Land Reform"]. *Közgazdasági Szemle* 65 (84): 16–29.

———. 1941b. "Többtermelés és Földtulajdon" ["Overproduction and Landed Property"] *Közgazdasági Szemle* 65 (84): 705–16.

———. 1940. "Agrárlakosságunk Foglalkozási Tagozódása" ["The Occupational Structure of Our Agrarian Population"]. *Közgazdasági Szemle* 64 (83): 576–89.

———. 1937. *Cselédsor. A Mezőgazdasági Cselédek Helyzete 1935-ben, Különös Tekintettel a Székesfehérvári Járásra* [*Servanthood: The Situation of Agricultural Servants in 1935, with Special Consideration of the District of Székesfehérvár*]. Budapest: A Szent-István Társulat.

———. 1936. "A Gazdasági Cselédek Gyermekeinek Táplálkozása és Szociális Helyzete" ["The Nutrition and Social Situation of the Children of Manorial Servants"]. *Népegészségügy* 17 (10): 455–60.

Henisz, Witold J., Bennet A. Zelner, and Mauro F. Guillén. 2005. "The Worldwide Diffusion of Market-Oriented Infrastructure Reform." *American Sociological Review* 70 (6): 871–97.

Henke, Chris. 2000. "Making a Place for Science: The Field Trial." *Social Studies of Science* 30: 482–512.

Hensch, Árpád. 1901. *Mezőgazdasági Üzemtan* [*Agricultural Firm Studies*]. Második Javított és Bővített Kiadás, vol. I-II. Magyar-Óvár: Czeh Sándor-Féle Könyvnyomda.

Hensch, Árpád János. 1941. "A Mezőgazdasági Kísérletügy Agrárpolitikai Jelentősége" ["The

Agrarian Policy Significance of Agricultural Experimentation"]. PhD. diss., M. Kir. József Nádor Műszaki és Gazdaságtudományi Egyetem Mezőgazdasági Osztálya.

Herf, Jeffrey. 1984. *Reactionary Modernism: Technology, Culture, and Politics in Weimar and the Third Reich.* Cambridge: Cambridge University Press.

Holtfrerich, Carl-Ludwig. 1986. *The German Inflation, 1914–1923: Causes and Effects in International Perspective.* Berlin; New York: De Gruyter.

———. 1985. "Germany and Other European Countries in the 1920s." In *Inflation and Indexation: Argentina, Brazil, and Israel,* ed. John Williamson, 123–40. Washington, DC: Institute for International Economics.

Horn, David G. 1994. *Social Bodies: Science, Reproduction, and Italian Modernity.* Princeton: Princeton University Press.

Horowitz, Irving Louis. 1964. "Sociological and Ideological Conceptions of Industrial Development." *American Journal of Economics and Sociology* 23 (4): 351–74.

Hoskin, Keith, and Richard Macve. 1986. "Accounting and the Examination: A Genealogy of Disciplinary Power." *Accounting, Organizations and Society* 11 (2): 105–13.

Howard, Albert. 1931. "An Experiment in the Management of Indian Labour." *International Labour Review* 23 (5): 636–43.

Ihrig, Károly. 1947. "Összhangban Kell Hozni a Szövetkezeti Gazdálkodás és Egyéni Birtoklás Előnyeit" ["We Must Harmonize the Advantages of Cooperative Farming and Private Property"]. *Szövetkezés* 1 (44):1–2.

———. 1935. "A Mezőgazdaság Irányítása" ["The Management of Agriculture"]. *Magyar Szemle* 23: 123–31.

Illyés, Gyula. 1936. *Puszták Népe* [*The People of the Puszta*]. Budapest: Nyugat Kiadó és Irodalmi R.T.

International Labour Office. 1932. "The Collectivisation of Agriculture in the USSR." *International Labour Review* 16 (3): 386–409.

———. 1930. "An Enquiry into Conditions of Work and Wages of Agricultural Workers in Czechoslovakia." *International Labour Review* 21 (6): 855–67.

———. 1928. "Labour Cost in Agriculture in England and in Illinois." *International Labour Review* 17 (2): 240–53.

———. 1927b. "Scientific Management in Agriculture." *International Labour Review* 16 (5): 703–6.

———. 1927a. "The Science of Farm Labour: Scientific Management and German Agriculture." *International Labour Review* 15 (3): 379–413.

———. 1926. *The Relation of Labour Cost to Total Costs of Production in Agriculture.* International Economic Conference Documentation.

Iordachi, Constantin, and Arnd Bauerkämper, eds. 2014. *The Collectivization of Agriculture in Communist Eastern Europe: Comparison and Entanglements.* Budapest: Central European University Press.

Iordachi, Constantin, and Dorin Dobrincu, eds. 2009. *Transforming Peasants, Property, and Power: The Collectivization of Agriculture in Romania, 1949–1962.* Budapest: Central European University Press.

Izsák, Lajos, and Miklós Kun, eds. 1944. *Moszkvának Jelentjük . . . Titkos Dokumentumok 1944–1948* [*We Report to Moscow . . . Secret Documents, 1944–1948*]. Budapest: Századvég.

Jain, S. 2006. *Injury: The Politics of Product Design and Safety Law in the United States.* Princeton: Princeton University Press.

BIBLIOGRAPHY

Jánossy, Andor. 1933. "A Szovjet Agrárpolitikája" ["Soviet Agrarian Policy"]. PhD. diss., Budapesti Királyi Magyar Egyetemi Közgazdaságtudományi Kar.

Jarausch, Konrad H., ed. 1999. *Dictatorship as Experience: Towards a Socio-Cultural History of the GDR.* New York: Berghahn.

Jarosz, Dariusz. 2014. "The Collectivization of Agriculture in Poland: Causes of Defeat." In *The Collectivization of Agriculture in Communist Eastern Europe*, ed. Constantin Iordachi and Arnd Bauerkämper, 137–46. Budapest: Central European University.

Jasanoff, Sheila. 2004. *States of Knowledge: The Co-Production of Science and the Social Order.* New Edition. London: Routledge.

Johnson, Björn, Edward Lorenz, and Bengt-Åke Lundvall. 2002. "Why All this Fuss about Codified and Tacit Knowledge?" *Industrial and Corporate Change* 11 (2): 245–62.

Juhász, Gyula. 1983. *Uralkodó Eszmék Magyarországon, 1939–1944* [*The Ruling Ideas in Hungary, 1939–1944*]. Budapest: Kossuth Könyvkiadó.

Kadvany, John David. 2001. *Imre Lakatos and the Guises of Reason.* Durham, NC: Duke University Press.

Káli, Csaba. 1999. "Politikai Gazdasági és Társadalmi Átalakulás Zala Megyében 1947–1956" ["The Political, Economic, and Social Transformation in Zala County, 1947–1956"]. In *Dokumentumok Zala Megye Történetéből 1947–1956* [*Documents from the History of Zala County, 1947–1956*], ed. Csaba Káli, 23–39. Zalaegerszeg: Zala Megyei Levéltár.

Kalocsay, Ferenc. 1970. *A Munkadíjazási Formák a Tszekben* [*The Forms of Wages in Cooperative Farms*]. Budapest: Agrárgazdasági Kutató Intézet.

———. 1968. *Munkanormák Kidolgozása és Alkalmazása a Mezőgazdaságban* [*Drawing Up and Using Work Norms in Agriculture*]. Budapest: Közgazdasági és Jogi Könyvkiadó.

———. 1961. *A Munka és Munkaidő Vizsgálatának Egyes Módszerei* [*Several Methods of Studying Work and Labor Time*]. Budapest: Közgazdasági és Jogi Könyvkiadó.

Károly, Rezső. 1926. "A Mezőgazdasági Számtartás Ugye Németországban" ["The Issue of Agricultural Accounting in Germany"]. *Mezőgazdasági Üzemtani Közlemények* 4 (4): 49–55.

———. 1925. "A Mezőgazdasági Könyvelés Szerepe és Szervei" ["The Role and Organs of Agrarian Accounting"]. *Mezőgazdasági Üzemtani Közlemények* 3 (1–2): 1–11.

———. 1924. "A Munkaszervezeti Javítása és a Taylor-rendszer a Mezőgazdaságban" ["Fixing Labor Organization and the Taylor System in Agriculture"]. *A Mezőgazdasági Üzemtani Közlemények* 2 (3–4): 75–86.

———. 1923. "Előszó: Mezőgazdasági Üzemtani Közlemények Alapítása" ["Foreword: Establishing the Proceedings of Agricultural Economics"]. *Mezőgazdasági Üzemtani Közlemények* 1 (1): 1–3.

Károlyi, Mihály. 1977. *Hit, Illúziók Nélkül* [*Belief without Illusions*]. Budapest: Magvető.

Karpik, Lucien. 2010. *Valuing the Unique: The Economics of Singularities.* Princeton: Princeton University Press.

Kemptner, Ernő. 1939. "A Mezőgazdasági Munka Célszerüsítése és a Szociális Helyzet Javítása" ["Rationalizing Agricultural Work and the Improvement of Social Conditions"]. *Köztelek Zsebnaptár* 45. évf. (II. kötet): 27–36.

Kenessey, Kálmán. 1868. *Mezőgazdasági Munkaerő-calamitás* [*Calamity of Agricultural Labor Power*]. Pest: Ráth Mór.

———. 1858. *A Szántóvető Aranyszabályai. Rövid Utasítás a Kis Gazdák Számára* [*The Golden Rules of the Ploughman: Brief Instructions for Small Farmers*]. Pest: Herz János.

Kenez, Peter. 2006. *Hungary from the Nazis to the Soviets: The Establishment of the Communist Regime in Hungary, 1944–1948.* Cambridge: Cambridge University Press.

Kenney, Padraic. 1997. *Rebuilding Poland: Workers and Communists, 1945–1950*. Ithaca: Cornell University Press.

Kerék, Mihály. 1942. *A Földreform Útja* [*The Road to Land Reform*]. Budapest: Magyar Élet Kiadása.

———. 1939. *A Magyar Földkérdés* [*The Hungarian Land Question*]. Budapest: Mefhosz Könyvkiadó.

———. 1937. "Az Agrárproletariátus Jövője" ["The Future of the Agrarian Proletariat"]. *Magyar Szemle* 31 (szept.): 44–52.

Keszler, Gyula. 1941. "A Teljesítménnyel Arányos Munkabérfizetés" ["Payment of Wages Proportionate to Output"]. *Közgazdasági Szemle* 65 (84): 869–93.

Kesztyűs, Lajos. 1943. "Számtartásstatisztika és Termelésirányítás a Németbirodalomban" ["Statistics of Accounting and the Guidance of Production in the German Reich"]. *Mezőgazdasági Közlöny* 7: 1–11.

———. 1942. *Mezőgazdasági Üzemi Tanácsadás a Németbirodalomban* [*Agricultural Economics Advising in the German Reich*]. Debrecen: Nagy Károly Grafikai Műintézet.

———. 1932. "A Mezőgazdasági Üzemtani Tudományok és Különösen a Gazdasági Számtartás Főiskolai Tanítása" ["The Sciences of Agricultural Economics and in Particular the Technical School Teaching of Farm Accounting"]. *Mezőgazdasági Üzem És Számtartás* 4 (125): 130, 145–52.

———. 1929b. "A Magyar Mezőgazdasági Számtartás Elterjesztése és Üzemtudomány, Valamint Agrárpolitikai Fontosságának Elismertetése" ["The Dissemination of Hungarian Agricultural Accounting and Economics, and the Recognition of its Importance for Agrarian Policy"]. *Mezőgazdasági Üzem És Számtartás* 1: 49–55.

———. 1929a. "A Magyar Mezőgazdasági Számtartás Elterjesztése és Üzemtudomány, Valamint Agrárpolitikai Fontosságának Elismertetése" ["The Dissemination of Hungarian Agricultural Accounting and Economics, and the Recognition of its Importance for Agrarian Policy"]. *Mezőgazdasági Üzem és Számtartás* 1: 14–21.

Király, Béla, Barbara Lotze, and Nándor F. Driesinger, eds. 1984. *The First War between Socialist States: The Hungarian Revolution of 1956 and Its Impact*. New York: Social Science Monographs.

Kirsch, Michael. 2006. "Pay-for-Performance Medicine—Quality or Quagmire?" *American Journal of Gastroenterology* 101 (11): 2453–55.

Klein, Judy L., and Mary S. Morgan, eds. 2001. *The Age of Measurement*. Durham: Duke.

Kligman, Gail, and Katherine Verdery. 2011. *Peasants under Siege: The Collectivization of Romanian Agriculture, 1949–1962*. Princeton: Princeton University Press.

Kodar, Jenő. 1944. "Üzemvezetés és Psychológia" ["Firm Management and Psychology"]. *1944 Január 10-Én a Weiss Manfréd Vállalatok Kulturális-Szociális Osztályának Szabad Egyetemén Tartott Előadása*: 27.

Kölber, László. 1932. "Munkatanulmányok a Dunántúli Tengeritermelés köréből" ["Work Studies from the Domain of Corn Cultivation in Dunántúl"].Kir. M. Tudományegyetemi Közgazdaságtudományi Kar Mezőgazdasági Üzemtan Intézetéből.

Kosáry, Domokos G. 1983. *Művelődés a XVIII. Századi Magyarországon* [*Culture in Eighteenth Century Hungary*]. Második, változatlan Kiad ed. Budapest: Akadémiai Kiadó.

Kotkin, Stephen. 1995. *Magnetic Mountain: Stalinism as Civilization*. Berkeley: University of California Press.

Kotsonis, Yanni. 1999. *Making Peasants Backward: Agricultural Cooperatives and the Agrarian Question in Russia, 1861–1914*. New York: St. Martin's Press.

Kovács, Imre. 1941. "A Nyilások Agrárprogrammja" ["The Agrarian Program of the Arrow Cross Party"]. *Magyar Szemle* 41 (jul.): 6–13.

———. 1940. *Szovjetoroszország Agrárpolitikája* [*The Agrarian Policies of Soviet Russia*]. Budapest: Cserépfalvi Kiadó.

———. 1937. *A Néma Forradalom* [*The Silent Revolution*]. Budapest: Cserépfalvi.

———. 1935. "A Gazdasági Cselédek Kereseti és Megélhetési Viszonyai" ["The Income and Livelihood Conditions of Manorial Servants"]. *Magyar Szemle* 35 (99): 211–22.

Köver, György. 1998. *Losonczy Géza 1917–1957* [*Géza Losonczy, 1917–1957*]. Budapest: 1956-os Intézet.

Kovrig, Bennett. 1979. *Communism in Hungary: From Kun to Kádár*. Palo Alto: Hoover Institute.

Kuhn, Thomas. 2012. *The Structure of Scientific Revolutions: 50th Anniversary Edition*. Fourth ed. Chicago: University of Chicago Press.

Kulcsár, Aurél. 1947. "A Szovjet 'Könyvszámvitel' Elmélete" ["The Theory of 'Soviet' Accounting"]. *Gazdaság* 10: 438–43.

Kulin, Sándor. 1947. *A Dunántúli Parasztgazdaságok Számokban és Ábrákban* [*Peasant Farms of Dunántúl in Numbers and Figures*]. Kaposvár: Az Alsódunántúli Földmívelésügyi Tanács kiadványa.

Kuszenda, Sándor. 1948. "A Teljesítménybérek Meghatározása" ["Determining Output Wages"]. *Közgazdasági Szemle*: 106–9.

Lampland, Martha. 2011. "The Technopolitial Lineage of State Planning in Mid-Century Hungary (1930–1956)." In *Entangled Geographies: Empire and Technopolitics in the Global Cold War*, ed. Gabrielle Hecht, 155–84. Cambridge: MIT Press.

———. 2010. "False Numbers as Formalizing Practices." *Social Studies of Science* 40 (3): 377–404.

———. 2009. "Classifying Laborers: Instinct, Property, and the Psychology of Productivity in Hungary (1920–1956)." In *Standards and their Stories: How Quantifying, Classifying, and Formalizing Practices Shape Everyday Life*, ed. Martha Lampland and Susan Leigh Star, 123–42. Ithaca: Cornell University.

———. 2002. "The Advantages of Being Collectivized: Cooperative Farm Managers in the Postsocialist Economy." In *Postsocialism: Ideas, Ideologies, and Practices in Europe and Asia*, ed. Chris Hann, 72–123. London: Routledge.

———. 1998. "Corvée, Maps, and Contracts: Agricultural Policy and the Rise of the Modern State in Hungary during the 19th Century." *Irish Journal of Anthropology* 3: 7–40.

———. 1995. *The Object of Labor: Commodification in Socialist Hungary*. Chicago: University of Chicago Press.

Lampland, Martha, and Susan Leigh Star. 2009. *Standards and their Stories: How Quantifying, Classifying, and Formalizing Practices Shape Everyday Life*. Ithaca: Cornell University Press.

Landsman, Mark. 2005. *Dictatorship and Demand: The Politics of Consumerism in East Germany*. Cambridge: Harvard University Press.

Larkin, Brian. 2013. "The Politics and Poetics of Infrastructure." *Annual Review of Anthropology* 42: 327–43.

Latour, Bruno. 1988. *Science in Action: How to Follow Scientists and Engineers through Society*. Cambridge: Harvard University Press.

Laur, Ernst. 1930. *Einführung in die Wirtschaftslehre des Landbaus, unter Besonderer Berücksichtigung der Landarbeitslehre* [*Introduction to the Economics of Agriculture, with Particular Attention to Farm Labor*]. Zweite, neubearbeitete Auflage ed. Berlin: P. Parey.

Lautenberg, István. 1933. "Üzemgazdasági Statisztika és Könyvitel" ["Statistics and Bookkeeping for Firm Studies"]. *Mezőgazdasági Üzem és Számtartás* 5: 28–32.

Lave, Jean. 1988. *Cognition in Practice: Mind, Mathematics, and Culture in Everyday Life.* Cambridge: Cambridge University Press.

Law, John. 1996. "Organizing Accountabilities: Ontology and the Mode of Accounting." In *Accountability: Power, Ethos, and the Technologies of Managing*, ed. Rolland Munro and Jan Mouritsen, 283–306. London: International Thomson Business Press.

Leányfalusi, Károly, and Ádám Nagy. 1997. *Magyarország Fém- és Papírpénzei 1892–1925* [*Hungary's Coins and Paper Money, 1892–1925*]. 3 kiad ed. Csongrád: Magyar Éremgyűjtők Egyesülete.

Lebow, Katherine A. 2013. *Unfinished Utopia: Nowa Huta, Stalinism, and Polish Society, 1949–1956.* Ithaca: Cornell University Press.

Lénárd, Ödön. 1946. "Munka, Bér, Tulajdon" ["Work, Wage, Property"]. Actio Catholica Országos Elnöksége. Budapest: Korda R. T. Nyomda.

Lencsés, Ferenc. 1982. *Mezőgazdasági Idénymunkások a Negyvenes Években* [*Seasonal Workers in Agriculture during the 1940s*]. Budapest: Akadémiai Kiadó.

Lenin, Vladimir Ilyics. 1975. *The State and Revolution: The Marxist Theory of the State and the Tasks of the Proletariat in the Revolution.* Moscow: Progress Publishers.

Leopold, Gusztáv. 1911. "Kapitalisztikus Mezőgazdaság Magyarországon" ["Capitalist Agriculture in Hungary"]. *Huszadik Század* 12 évf., 24 kötet (5. szám): 533–54.

Leopold, Lajos. 1934. "Munkabér és Nagybirtok" ["Wage Labor and Large Estates"]. *Közgazdasági Szemle* 58 (77): 445–79.

Lett, Miklós, Róbert Vértes, and Gábor Székely. 2002. *Magyarországi Zsidótörvények és Rendeletek 1938–1945* [*Hungarian Jewish Laws and Regulations, 1938–1945*] 2. változatlan kiadás ed. Budapest: Polgár.

Levy, Robert. 2001. *Ana Pauker: The Rise and Fall of a Jewish Communist.* Berkeley: University of California Press.

Lewin, Moshe. 1974. *Political Undercurrents in Soviet Economic Debates: From Bukharin to Modern Reformers.* Princeton: Princeton University Press.

Liszka, Jenő. 1943. "Teljesítményprobléma a Kertészetben" ["Performance Problems in Horticulture"]. *Mezőgazdasági Munkatudomány* (jul-aug): 336–40.

Livingstone, David N. 2003. *Putting Science in its Place: Geographies of Scientific Knowledge.* Chicago: University of Chicago Press.

Loft, Anne. 1986. "Towards a Critical Understanding of Accounting: The Case of Cost Accounting in the UK, 1914–1925." *Accounting, Organizations and Society* 11 (2): 137–69.

Lucy, John, ed. 1993. *Reflexive Language, Reported Speech, and Metapragmatics.* New York: Cambridge University Press.

Lynch, Michael. 1991. "Method: Measurement—Ordinary and Scientific Measurement as Ethnomethodological Phenomena." In *Ethnomethodology and the Human Sciences: A Foundational Reconstruction*, ed. G. Button, 77–108. Cambridge: Cambridge University Press.

Lytle, Charles. 1942. *Wage Incentive Methods: The Selection, Installation, and Operation.* New York: Ronald Press Company.

Maasen, Sabine, and P. Weingart. 2006. *Democratization of Expertise? Exploring Novel Forms of Scientific Advice in Political Decision-Making.* New York: Springer Science & Business Media.

Macara, Chas W. 1926. *Modern Industrial Tendencies.* Manchester: Sherrat and Hughes.

MacFarquhar, Roderick, and Michael Schoenhals. 2006. *Mao's Last Revolution.* Cambridge: Harvard University Press.

MacKenzie, Donald A. 2006. *An Engine, Not a Camera: How Financial Models Shape Markets.* Cambridge: MIT Press.

Macy, Loring K., Lloyd E. Arnold, and Eugene G. McKibben. 1938. *Changes in Technology and Labor Requirements in Crop Production: Corn.* Works Progress Administration, National Research Project Report, no. A5.

Macy, Loring K., Lloyd E. Arnold, Eugene G. McKibben, and Edmund J. Stone. 1937. *Changes in Technology and Labor Requirements in Crop Production: Sugar Beets.* Works Progress Administration, National Research Project Report, no. A1.

Magyary, Zoltán. 1933. *Scientific Management in Public Administration.* Hungarian Institute of Public Administration, no. 12. Budapest: Royal Hungarian University Press.

———. 1930. *A Magyar Közigazgatás Racionalizálása* [*The Rationalization of Hungarian Public Administration*]. A Debreceni Tisza István Tudományos Társaság I. Osztályának Kiadványai. III. kötet, 3. sz. ed. Debrecen: Debrecen Sz.Kir. Város és a Tiszántúli Református Egyházkerület Könyvnyomdája-Vállalata.

Maier, Charles S. 1975. *Recasting Bourgeois Europe.* Princeton: Princeton University Press.

Major, Patrick, and Jonathan Osmond, eds. 2002. *The Workers' and Peasants' State. Communism and Society in East Germany under Ulbricht, 1945–1971.* Manchester: Manchester University Press.

Mark, James. 2011. *The Unfinished Revolution: Making Sense of the Communist Past in Central-Eastern Europe.* New Haven: Yale University Press.

———. 2005b. "Society, Resistance, and Revolution: The Budapest Middle Class and the Hungarian Communist State, 1948–1956." *English Historical Review* 488: 963–86.

———. 2005a. "Discrimination, Opportunity, and Middle-Class Success in Early Communist Hungary." *Historical Journal* 48 (2): 499–521.

Markó, Lajos. 1960. "A Termelőszövetkezeti Jövedelemelosztás Vizsgálata az Anyagi Érdekeltség Szempontjából Különös Tekintettel a Vezetők Javadalmazására" ["The Examination of the Distribution of Cooperative Farm Income from the Perspective of Material Incentives with a Specific Attention to Managers' Salaries"]. *Jelenetés a Mezőgazdasági Szervezési Intézet 1960. Évi Kutatói Munkájáról*: 66–70.

Marosi, Miklós. 1947. "A Szocialista Tervgazdaság Üzemgazdasági Problémái" ["The Problems of Firm Operations of a Socialist Planned Economy"]. *Gazdaság* 13: 600–602.

Mártonfi, Rudolf. 1949. *Munka- és Időelemzés. Teljesítménynormák Szakszerű Megállapítása* [*Labor and Time Studies: The Technical Determination of Output Norms*]. Budapest: Munkatudományi és Racionalizálási Intézet.

———. 1946. "Munkateljesítmény, Munkabérek, Bérrendszerek" ["Work Output, Wages, and Wage Systems"]. Üzemi Bizottsági Tanfolyam III. füzet. Budapest: Szakszervezeti Tanács Kiadása.

Marx, Karl, Ben Fowkes, and Ernest Mandel. 1976; 1981. *Capital: A Critique of Political Economy.* Pelican Marx Library. Harmondsworth, UK: Penguin Books, in association with New Left Review.

Matolcsy, Mátyás. 1943. "Terménybeszolgáltatási Módszerek" ["Methods of Requisitioning Produce"]. *Új Európa* 1 (1): 18–23.

———. 1942. "Új Európa" ["New Europe"]. *Új Európa* 1 (1): 1–6.

Maurer, Bill. 2006. "The Anthropology of Money." *Annual Review of Anthropology* 35: 15–36.

McCagg, William O. 1972. *Jewish Nobles and Geniuses in Modern Hungary.* East European Monographs, vol. 3. Boulder, CO: East European Quarterly.

Mehos, Donna, and Suzanne Moon. 2011. "The Uses of Portability: Circulating Experts in the

Technopolitics of Cold War and Decolonization." In *Entangled Geographies: Empire and Technopolitics in the Global Cold War*, ed. Gabrielle Hecht, 43–74. Cambridge: MIT Press.

Miller, Daniel. 2005. "Reply to Michel Callon." *Economic Sociology: European Electronic Newsletter* 6 (3): 3–13.

———. 2002. "Turning Callon the Right Way Up." *Economy and Society* 31 (2): 218–33.

Miller, Peter. 1994. "Accounting as Social and Institutional Practice: An Introduction." In *Accounting as Social and Institutional Practice*, eds. Anthony G. Hopwood and Peter Miller, 1–39. Cambridge: Cambridge University Press.

———. 1992. "Accounting and Objectivity: The Invention of Calculating Selves and Calculable Spaces." *Annals of Scholarship* 9: 61–86.

Miller, Peter, Trevor Hopper, and Richard Laughlin. 1991. "The New Accounting History: An Introduction." *Accounting, Organizations, and Society* 16 (5/6): 395–403.

Miller, Peter, and Nikolas Rose. 1990. "Governing Economic Life." *Economy and Society* 19 (1): 1–31.

Mirowski, Philip. 2002. *Machine Dreams: Economics Becomes a Cyborg Science*. Cambridge: Cambridge University Press.

———. 1989. *More Heat than Light: Economics as Social Physics, Physics as Nature's Economics*. New York: Cambridge University Press.

Mirowski, Philip, and Edward Nik-Khah. 2007. "Markets Made Flesh: Performativity, and a Problem in Science Studies, Augmented with Consideration of the FCC Auctions." In *Do Economists Make Markets? On the Performativity of Economics*, eds. Donald MacKenzie, Fabian Muneisa, and Lucia Siu, 190–224. Princeton: Princeton University Press.

Mitchell, David, and Sharon Snyder. 2003. "The Eugenic Atlantic: Race, Disability, and the Making of an International Eugenic Science, 1800–1945." *Disability and Society* 18 (7): 843–64.

Mitchell, Timothy. 2002. *Rule of Experts: Egypt, Techno-Politics, Modernity*. Berkeley: University of California Press.

Morgan, Mary S. 2003. "Economics." In *The Cambridge History of History of Science: The Modern Social Sciences*, ed. T. Porter and D. Ross, 275–305. Cambridge: Cambridge University Press.

Morris-Suzuki, Tessa. 2000. "Ethnic Engineering: Scientific Racism and Public Opinion Surveys in Midcentury Japan." *Positions* 8 (2): 499–529.

Muneisa, Fabian. 2007. "Market Technologies and the Pragmatics of Prices." *Economy and Society* 36 (3): 377–95.

Muneisa, Fabian, and Michel Callon. 2007. "Economic Experiments and the Construction of Markets." In *Do Economists Make Markets? On the Performativity of Economics*, ed. Donald MacKenzie, Fabian Muneisa, and Lucia Siu, 163–89. Princeton: Princeton University Press.

Munn, Nancy. 1986. *The Fame of Gawa: A Symbolic Study of Value Transformation in a Massim (Papua New Guinea) Society*. Durham: Duke University Press.

Nagengast, Carole. 1991. *Reluctant Socialists, Rural Entrepreneurs: Class, Culture, and the Polish State*. Boulder, CO: Westview Press.

Naimark, Norman. 1997. *The Russians in Germany: A History of the Soviet Zone of Occupation, 1945–1949*. Cambridge: Harvard University Press.

Nelson, Daniel. 1992. *A Mental Revolution: Scientific Management since Taylor*. Columbus: Ohio State Press.

———. 1980. *Frederick W. Taylor and the Rise of Scientific Management*. Madison: University of Wisconsin.

BIBLIOGRAPHY 319

Niemelä, Raimo, Mika Hannula, Sari Rautio, Kari Reijula, and Jorma Railio. 2002. "The Effect of Air Temperature on Labour Productivity in Call Centres—A Case Study." *Energy & Buildings* 34 (8) (09): 759.

Nogaro, Bertrand. 1948. "Hungary's Recent Monetary Crisis and its Theoretical Meaning." *American Economic Review* (4): 526–42.

Nolan, Mary. 1994. *Visions of Modernity*. New York: Oxford University Press.

Nötel, R. 1986. "International Finance and Monetary Reforms." In *The Economic History of Eastern Europe, 1919–1975*, ed. M. C. Kaser and E. A. Radice, vol. 2, 520–63. Oxford: Clarendon Press.

Oldenziel, Ruth. 2000. "Gender and Scientific Management: Women and the History of the International Institute for Industrial Relations, 1922–1946." *Journal of Management History* 6 (7): 323–42.

Orbán, Sándor. 1972. *Két Agrárforradalom Magyarországon. Demokratikus és Szocialista Agrárátalakulás 1945–1961* [*Two Agrarian Revolutions in Hungary: Democratic and Socialist Agrarian Transformation, 1945–1961*]. Budapest: Akadémiai Kiadó.

Oreskes, Naomi. 2003. "The Role of Quantitative Models in Science." In *Models in Ecosystem Science*, eds. C. D. Canham, J. J. Cole, and W. K. Laurenroth, 13–32. Princeton: Princeton University Press.

Orosz, István. 1962. "Széchenyi és a Jobbágykérdés" ["Széchenyi and the Serf Question"]. *Agrártörténeti Szemle* 4: 52–90.

Országh, László. 1974. *Magyar-angol Szótár* [*Hungarian-English Dictionary*]. Fourth unrevised ed. Vol. 1–2. Budapest: Akadémiai Kiadó.

Országos Magyar Gazdasági Egyesület. 1942. "Tudományos Intézetek és Kísérleti Állomások. Az Országos Mezőgazdasági Üzemi és Termelési Költségvizsgáló Intézet" ["Scientific Institutes and Experimental Stations: The National Institute for the Study of Agricultural Organization and Production Costs"]. *Köztelek Zsebnaptár* 48 évf. (II. kötet): 546–50.

———. 1922. "Mezőgazdasági Szakoktatási és Kísérletügyi Intézetek" ["Agricultural Vocational and Experimental Institutes"]. *Köztelek Zsebnaptár* 28 évf.: 263–66.

Palasik, Mária. 2000. *A Jogállamiság Megteremtésének Kísérlete és Kudarca Magyarországon 1944–1949* [*The Attempt and Failure to Establish a Constitutional State in Hungary, 1944–1949*]. Budapest: Napvilág Kiadó.

Pardo-Guerra, Juan Pablo. 2013. "Priceless Calculations: Reappraising the Sociotechnical Appendages of Art." *Euopean Societies* 15 (2): 196–211.

Parmentier, Richard. 1994. "Pierce Divested for Nonintimates." In *Signs in Society: Studies in Semiotic Anthropoloogy*, ed. Richard Parmentier, 3–22. Bloomington: Indiana University Press.

Perneczky, Béla. 1943. *Koreszméink* [*The Ideas of Our Age*]. Budapest: Királyi Magyar Egyetemi Nyomda.

Péteri, György. 1998. "A Fordulat a Magyar Közgazdaság-tudományban. A Magyar Gazdaságkutató Intézettől a Közgazdaság-tudományi Intézetig" ["The Change in Hungarian Economic Sciences: From the Hungarian Economics Research Institute to the Institute for Economic Science"]. In *A Fordulat Évei. 1947–1949*, ed. Éva Standeisky, Gyula Kozák, Gábor Pataki, and M. János Rainer, 185–201. Budapest: 1956-os Intézet.

———. 1997. "New Course Economics: The Field of Economic Research in Hungary after Stalin, 1953–56." *Contemporary European History* 6 (3): 295–327.

———. 1993. "The Politics of Statistical Information and Economic Research in Communist Hungary, 1949–56." *Contemporary European History* 2 (2): 149–67.

———. 1991. "Academic Elite into Scientific Cadres: A Statistical Contribution to the History of the Hungarian Academy of Sciences, 1945–49." *Soviet Studies* 43 (2): 281–99.

———. 1989. "Engineer Utopia: On the Position of Technostructure in Hungary's War Communism, 1919." *International Studies of Management and Organization* 19 (3): 82–102.

Pető, Iván, and Sándor Szakács. 1985. *A Hazai Gazdaság Négy Évtizedének Története. 1945–1985* [*Four Decades of the Domestic Economy, 1945–1985*]. Budapest: Közgazdasági és Jogi Könyvkiadó.

Peirce, Charles S. 1972. *The Essential Writings*, ed. Edward C. Moore. New York: Harper and Row.

Pintér, János. 1990–1991. "Adalékok a Mezőhegyesi Állami Gazdaság Történetéhez (1945–1948)" ["Contributions to the History of the State Farm of Mezőhegyes (1945–1948)"]. *A Magyar Mezőgazdasági Múzeum Közleményei*: 243–62.

Pittaway, Mark. 2012. *The Workers State: Industrial Labor and the Making of Socialist Hungary, 1944–1958*. Pittsburgh: University of Pittsburgh Press.

Polich, Simon. 1930. "A Gazdasági Számvitel Tanításáról" ["Teaching Business Accounting"]. *Mezőgazdasági Üzem És Számtartás* 2 (4–5): 115–6.

Pollock, Ethan. 2006. *Stalin and the Soviet Science Wars*. Princeton: Princeton University Press.

Poovey, Mary. 1998. *A History of the Modern Fact: Problems of Knowledge in the Sciences of Wealth and Society*. Chicago: University of Chicago Press.

Porter, Theodore M. 1995. *Trust in Numbers*. Princeton: Princeton University Press.

Postone, Moishe. 1993. *Time, Labor, and Social Domination: A Reinterpretation of Marx's Critical Theory*. Cambridge: Cambridge University Press.

Powell, Richard C. 2007. "Geographies of Science: Histories, Localities, Practices, Futures." *Progress in Human Geography* 31 (3): 309–29.

Quattrone, Paolo. 2004. "Accounting for God: Accounting and Accountability Practices in the Society of Jesus (Italy, XIV-XVII Centuries)." *Accounting, Organizations, and Society* 29: 647–83.

Rabinbach, Anson. 1992. *The Human Motor: Energy, Fatigue, and the Origins of Modernity*. Berkeley: University of California Press.

Rádóczy, Gyula. 1984. *A Legújabb Kori Magyar Pénzek (1892–1981)* [*Hungarian Monies from the Most Recent Period, 1892–1981*]. Budapest: Corvina.

Rainer, János M. 1999. *Nagy Imre. Politikai Életrajz II. 1953–1958* [*Imre Nagy: Political Biography, Vol. II, 1953–1958*]. Budapest: 1956-os Intézet.

Raith, Tivadar. 1930. "A Racionalizálásról" ["About Rationalization"]. *Magyar Szemle* 10 (11): 248–58.

Reichenbach, Béla. 1944. "Gazdálkodjunk Előirányzattal!" ["Let's Farm Using Estimates!"]. *Köztelek Zsebkönyv* 50 (2): 28–34.

———. 1930. *A Mezőgazdasági Üzem Berendezése és Szervezése. A Mezőgazdasági Termelés Alapkellékei és ezek Üzemszervezési Vonatkozásai. Első kötet* [*The Organization and Arrangement of the Agricultural Firm: Basic Requirements of Agricultural Production and the Conditions of their Firm Organization, Vol. 1*]. Budapest: "Pátria" Irodalmi Vállalat és Nyomdai Részvénytársaság.

———. 1925. *A Munka Eredményének Fokozása a Mezőgazdaságban* [*The Intensification of the Results of Labor in Agriculture*]. Budapest: "Pátria" Irodalmi Vállalat és Nyomdai Részvénytársaság.

Reitzer, Béla. 1941. "A Munkapiac Helyzete a Munkaközvetítés Reformja Előtt" ["The Situation of the Labor Market before the Reform of the Labor Exchange"]. *Közgazdasági Szemle* 65 (84): 996–1013.

Rege, Károly. 1929. "A Nemzetközi Mezőgazdasági Számtartási Statisztikai Irányelvei" ["The International Guiding Principles of Statistics for Agricultural Accounting"]. *Mezőgazdasági Üzem és Számtartás* 1: 1–11.

Révay, József. 1946. "A Munka és a Szórakozás Lélektana" ["The Psychology of Work and Leisure"]. *Lélektani Tanulmányok* 8: 86–107.

Rézler, Gyula. 1944. "A Magyar Ipari Munkatudományi Intézet" ["The Hungarian Industrial Work Science Institute"]. Budapest: A Magyar Ipari Munkatudományi Intézet.

———. 1940. *A Paraszt Szovjetoroszországban* [*The Peasant in Soviet Russia*]. Budapest: Stádium Sajtóvállalat.

Ries, L. W. 1930. *Die Menschlichen Arbeitskräfte der Landgutswirtschaft* [*The Human Workforce of the Estate Economy*]. Berlin: Parey.

———. 1924. *Leistungen und Lohn in der Landwirtschaft* [*Productivity and Wages in Agriculture*]. Berlin: Parey.

Robson, Keith. 1992. "Accounting Numbers as 'Inscription': Action at a Distance and the Development of Accounting." *Accounting, Organizations, and Society* 17 (7): 685–708.

Roediger, David, and Elizabeth D. Esch. 2012. *The Production of Difference: Race and the Management of Labor in US History*. New York: Oxford University Press.

Rogger, Hans. 1981. "*Amerikanizm* and the Economic Development of Russia." *Comparative Studies in Society and History* 23 (3): 382–420.

Roitman, Janet. 2005. *Fiscal Disobedience: An Anthropology of Economic Regulation in Central America*. Princeton: Princeton University Press.

Roman, Eric. 1996. *Hungary and the Victor Powers, 1945–1950*. First ed. New York: St. Martin's Press.

Romanov, R. Sz. 1948. "A Munka Termelékenység Fokozására Irányuló Intézkedések a Szovjet-Unióban" ["Measures Directed to Increase the Productivity of Labor in the Soviet Union"]. *Közgazdasági Szemle*: 283–90.

Romány, Pál. 1996. "A Nagyüzemi Átszervezés és a Mezőgazdaság Integrációja" ["The Integration of Agriculture and Large-Scale Firm Reorganization"]. In *Magyarország Agrártörténete. Agrártörténeti Tanulmányok*, ed. István Orosz, Lajos Für, and Pál Romány, 503–30. Budapest: Mezőgazda Kiadó.

Romsics, Ignác. 1999. *Magyarország Története a XX. Században* [*The History of Hungary in the Twentieth Century*]. Budapest: Osiris.

Róna-Tas, Ákos. 1997. *The Great Surprise of the Small Transformation: the Demise of Communism and the Rise of the Private Sector in Hungary*. Ann Arbor: University of Michigan Press.

Rotman, Brian. 2000. *Mathematics as Sign: Writing, Imagining, Counting*, eds. Timothy Lenoir and Hans Ulrich Gumbrecht. Writing Science ed. Stanford: Stanford University Press.

Roth, Louise Marie. 2006. "Because I'm Worth it? Understanding Inequality in a Performance-Based Pay System." *Sociological Inquiry* 76 (1) (02): 116–39.

Rubin, Eli. 2008. *Synthetic Socialism: Plastics and Dictatorship in the German Democratic Republic*. Chapel Hill: University of North Carolina Press.

Santos, Ana, and João Rodrigues. 2009. "Economics as Social Engineering? Questioning the Performativity Thesis." *Cambridge Journal of Economics* 33: 985–1000.

Schandl, Károly. 1941. "A Szövetkezeti Mozgalom Fundamentuma" ["The Fundamentals of the Cooperative Movement"]. In *Rerum Novarum. XIII. Pápa Szociális és Társadalomújító*

Szózatának Hatása Szent István Magyar Birodalmában 1891–1941, ed. Rerum Novarum Emlékbizottság, 48–50. Budapest: Rerum Novarum Emlékbizottság.

Scherer, Péter Pál. 1943. "Munkabér a Mezőgazdasági Számvitelben" ["Wages in Agricultural Accounting"]. *Mezőgazdasági Munkatudomány* (szept.-okt.): 441–48.

———. 1941. "Mezőgazdasági Nagyüzemek Szociális Gondossága és Feladata" ["The Social Care and Task of Agricultural Large Enterprises"]. In *Rerum novarum. XIII. Leo Pápa Szociális és Társadalomújitó Szózatának Hatása Szent István Magyar Birodalmában 1891–1941*, ed. Rerum Novarum Emlékbizottság, 51–53. Budapest: Rerum Novarum Emlékbizottság.

———. 1940. "A Nagybirtok Kérdése" ["The Question of Large Estates"]. In *A Mezőgazdasági Haladás Legujabb Vívmányai. az 1940. Évi Gazdatisti Továbbképző Tanfolyam Előadásai*, ed. György Faber, 67–73. Budapest: Magyar Gazdatisztek Országo Egyesülete.

Schlesinger, Georg. 1949. *The Factory: Fundamental Problems of Materials, Labour, Overhead, Plant, Manufacture, Management, and Economic Control*. London: Pitman.

Schmid, Sonja D. 2011. "Nuclear Colonization?: Soviet Technopolitics in the Second World." In *Entangled Geographies: Empire and Technopolitics in the Global Cold War*, ed. Gabrielle Hecht, 125–54. Cambridge: MIT Press.

Schöne, Jens. 2014. "Ideology and Asymmetrical Entanglements: Collectivization in the German Democratic Republic." In *The Collectivization of Agriculture in Communist Eastern Europe*, ed. Constantin Iordachi and Arnd Bauerkämper, 147–80. Budapest: Central European University.

Schranz, András. 1936. *A Német Üzemgazdaságtan* [*German Industrial Economics*]. Budapest: Czeman és Sövegjártó.

———. 1930. "A Német Racionalizáló Mozgalom" ["The German Rationalization Movement"]. *Mezőgazdasági Üzem És Számtartás* 2 (4–5): 119–23.

Schrikker, Sándor. 1942. "A Sommások Alkalmazásának és Kereseti Viszonyainak Üzemi Vonatkozásai" ["The Connections between the Employment of Migrant Workers and the Conditions of their Income"]. Budapest: M. Kir. József Nádor Műszaki és Gazdaságtudományi Egyetem Mezőgazdasági Osztályának Mezőgazdasági Üzemtan Intézete.

Scott, James. 1999. *Seeing Like a State: How Certain Schemes to Improve the Human Condition Have Failed*. New Haven: Yale University Press.

Seedorf, Wilhelm. 1919. *Die Vervollkommnug der Landarbeit und die Bessere Ausbildung der Landarbeiter unter Besonderer Berücksichtigung des Taylor-Systems* [*The Improvement of Agrarian Labor and Better Training of Agrarian Workers with Particular Attention to the Taylor System*]. Berlin: Deutsche Landbuchhandlung.

Shearer, David R. 1996. *Industry, State, and Society in Stalin's Russia, 1926–1934*. Ithaca: Cornell University Press.

Shenhav, Yehouda. 1999. *Manufacturing Rationality: The Engineering Foundations of the Managerial Revolution*. Oxford: Oxford University Press.

———. 1995. "From Chaos to Systems: The Engineering Foundations of Organization Theory." *Administrative Science Quarterly* 40 (4): 557–85.

Siegelbaum, Lewis. 1988. *Stakhanovism and the Politics of Productivity in the USSR, 1935–1941*. Cambridge: Cambridge University Press.

Siklos, Pierre L. 1991. *War Finance, Reconstruction, Hyperinflation, and Stabilization in Hungary, 1938–48*. St Antony's/Macmillan series. Basingstoke: Macmillan / St Antony's College, Oxford.

———. 1989. "The End of the Hungarian Hyperinflation of 1945–46." *Journal of Money, Credit, and Banking* 21 (2): 135–47.

Simmel, Georg, and David Frisby. 1990. *The Philosophy of Money [Philosophie des Geldes]*. 2nd ed. London; New York: Routledge.

Simon, Péter. 1971. "Termelőszövetkezeti Mozgalmunk az Ellenforradalmi Válság Idején" ["Our Cooperative Movement during the Counterrevolutionary Crisis"]. *Párttörténeti Közlemények* 2: 81, 107.

Slonim, S. 1922. "Russian Scientists in Quest of American Efficiency." *Bulletin of the Taylor Society* 7 (5): 194–97.

Smith, Crosbie, and Jon Agar, eds. 1998. *Making Space for Science: Territorial Themes in the Shaping of Knowledge*. New York: St. Martin's Press.

Solomons, David. 1968. *Studies of Cost Accounting*. Homewood, IL: R. D. Irwin.

Somlyai, Magda. 1985. "A Közigazgatási Hivatalok Helye, Szerepe a Felszabadulás Időszkában" ["The Situation and Role of Administrative Offices at the Time of the Liberation"] In *Tanulmányok a Magyar Népi Demokrácia Negyven Évéről*, ed. János Molnár, Sándor Orbán, and Károly Urbán, 65–82. Budapest: Kossuth Kiadó.

Somogyi, Imre. 1942. *Kertmagyarország felé [Toward Garden Hungary]*. Budapest: Magyar Élet Kiadása.

Soproni (Schmidt), Elek. 1935. *A Mezőgazdaság Irányítása a Németbirodalomban [The Governance of Agriculture in the German Reich]*. Budapest: Királyi Magyar Egyetemi Nyomda.

Sorokin, Pitirim. 1960. "Mutual Convergence of the United States and the USSR in Mixed Sociocultural Type." *International Journal of Comparative Sociology* 1: 143–76.

Stapleford, Thomas A. 2009. *The Cost of Living in America: A Political History of Economic Statistics, 1880–2000*. New York: Cambridge.

Star, Susan Leigh, and Geoffrey Bowker. 2006. "How to Infrastructure." In *Handbook of New Media: Social Shaping and Social Consequences of ICTs*, ed. Leah A. Lievrouw and Sonia Livingstone, 230–45. London: Sage.

Stark, David, and László Bruszt. 1998. *Postsocialist Pathways: Transforming Politics and Property in East Center Europe*. New York: Cambridge University Press.

Stein, Lajos. 1930. "A Számvitel Aláértékelése a Közhiedelemben" ["The Undervaluation of Accounting in Popular Belief"]. *Mezőgazdasági Üzem És Számtartás* 2 (10): 221–28.

Stern, Alexandra M. 2005. *Eugenic Nation: Faults and Frontiers of Better Breeding in Modern America*. Berkeley: University of California Press.

Steuer, György. 1938. "A Legkisebb Földmunkabérek Megállapítása" ["Determining Minimum Field-Work Wages"]. *Katolikus Szemle* 3 (2): 601–11.

Stewart, Michael. 1993. "Gypsies, the Work Ethic, and Hungarian Socialism." In *Socialism: Ideals, Ideologies, and Local Practice*, ed. Chris Hann, 187–203. London: Routledge.

Stone, Randall W. 1996. *Satellites and Commissars: Strategy and Conflict in the Politics of Soviet-Bloc Trade*. Princeton Studies in International History and Politics. Princeton: Princeton University Press.

Strauss, Julia C. 2002. "Paternalist Terror: The Campaign to Suppress Counterrevolutionaries and Regime Consolidation in the People's Republic of China, 1950–1953." *Comparative Studies in Society and History* 44 (1): 80–105.

Swain, Nigel. 1992. *Hungary: The Rise and Fall of Feasible Socialism*. London: Verso Books.

Szabó, Bálint. 1986. *Az "Ötvenes Évek." Elmélet és Politika a Szocialista Építés Első Szakaszában Magyarországon 1948–1957 [The '50s: Theory and Policy in the First Period of Constructing Socialism in Hungary, 1948–1957]*. Budapest: Kossuth Könyvkiadó.

Szabó, Károly, and László Virágh. 1984. "A Begyűjtés 'Klasszikus' Formája Magyarországon

(1950–1953)" ["The 'Classical' Form of Procurement in Hungary (1950–1953)"]. *Medvetánc* 2–3: 159–79.

Szabó, Zoltán. 1937. *A Tardi Helyzet* [*The Situation in Tard*]. Budapest: Cserépfalvi Kiadása.

———. 1935. "Hét Falu Táplálkozási Viszonyai" ["The Nutritional Conditions of Seven Villages"]. *Magyar Szemle* 35 (98): 132–41.

Szakáll, Sándor. 1944. "A Cukorrépaszedés Ökonómiája és Célszerűsítésének Irányelvei II" ["The Guiding Principles of the Economics of Harvesting Sugar Beets and Its Rationalization II"]. *Mezőgazdasági Munkatudomány* (4–5):244–63.

———. 1943b. "Irányelvek a Kapás Normális Munkájának Megállapítására" ["Guiding Principles for Determining the Normal Work of Hoeing"]. *Mezőgazdasági Munkatudomány* (szept-okt): 421–31.

———. 1943a. "A Magas Hőmérsékleten Végzett Munka Hatása az Emberi Szervezetre" ["The Impact of Work Conducted at High Temperatures on the Human Constitution"]. *Mezőgazdasági Munkatudomány* (máj-jun): 249–56.

Szalay, István. 1931. *A Mezőgazdasági Üzemek Kézimunkaszükséglete és Kézimunkaköltsége 121 Magyar Közép- és Nagybirtok Üzemadta Alapján Tanulmányozva* [*Requirements and Costs of Work by Hand at Agricultural Firms Based on a Study of 121 Hungarian Medium and Large Estates*]. Budapest: "Pátria" Irodalmi Vállalat és Nyomdai Részvénytársaság.

Széchenyi, István. 1991 [1830]. *Hitel* [*Credit*]. Budapest: Közgazdasági és Jogi Könyvkiadó.

Szeibert, János. 1939. *Munkanélküliség és Napszámbér a Mezőgazdaságban* [*Unemployment and Daily Wages in Agriculture*]. Budapest: Hornyánszky Viktor R. T.

———. 1938. "Családi Munkabér a Mezőgazdaságban" ["Family Wages in Agriculture"]. *Katolikus Szemle* 3 (jul.): 391–411.

Szulanski, Gabriel. 2000. "The Process of Knowledge Transfer: A Diachronic Analysis of Stickiness." *Organizational Behavior and Human Decision Processes* 82 (1): 9–27.

Tassey, Gregory. 1991. "The Functions of Technology Infrastructure in a Competitive Economy." *Research Policy* 20 (4): 345–61.

Thévenot, Laurent. 1984. "Rules and Implements: Investment in Forms." *Social Science Information* 23 (1): 1–45.

Thomas, Robert. 1994. *What Machines Can't Do: Politics and Technology in the Industrial Enterprise*. University of California Press: Berkeley.

Thompson, Grahame. 1991. "Is Accounting Rhetorical? Methodology, Luca Pacioli, and Printing." *Accounting, Organizations, and Society* 16 (5/6): 572–99.

———. 1987. "Inflation Accounting in a Theory of Calculation." *Accounting, Organizations, and Society* 12 (5): 523–43.

Thurzó, Jenő. 1934. "Psychológiai és Bioszociológiai Észrevételek az Amerikai Embertypusról és az 'Amerikanizmus'-ról" ["Psychological and Biosociological Observations on the American Type of Person and 'Americanism'"]. *Orvosok és Gyógyszerészek Lapja*: 1–22.

Tihanyi, Laszlo, and Anthony S. Roath. 2002. "Technology Transfer and Institutional Development in Central and Eastern Europe." *Journal of World Business* 37 (3): 188–98.

Tipps, Dean C. 1973. "Modernization Theory and the Comparative Study of National Societies: A Critical Perspective." *Comparative Studies in Society and History* 15 (2): 199–226.

Tomlinson, Jim. 1994. "The Politics of Economic Measurement: The Rise of the 'Productivity Problem' in the 1940s." In *Accounting as Social and Institutional Practice*, ed. Anthony G. Hopwood and Peter Miller, 168–89. Cambridge: Cambridge University Press.

Tóth, Pál Péter. 1984. *Agrárszociológiai Írások Magyarországon* [*Agrarian Sociology Writings in Hungary*]. Budapest: Kossuth Kiadó.

Trotsky, Leon. 1937. *The Revolution Betrayed: What is the Soviet Union and Where is It Going?* Trans. and ed. Max Eastman. New York: Doubleday.

Turda, Marius, and Paul Weindling, eds. 2007. *Blood and Homeland: Eugenics and Racial Nationalism in Central and Southeast Europe, 1900–1940*. Budapest: Central European University.

Ujlaki Nagy, Árpád. 1943. "A Mezőgazdasági Munka Mechanikai, Termodinamikai és Biologiai Egyenértéke" ["The Mechanical, Thermodynamic, and Biological Exchange Value of Agricultural Work"]. *Mezőgazdasági Munkatudomány* (jul/aug): 316–27.

Valuch, Tibor. 1996. "Agrárpolitika, Döntéshozatal És Agrárintézmények Magyarországon 1944–1956 Között" ["Agrarian Policy, Decision Making, and Agrarian Institutions in Hungary, 1944–1945"]. Unpublished manuscript.

van de Grift, Liesbeth. 2012. *Securing the Communist State: The Reconstruction of Coercive Institutions in the Soviet Zone of Germany and Romania, 1944–1948*. The Harvard Cold War Studies Book Series, ed. Mark Kramer. Lanham: Lexington Books.

van Walré de Bordes, J. 1924. *The Austrian Crown, its Depreciation and Stabilization*. London: P. S. King.

Varga, Zsuzsanna. 2014. "The Appropriation and Modification of the 'Soviet Model' of Collectivization: The Case of Hungary." In *The Collectivization of Agriculture in Communist Eastern Europe*, ed. Constantin Iordachi and Arnd Bauerkämper, 433–65. Budapest: Central European University.

———. 2013. *Az Agrárlobbi Tündöklése és Bukása az Államszocializmus Időszakában* [*The Rise and Fall of the Agrarian Lobby during State Socialism*]. Budapest: Gondolat Kiadó.

———. 2010. "Erdei Ferenc és az Agrárlobbi" ["Ferenc Erdei and the Agrarian Lobby"]. In *Magyar Agrárpolitikusok a XIX. és a XX. Században* [*Hungarian Agrarian Politicians in the Nineteenth and Twentieth Centuries*], ed. Levente Sipos, 223–47. Budapest: Napvilág Kiadó.

Velthuis, Olav. 2005. *Talking Prices: Symbolic Meanings of Prices on the Market for Contemporary Art*. Princeton: Princeton University Press.

Verdery, Katherine. 2003. *The Vanishing Hectare: Property and Value in Postsocialist Transylvania*. Ithaca: Cornell University Press.

Vértes, Róbert. 1997. *Magyarországi Zsidótörvények és Rendeletek 1938-1945* [*Hungary's Jewish Laws and Regulations, 1938–1945*]. Budapest: Polgár Kiadó.

V. Fodor, Zsigmond. 1925. "A Taylor-Rendszer és a Mezőgazdaság" ["The Taylor System and Agriculture"]. *Mezőgazdasági Üzemtani Közlemények* 3 (3–4): 42–49.

Világhy, Károly. 1930. *A Mezőgazdasági Üzemtan Alapjai* [*The Foundations of Agricultural Economics*]. Budapest: "Pátria" Irodalmi Vállalat és Nyomdai R. T.

Viola, Lynne. 1999. *Peasant Rebels under Stalin: Collectivization and the Culture of Peasant Resistance*. New York: Oxford University Press.

Vörös, Antal. 1976. "A Magyar Mezőgazdaság a Kapitalista Átalakulás Útján (1849–1890)" ["Hungarian Agriculture on the Road to Capitalist Transformation (1849–1890)"]. In *A Magyar Mezőgazdaság a XIX.-XX. Században (1849–1949)*, ed. Péter Gunst and Tamás Hoffmann, 11–152. Budapest: Corvina Kiadó.

Vutskits, György. 1934. "Az Irányított Termelés Üzemi Előfeltételei és Várható Üzemi Következményei." *Mezőgazdasági Közlöny* 6 (7–9): 81–88.

Wädekin, Karl Eugen, and Everett M. Jacobs. 1982. *Agrarian Policies in Communist Europe: A Critical Introduction*. Studies in East European and Soviet Russian Agrarian Policy, vol. 1. Totowa, NJ: Allanheld, Osmun.

Walleshausen, Gyula. 1993. *A Magyaróvári Agrárfelsőoktatás 175 éve (1818–1993)* [*175 Years of*

Agrarian Higher Education at Magyaróvár, 1818–1993]. Mosonmagyaróvár: Pannon Agrártudományi Egyetem Mezőgazdaságtudományi Kar.

Weber, Max. 1978. *Economy and Society*, eds. Guerther Roth and Claus Wittich. Vols. 1–2. Berkeley: University of California Press.

Weill, Peter, and Marianne Broadbent. 1998. *Leveraging the New Infrastructure: How Market Leads Capitalize on Information Technology*. Cambridge: Harvard Business Press.

Weinberg, Ian. 1969. "The Problem of Convergence of Industrial Societies: A Critical Look at the State of a Theory." *Comparative Studies in Society and History* 11 (1): 1–15.

Weistroffer, H. Roland, Michael A. Spinelli, George C. Canavos, and F. Paul Fuhs. 2001. "A Merit Pay Allocation Model for College Faculty Based on Performance Quality and Quantity." *Economics of Education Review* 20 (1): 41–9.

Wiggers, Richard Dominic. 2003. "The United States and the Refusal to Feed German Civilians after World War II." In *Ethnic Cleansing in Twentieth-Century Europe*, ed. S. B. Várdy, T. H. Tooley, and A. H. Várdy, 441–66. New York: Columbia University Press.

Wise, M. Norton, ed. 1995. *The Values of Precision*. Princeton: Princeton University Press.

Wise, M. Norton, and Crosbie Smith. 1990. "Work and Waste: Political Economy and Natural Philosophy in Nineteenth Century Britain (III)." *History of Science* 27: 221–61.

———. 1989b. "Work and Waste: Political Economy and Natural Philosophy in Nineteenth Century Britain (II)." *History of Science* 27: 391–499.

———. 1989a. "Work and Waste: Political Economy and Natural Philosophy in Nineteenth Century Britain (I)." *History of Science* 27: 263–301.

Yates, JoAnne. 1989. *Control through Communication*. Studies in Industry and Society. Baltimore: Johns Hopkins University Press.

Zala, Ferenc. 1948. "Munkabérpolitika" ["Wage Policy"]. *Gazdaság* 7: 402–9.

Zarecor, Kimberly Elman. 2011. *Manufacturing a Socialist Modernity: Housing in Czechoslovakia, 1945–1960*. Pittsburgh: University of Pittsburgh Press.

Zarnowski, C. Frank. 2004. "Working at Play: The Phenomenon of 19th-Century Worker Competitions." *Journal of Leisure Research* 36 (2): 257–81.

Zavada, Pál. 1986. *Kulákprés. Dokumentumok és Kommentárok egy Parasztgazdaság Történetéhez* [*Kulak Press: Documents and Commentaries about the History of Peasant Economy*]. Budapest: Művelődéskutató Intézet.

———. 1984. "Teljes Erővel: Agrárpolitika, 1949–1953" ["With Full Force: Agrarian Policies, 1949–1953"]. *Medvetánc* 2–3: 137–58.

Index

abolition of serfdom, 109–10, 112–13, 122, 144, 286n4
Academy of Sciences, 22–23, 231, 260, 271, 277, 287n4, 298n22
accounting, 8–9, 52–53, 61–62, 64, 66–75, 195–96, 238–43, 249–52, 259–60
administration, 91, 168, 190, 219, 239, 257. *See also* county administration; government administration
Aereboe, Friedrich, 16, 54–56, 90, 94, 97–98, 100
agricultural economics, 6, 10, 15–17, 21–22, 29–30, 33, 35–37, 43, 49–71, 75–76, 155–56, 227–28, 231, 267–68, 293n6. *See also* business economics; economics
Albania, 170
animal husbandry. *See* livestock
anti-Semitism, 46–47, 110. *See also* Jews
archives, 15, 18–24, 268, 275–77
Austria, 59, 95, 124, 282n3, 290n46, 301n131
Austro-Hungarian Empire, collapse of, 30, 123, 285n2

Badics, József, 88–89, 106–7, 284n10, 297n20
banknotes, 123–27, 131, 135, 287n23
banks and banking, 10, 63, 71, 110, 122–26, 128, 130–31, 159, 162, 170, 239, 242, 251
birthrate, 43–44
bodies, 78, 81, 283n8; as metaphors, 57–58; sweat and tears, 86–87, 269
bookkeeping. *See* accounting
brigades, 3, 170, 185, 187, 194, 201–2, 224, 238, 252–54, 258, 263, 293n33
Bulgaria, 170, 172, 180, 186, 261, 293n10
bureaucracy, 19–20, 22–25, 148, 190, 192, 216, 222, 226, 262, 270–71

business economics, 10, 16, 21–22, 32, 50–62, 65–79, 82, 87–88, 90, 96–97, 102, 106, 109–10, 135, 164, 261, 269. *See also* agricultural economics; economics

calculations. *See* formulae; mathematics
Callon, Michael, 6–9, 279n5
calorie money, 128–30
Cartwright, Nancy, 12
Catholic Church, 25, 30, 32, 41–42, 47, 110, 120, 139, 151, 189, 209, 215, 269, 281n4, 282n13
children, 38–40, 44, 47, 84, 94, 116, 128, 195, 266
Christianity. *See* Catholic Church
class, 19, 36, 42, 44, 59, 77, 91, 93–94, 97, 113, 115–17, 142, 145, 150, 153, 155, 166, 170–73, 180, 194, 208–17, 224–29, 242–49, 263, 266, 269, 273; class enemies, 140, 171, 197, 214–16; class warfare, 187, 189–90, 203–11, 212, 217, 221, 224–25, 247, 269n86; rehabilitation of class enemies, 230–31
cliques, 194
Cold War, shift away from, 18
collectivization, 3, 19, 21, 25, 29, 54–55, 170–71, 173–74, 177–81, 187–90, 217, 222, 252, 267–69, 291n21, 292n1, 293n10; failures of, 195; fears of, 98; resistance to, 194–95. *See also* cooperative farms; state farm
commensuration, 1, 8, 11, 14, 25, 30, 78–79, 83, 87, 90, 108, 187, 27, 279n1
commerce, 59, 66, 70, 115–17, 120, 122, 283n14, 285n1
commodification, 1, 4–6, 9, 265, 272–74
competition, 95, 197–203
complaints, 3, 24, 192, 203, 213, 239, 248–51, 296n96

confusion, 25, 70–71, 118, 131, 156, 201, 207–9, 241–42, 263, 268
context, importance of, 14. *See also* local situation
continuity, 140–41, 182
contracts, 25–26, 31–32, 35–36, 78–79, 81–82, 102–3, 118, 176–77, 244, 254, 256
cooperative farms, 2–3, 10, 17, 20, 23, 25–26, 64, 152, 156, 165, 170, 173–82, 185–89, 192–228, 236–74; dismantling of, 223–24, 243–44, 246–48, 250–51, 263; type I, type II, and type III, 177, 179. *See also* collectivization; state farm
county administration, 15, 17, 22–24, 144–46, 159, 189, 192–96, 202, 209, 217–18, 221, 224, 227–30, 238–43, 246, 249–50, 253, 263, 271, 300n81
credit. *See* debt
crisis, 30, 43, 66, 131, 190, 256, 286n8
Czechoslovakia, 18, 30, 53–54, 99, 261, 282n3
Czettler, Jenő, 34, 41–42

debt, 113, 124, 126, 185, 195, 220, 243, 246, 250–52
decimals, 184, 186, 236–37, 267
demographics, 30, 36, 38, 43–44, 77
Denmark, 59, 68
documentation, 4, 11, 15, 17, 19–24
Donáth, Ferenc, 151, 174, 178–79, 205, 207, 244, 247, 257, 261, 266, 288n15, 301n129

Éber, Ernő, 37–38, 40–41, 44
economic planning, 15, 140–41, 158–60, 164–65, 171, 198, 219, 270
economics, 49, 67, 153–56, 235–36; influencing economic policy, 6–8; new departments of, 10, 59; performativity of, 7–8. *See also* agricultural economics; business economics
Economics University, 58, 153–54, 237, 262
education, 21, 25, 55, 58–59, 146, 151–52, 157, 162, 168, 271; technical school, 21, 61, 64, 83–84, 283n10, 290n9; training, 6, 35, 47, 51, 55, 58–63, 71, 76, 92, 96, 142, 152–53, 155–57, 163–64, 219, 226, 254, 271, 289n30; university, 10, 20, 22, 51, 58–62, 76, 142, 152–55, 231, 262, 271. *See also* Economics University; research and research institutions
elders, 3, 181, 205, 207, 245
elections, 29, 143, 147–48, 150, 170
Erdei, Ferenc, 93, 156, 177, 179, 192, 205, 208, 220–21, 228, 246, 257–58, 292n34, 293n32, 295n58, 296n94, 300n114
Erdmann, Gyula, 174–77, 204, 209, 291nn23–26
Espeland, Wendy, 8–9
ethnicity, 43, 46, 93, 184–85; traits of Hungarians, 57. *See also* racial purity
ethnography, 19, 22, 93
experts and expertise, 16, 25, 49, 58, 63–64, 69, 71–72, 76–77, 79, 87, 140–42, 148–49, 157, 167, 203, 214, 223, 226–28, 231, 244, 262

Finland, 53, 282n3
firm studies, 16, 56–57, 64, 67, 77, 90–91
Fischer, Andor, 81, 86, 97–98, 100–101, 104, 106
food supply, 38–40, 128–30, 148, 150, 170, 203, 207, 219–20, 246
force, willingness to use, 5, 17, 219, 223, 225–26, 262, 295n79. *See also* terror; violence
Fordism, 50, 121
forint, 116, 132, 134, 161
formalizing practices, 5, 8, 11–12, 14, 26, 77–108
formulae, 11, 13–14, 90–91, 184
Foucault, Michel, 8–9
Fourcade, Marion, 6, 8
France, 53, 140, 154, 282n3
Friss, István, 219, 235–36, 257
futures. *See* speculation

gender, 84, 94, 105–6, 194, 239, 242, 245
generation gap, 3
Germany (including GDR / East Germany), 16, 18, 19, 21, 46, 49–55, 67–68, 75, 77, 85–87, 96, 100, 105, 126, 146, 155–56, 160–61, 169, 187, 261, 267
Gerő, Ernő, 171, 220–21, 246
government administration, 59, 145–46, 148, 162, 190, 218–19, 231, 239, 257. *See also* county administration
greed, 113, 115, 150, 250

health and healthcare, 37–38, 58, 121, 245
Heller, András, 21, 35–40, 43–47, 55, 155, 281n5, 282n8, 284n3, 297n20
historiography, 15, 24
household, 32, 88, 128, 175, 207–8, 248, 292n35
housing, 32, 36–37, 82, 87, 89, 116, 281n6, 299n62

Ihrig, Károly, 159, 167, 231
immorality. *See* moral decay and immorality
implementation, 13, 22–24, 188, 192–94, 198, 216–18, 268–70, 273
incentives, 6, 25, 52, 91, 100–101, 104, 106, 120, 196, 201, 238
India, 54
indifference, 25, 51, 197, 201, 218, 247, 254; to accounting, 70, 238; to reforms, 56–57, 201, 269. *See also* resistance; skepticism
inflation, 10, 34, 110–11, 123–31, 252
infrastructure, 2, 7–10, 13, 17, 49–50, 52, 76, 135, 141, 158, 189–90, 249, 265, 271–74
in-kind compensation, 24–25, 37, 74, 83, 85, 87–89, 105–8, 114, 130–31, 177, 245–46, 251, 255. *See also* sharecroppers and sharecropping
interviews, 4, 23, 155, 169
interwar period, 16–17, 29–30
Italy, 21, 55, 59, 86, 95, 161, 282n3

INDEX

Jews, 45–47, 117, 282n15, 283n11, 285n1, 286n7, 287n6. *See also* anti-Semitism

Kalocsay, Ferenc, 155, 168–69, 181, 184, 290n3
Károly, Rezső, 21, 56, 59, 60–61, 68–70, 83–85, 95–105
Kerék, Mihály, 30, 31, 33, 46–47
Kesztyűs, Lajos, 21, 35, 56–61, 65, 67–70, 75, 84
Kingdom of Serbs, Croats, and Slovenes, 30, 124
knowledge transfer, 5
Kölber, László, 86, 94, 98, 100, 102, 105
Kovács, Imre, 21, 32, 39–40, 55, 93, 107, 282, 293
kulaks (wealthy peasants), 149–50, 153, 171–77, 205–16, 221, 229, 244, 249
Kuzmiak, Mihály, 181, 257, 292n42

labor markets, 2–4, 10, 16, 32, 36, 45, 50, 108, 187
labor power, 80
landowners, 10, 29, 31–38, 41–45, 51, 55, 58, 62–64, 112–23, 179, 225, 248, 262, 269
land reform, 17, 21, 29–34, 36, 40–42, 45–46, 151, 172
Latkovics, György, 184, 290n13, 292n39, 301n114
law and legal matters, 24, 52–53, 59, 63, 73, 118, 130, 142, 145, 147, 150, 158, 174, 180, 186, 225, 229, 262–63, 281n5, 287n1, 288n15, 299n75
laziness, 32, 93, 95–96, 113, 115, 184, 194, 247, 285n20
legitimacy, 13, 24, 30, 49, 79, 207, 235, 242–43, 250–51, 254, 263, 288n9, 302n3; of money, 125–26
Lenin, Vladimir, 18, 157, 168, 209
livestock, 2, 98 101–3, 107, 122, 180, 186, 219, 246, 250
local situation, 22–24, 189, 202, 208, 217–18, 229–30. *See also* context

MacKenzie, Donald, 7–8, 273
manorial estates, 29–46, 62–64, 66, 80–82, 88–89, 94, 98, 106–7, 115, 151, 181–83, 186, 222, 227–29, 263, 269, 281nn4–6
Mark, James, 19, 269
marriage, 33, 43–44, 107–8, 150, 178, 209, 282n15
Mártonfi, Rudolf, 166–67, 290n2
Marx, Karl, 78, 109–11, 134–35, 279n1; *Capital*, 1; commodity fetishism, 8
Marxism-Leninism, 16, 154, 163, 165, 172–73, 193
materiality, 13, 20, 111
mathematics, 11, 13–14, 90, 131, 184, 187, 199–200, 236–37
media, 4, 72, 139–40, 186
middle peasants, 172, 244, 248–49
migrant workers, 31–32, 36, 44, 82, 93–94, 98, 103, 160, 179, 183, 201, 222, 264, 284n4, 292n45, 299n62

military, 10, 92, 126, 153, 160, 215
Miller, Daniel, 7
Miller, Peter, 8–9
Ministry of Agriculture, 20, 37, 60, 63, 66, 69, 75, 95, 149–56, 159, 169, 180–82, 186, 190–92, 202–4, 219–21, 231–34, 237, 249–57; attack on, 139–40
Mirowski, Philip, 6–7, 79
models, scientific, 12–15, 273, 280nn9–10
money, 2–3, 25, 34, 79, 87, 89, 107–8, 109–35, 191, 195, 210, 240–41, 247, 264, 266; inability of people to use wisely, 107, 115, 240–41, 243, 285n20; unreliability of, 110–11, 127
mood, 91, 211, 220, 256. *See also* psychology
moral decay and immorality, 43–45, 114–15, 218
morality, 47, 93, 114, 119
Muneisa, Fabian, 7, 280n12

Nagy, Imre, 134, 154, 157, 171–73, 220, 223–26, 230, 235, 244, 247, 257, 261, 263, 290n14, 291n33, 292n34, 298n32
nationalization, 17, 134, 162, 170–71
National Peasant Party, 134, 143, 145, 296n94
New Phase, 223, 250
norms. *See* standards and standardization
nutrition. *See* food supply

Object of Labor, The, 3

paperwork, 192, 222, 241, 249, 257. *See also* accounting
performativity, of economics, 7–8, 272
periodization, 15–16, 267
Perneczky, Béla, 101, 151
personnel, 23, 64, 85, 97, 99, 140–42, 146–48, 168, 216, 218, 232–34, 238, 256, 260; lack of qualified, 17, 155–57, 162, 191, 207, 227–30, 269, 281n18
Péteri, György, 154–55, 227, 235, 287n4, 298n22
Pittaway, Mark, 19, 129, 272, 286–87, 294n34
planned economy. *See* economic planning
Poland, 18, 19, 45, 54, 173, 261, 282n3
policy and policy making, 5 6, 8, 13 25, 35–36, 77, 225–26
precision, 78–79, 100, 164, 167, 185–86
premiums, 105–6, 259–60
price fluctuations, 74, 88
private farms, 172–78, 186–88, 204–5, 207, 245–48
productivity, 86, 90–91, 95, 102–5, 119–21, 220, 258
propaganda, 4, 32, 99, 139–40, 202–4, 215
psychology, 79, 91–92, 96, 100, 104, 108, 211–15. *See also* mood
public shaming, 200, 205, 210, 214
purges. *See* screening of individuals

racial purity, 43, 46, 93, 117
Rákosi, Mátyás, 132, 134, 152, 172–73, 178–79, 190, 192, 194, 209, 213, 223, 261, 268, 292n34, 296n94, 297n1, 298n22
rationalization, 1, 13, 20, 29, 50–51, 53–54, 58, 60, 65, 81, 91, 166–68, 265, 285n14, 292n3
rationing, 128, 148, 176–77, 207, 291n26
Red Army, 45, 126, 128, 143–44, 147, 150, 161, 286n11
regulation, 4, 26, 37, 130, 140–42, 146, 152, 158–59, 176–77, 192, 198, 208–10, 220, 222, 236–37, 243, 256–58, 269, 282n15
Reichenbach, Béla, 57, 80, 82, 84–85, 87, 93–107, 181, 183, 231, 285n18, 292n44, 297nn19–20
religion. *See* Catholic Church; Jews
reparations, 124, 128–29, 143, 158, 161
requisitions, 46, 174–76, 178, 186, 198, 200, 203–4, 206, 208, 210, 218, 220, 244, 291nn23–27
research and research institutes, 23, 59–61, 168, 235–36, 260
resistance, 25, 46, 97, 173, 187, 268–69, 284n12
revolution (of 1956), 261–62
rhetoric, 12–13, 21, 140, 171
Romania, 18, 99, 194, 261

sabotage, 139, 205–6, 212–13, 215, 228
Scherer, Péter Pál, 37, 45, 68, 71–75, 281n6, 282n13
Schranz, András, 53, 60, 67
scientific management, 9–10, 273
scientific models, 12–13, 15, 79, 273, 280n9
screening of individuals, 147–52, 192
secrecy, importance of, 219
secret police, 139, 218–19, 281n20, 292n1
shaming. *See* public shaming
sharecroppers and sharecropping, 81–82, 94, 103, 106, 113–14, 183, 254–56, 259–60, 264. *See also* in-kind compensation
Siklos, Pierre, 126–27, 131, 134, 286n9
skepticism: about bookkeeping, 241; about brigades, 254; about experiments, 169; of experts, 228; about reforms, 61–62, 188, 193; about source material, 20, 22, 24. *See also* indifference; resistance
Smallholders' Party, 45, 132, 134, 139, 143, 145, 157
Social Democrats, 17, 21, 130, 134, 143, 145, 161, 234
sociographs, 32, 36, 38–39
source materials, 18, 19–24
Soviet bloc, 18, 165
Soviet influence, 5, 16, 21, 143–45, 157, 160–61, 170, 172–73, 180–82, 186–87, 207–8, 224–27, 236–37, 267–68
Soviet military. *See* Red Army
Soviet Republic (of 1919), 98–99, 124–25
Soviet Union, 53–54, 185; travel to, 54, 196–97, 218, 261. *See also* Red Army; Soviet influence
speculation, 75, 116–17

Stalin, Joseph, 18, 26, 166, 190, 194, 197, 223, 268
Stalinist period, 5, 16–17, 174
standards and standardization, 79, 83–85, 166–69, 224
state farm, 23, 150, 157, 227, 229, 231, 251, 256. *See also* collectivization; cooperative farms
statistics, 10, 36, 65–68, 77, 89, 155–56, 219
Switzerland, 50
Szabó, Zoltán, 32, 38–40, 93
Szakáll, Sándor, 86, 284n3
Szalay, István, 68, 84, 88–90, 107, 281n2, 284n6, 301n125
Szeibert, János, 36, 43–44, 82–83, 87, 99, 102, 107

taxation, 122, 175–76, 185, 198, 223
Taylorism, 9–10, 13, 78, 84–85, 100, 103–4, 106
technical training. *See* education
terror, 5, 222, 295n79. *See also* force, willingness to use; violence
Thaw. *See* New Phase
Thévenot, Laurent, 9–10
time, as a way to measure labor, 83–85, 183, 191, 259, 267
Tito, Josip Broz, 173, 189, 206, 213
tools, 10, 14, 52, 56, 80–81, 85–86, 95, 111, 155, 167, 182, 203, 212; equipment, 250
trials, 139, 147, 150–52, 287n1, 297n20

uncertainty, 71, 105, 191–92, 219; in paperwork, 241
unemployment, 36, 38, 198
United Kingdom (Britain), 51, 53–54, 68, 139
United States, 18, 21, 50, 52–55, 68, 76, 78, 100, 139
unrest, 24, 44, 62, 78, 99, 118, 261
utilitarianism, 118–19

Varga, Zsuzsanna, 19
violence, 187, 204, 262, 295n79. *See also* force, willingness to use; terror
Voice of America, 206

wage scheme, 8, 11, 14, 100, 108, 164–68, 182; failure of, 17
wealthy peasants. *See* kulaks (wealthy peasants)
Western influence, 5, 17
women, 44, 84, 94, 105–6, 194, 239, 242, 245
work science, 4–6, 10, 14, 16, 20–21, 24–25, 49, 81–82, 164
work units, 2–3, 180–86, 192, 194–95, 224, 236–38, 252, 257–61, 264–66; failure of, 258

youth, 3, 42, 245–46. *See also* children
Yugoslavia, 170, 173. *See also* Kingdom of Serbs, Croats, and Slovenes

Zavada, Pál, 270, 291n23

www.ingramcontent.com/pod-product-compliance
Lightning Source LLC
Chambersburg PA
CBHW051349290426
44108CB00015B/1940